BACKGROUND TO SANS 10160

BASIS OF STRUCTURAL DESIGN AND ACTIONS FOR BUILDINGS AND INDUSTRIAL STRUCTURES

EDITORS | JV RETIEF & PE DUNAISKI

SUN PRESS

Background to SANS 10160

Published by SUN MeDIA Stellenbosch

First edition 2009

ISBN 978-1-920338-10-7

Cover design by Sonja Badenhorst
Typesetting: Author
Publisher: Liezel Meintjes

SUN PReSS is an imprint of SUN MeDIA Stellenbosch. Academic, professional and reference works are published under this imprint in print and electronic format. This publication may be ordered directly from www.sun-e-shop.co.za.

Packaging, printing and binding by SUN MeDIA Stellenbosch, Ryneveld Street, Stellenbosch 7600.

www.africansunmedia.co.za

www.sun-e-shop.co.za

CONTENTS

Part 1 – Basis of Structural Design

Parts 2, 7 & 8 – General actions:
Self-weight & Imposed Loads; Thermal Actions; Actions during Execution

Part 3 – Wind Actions

Part 4 – Seismic actions and general requirements for buildings

Part 5 – Basis for geotechnical design and actions

Part 6 – Actions induced by cranes & machinery

FOREWORD

The Draft South African Loading Code SANS 10160 *Basis for structural design and actions for buildings and industrial structures* represents a substantial revision of the present Standard SABS 0160:1989 (Amended 1993) *The general procedures and loading to be adopted in the design of buildings*. Proper substantiation of the changes and additions is therefore required. The purpose of this *Background Report* is to capture the main sources of reference; assessments; decisions and motivations applied in the formulation of SANS 10160 (Draft).

The background information should primarily be considered when SANS 10160 is evaluated for acceptance into design practice by the profession, and subsequently its approval for publication by SABS as a South African National Standard. The *Background Report* should also serve as the point of departure for the inevitable future revision and updating of SANS 10160. The lack of background to various sections of SABS 0160:1989 was in fact found to seriously impede its revision, particularly during early stages of the activities, some of which were carried through to the final stages!

The *Background Report* is not intended to serve as a commentary to the future use of SANS 10160 in design practice, since the function of a commentary would require substantially different contents and presentation in order to clarify the application and intentions of the various stipulations and procedures. However the background reported here should provide additional understanding as a complementary source in cases where critical consideration of design implications is required.

Due to the close link between the Loading Code and the respective materials-based standards for structural design, viz. structural concrete, steel, timber and masonry, the *Background Report* also serves to validate the use of SANS 10160 with the present materials standards through the demonstration of how consistency between SANS 10160 and SABS 0160 has been maintained. Background on the basis of design also provides important information on the future revision of materials-based design standards or even the introduction of new standards, particularly for related geotechnical design!

The development of SANS 10160 (Draft) took a decade to complete, from the initiation of the review process at the South African National Conference on Loading (SA-NCL) which was held in 1998! There are some explanations and justifications for this drawn out process, but also some positive consequences. Guidelines given at the Loading Conference were rather general and the technology base for the revision was not clear. Although responsibility for the revision process was taken appropriately by SAICE, particularly the Joint Structural and Geotechnical Divisions, the work was done on a voluntary basis, without resources being allocated, even for direct expenses. Strong motivation from practice for an updated South African Loading Code only emerged relatively recently, as the result of the strong upturn of the construction industry.

The extended duration of the revision allowed for some significant developments to take place, particularly the development of the Eurocode Standards for structural design. Interest in the voluntary Eurocode standard on geotechnical design ENV 1997:1995 initiated the SA-NCL due to incompatibilities between SABS 0160:1989 and Eurocode basis of design ENV 1991-1 at that stage. Significant further development of Eurocode subsequently took place in parallel with the South African review process, which removed these incompatibilities.

The conversion of the ENV Eurocode into normative European Standards as the EN Eurocode started in 1998. The first EN Eurocode Standards were published in July 2002. The last of the nine Parts related to the scope of the South African Loading Code was published as recently as September 2006, illustrating the parallel mode of development of SANS 10160 with the finalisation of the Eurocode reference standards. However the final step of implementation of these Standards and Parts by the publication of the related National Annex by the various Member States is not yet complete. It should be noted that the Eurocode Standards only become operational upon publication of the related National Annex. By the end of July 2008, in the UK the BS EN NA of only four Parts had been published.

Since Eurocode has evolved as an important reference for SANS 10160 (Draft), the rate of progress can be considered to be quite reasonable. In fact, direct interaction with Eurocode activities put the development on a fast tract, comparable to that of Member States by cutting short the lag that would have resulted from waiting for the ultimate publication of the Eurocode Parts.

The revision program also allowed the implementation of some academic research to take place. Not only did some useful insight derive from these investigations, but it also provided the base for 'voluntary participation' by academics in the activities.

ACKNOWLEDGEMENTS

SAICE WG MEMBERS

Membership of the SAICE WG on LC reflects a fair mixture of representatives from practice and academics (Some prominent members of the WG, being practitioners with a strong research base, or academics with extensive design experience, causes this classification to be rather fuzzy!). Needs for standardised design procedures and conditions are identified from practice. This function was fulfilled to a large extent by the SA-NCL. Acceptable requirements and procedures are identified and formulated by researchers and academics. The final formulation results from a joint decision making process where concept formulations are critically reviewed for sufficiency and clarity, also from an implementation point of view. A similar process was reflected in the preparation of concepts by champions for each topic which are supported by smaller groups.

WG Members: Peter Day; Prof Peter Dunaiski; John Duncan; Dr Adam Goliger; Dr Graham Grieve; Mike Hull; Dirk Loubser; Roy Mackenzie; Dr Alvin Masarira; Don Midgley; Dr Tony Paterson; Prof Johan Retief; Prof Chris Roth; Tim ter Haar; Prof Jan Wium: Nick Wright.

***Retired WG Members*:** Prof Alan Kemp; Dr Geoff Krige; Schalk Pienaar; John Lane; Tony Goldstein.

***Readers of WG Drafts*:** Victor Booth; Alan Berry; Keith Bokelman; Dr Hennie de Clercq; Karl Eschberger; Dr Bob Harrison, Ferdie Heymann; Nicol Labuschagne; Dr Irvin Luker; Prof Deneys Schreiner.

EUROCODE CEN TC250 SC1 & SC7

SANS 10160 can truly be considered to be an early conversion of Eurocode principles and procedures into operational design standards beyond the boundaries of the Eurocode Member States. The lag between the completion of Eurocode Standards and their application as National Standards is quite similar to that between Eurocode and the completion of SANS 10160. Although there are substantial differences in layout and format, scope of application and procedures, these differences are directly related to differences in institutional, regulatory, environmental and technical conditions of the two regions. Substantial harmonisation and consistency have however been maintained between SANS 10160 and Eurocode, basically with SANS 10160 that can be considered as a specific subset of Eurocode.

The high degree of harmonisation of SANS 10160 with Eurocode which is achieved with minimal lag between the development of SANS 10160, in comparison with the deployment of Eurocode, is a direct result of the high degree of accessibility and interaction provided by various individuals and Eurocode CEN TC 250 Sub-Committees. As previous chairperson of SC1 *Actions on structures*, Prof Haig Gulvanessian CBE played a pivotal role in the process of providing fast track access to developments which provided the motivation for referencing SANS 10160 to Eurocode. Continued interaction with SC 1 is maintained under the chair of Dr Nick Malakatas. Similar positive co-operation and support was received from SC7 *Geotechnical design*. Amongst the many colleagues who contributed to the exchanges at different levels, the interaction and co-operation with Prof Milan Holický should be mentioned specifically.

Part 1 – Basis of Structural Design

1-1 An Overview of the Revision of the South African Loading Code SANS 10160

Retief JV, Dunaiski PE & Day PW

1 INTRODUCTION

The current South African Loading Code SABS 0160:1989 (Amended 1993) has been substantially revised, updated and extended into the draft standard SANS 10160. The central role of the SA Loading Code in structural design practice has been further expanded in the draft standard, mainly through an extension of the provisions for the basis of structural design. The scope of the standard is also extended in terms of the range of actions for which specifications are provided. Provision for geotechnical design also represents a notable extension of the standard.

These developments require proper assessment before they are implemented in South African structural design practice. In South Africa, responsibility for structural design standards is taken primarily by the profession, supported by research institutions and materials-based institutions from industry. South African structural design standards therefore reflect the consensus of the profession on acceptable design practice for structural safety and economic performance of structures, as opposed to the dictates of the regulatory authorities.

In addition to the requirement of inviting public comment on any Draft South African Standard (DSS) before it is published as a South African National Standard (SANS), it is imperative that the profession reviews not only the contents of the new standard but also the basis on which it was formulated. There is also a need to record the considerations and assessments involved in drafting the standard for future reference, particularly during its inevitable future revision. In many cases, the development of SABS 0160 in the 1980's was severely restrained by the lack of background information. Proper understanding of the basis for the requirements and procedures of SANS 10160 will assist in its use in practice, particularly in difficult or marginal conditions.

This *Background Report* is therefore compiled to record the efforts made to gather related information, to establish the needs and requirements, to select appropriate procedures and then to implement them in the new SANS 10160. It also provides comparisons with existing practice. The various sections of the *Background Report* are generally compiled into three steps consisting of (a) a review of advances in standard practice within the context of South African conditions; (b) the selection of appropriate procedures, models and values for the requirements and stipulations; and (c) assessment of the implications of the stipulations in comparison to present practice. Where justified, the various steps are dealt with in separate chapters of this report. In the background to the basis of design, an overview is also given of the integral process of revision as it evolved with time and from which general principles have been distilled.

As the new draft standard does not yet have a date, it cannot be distinguished from the old code by its year of publication. For convenience and brevity the old standard is referred to as SABS 0160 and new standard as SANS 10160.

In this chapter, SANS 10160 is introduced, the revision process is summarised, the various Parts of the standard are outlined, and finally abstracts of the chapters of the *Background Report* are provided.

2 OUTLINE OF SANS 10160

The revised South African Loading Code SANS 10160 is best introduced by providing an outline of its structure and contents. Some reflection on its function and purpose is also required against which the revised contents can be considered.

2.1 Function of SA Loading Code

The function of the SA Loading Code is to provide the principles and design rules as well as the actions (loads) that need to be taken into account in the design of buildings and similar industrial structures. The basis of design for structural performance establishes the ability of structures to sustain actions and maintain their integrity and robustness. This basis of design applies not only to the assessment of actions and their effects on the structure, but also to the provision of sufficient resistance in accordance with the materials-based design standards, e.g. for structural concrete, steel, timber and masonry.

The scope of structures provided for in the old code SABS 0160, the general design procedures to be applied, the associated levels of reliability and the actions to be considered have generally been maintained as have the materials-based design standards which are intended to be used in conjunction with the new standard. Changes to the scope and contents of SABS 0160 are typically due to the incorporation of improved models and procedures, many of which are being implemented internationally.

The general basis of structural design uses limit states design procedures and partial factors to achieve appropriate levels of reliability for the safety and performance of structures. It is the intention that the code should specify all requirements for design which are independent of the specific structural materials used. Changes from the previous edition of this standard result mainly from an extension of design situations provided for. This should result in improved consistency of the reliability of structures by improving the reliability where necessary, but also by removing unwarranted conservatism.

The principle has been adopted that acceptable performance of structures designed according to existing procedures provides confirmation of sufficient levels of reliability. This provides the basis for the continued use of existing materials-based design standards together with this standard. The full potential of the extended reliability framework provided in the new code *vis-à-vis* the design of more efficient or advanced structures utilising modern structural materials will be realised when the materials-based design standards are also revised accordingly or when new compatible standards are introduced.

2.2 Structure of SANS 10160

For practical convenience, SANS 10160 will be published as a single document. However, due to the independent nature of the topics covered, it is divided into separate Parts. This separation into Parts allowed related material to be kept together whilst still complying with SABS requirements for layout and numbering (SANS 1-1:2003). The eight parts of SANS 10160 are as follows:

Part 1	*Basis of structural design*
Part 2	*Self-weight and imposed loads*
Part 3	*Wind actions*
Part 4	*Seismic actions and general requirements for buildings*
Part 5	*Basis of geotechnical design and actions*
Part 6	*Actions induced by cranes and machinery*
Part 7	*Thermal actions*
Part 8	*Actions during execution*

The *general procedures* given in SABS 0160 were upgraded into the separate Part 1 which provides the *basis of structural design*. The set of actions provided for in SABS 0160, viz. self-weight, imposed loads, wind actions and crane induced actions, were separated into individual Parts and the content was revised and updated. An important addition to the scope of SANS 10160 is the provision for *geotechnical design and actions* contained in Part5. Other additions include actions induced by stationary rotating machinery, thermal actions and actions during execution which include construction, maintenance and modification.

3 REVISION PROCESS

Although the emphasis in the *Background Report* is mainly on the technical contents of SANS 10160 including its motivation and implications, this cannot be separated from the development process. Structural design standards require a careful balance between the opposing needs for

- **standardisation** in order to set performance requirements and present procedures for use in practice that ensure compliance with these requirements;
- **development** to keep up with advances resulting from research and accumulation of experience and also improvements and innovation in structures and design procedures.

The revision of SABS 0160 was driven by the need both for development and for the correction of specific deficiencies identified in the old code. The most convenient way to provide an overview of the revision process is to recap the historical development of structural design practice as it relates to the development of the new code.

3.1 Historical Development

3.1.1 Limit states design

The practice of limit states design was introduced to South Africa with the design of concrete structures using the British concrete code CP110. In this code, the treatment of loading was characterised by the use of the partial load factors {1,4; 1,6} for dead and imposed loads respectively.

The next step in the evolution of reliability based, partial factor limit states design in South Africa was the publication of the SA Loading Code SABS 0160 in 1989. The general approach to applying reliability principles in this code was presented by Kemp *et al* [1987] whilst the calibration process was reported by Milford [1988]. Based on a probability of exceedance of load effects of 1%, the familiar partial factors for permanent and variable loads of {1,2; 1,6} were determined, with a check for dominant permanent load using the partial factors {1,5; 0}. Although this load combination scheme was developed independently, it was in agreement with the predecessor of ASCE-7, which followed from the seminal investigation on reliability based design by Ellingwood *et al* [1980].

By achieving a resistance fractile of 1% through the specifications of the materials-based standards, it was established that a target level of reliability as expressed through the reliability index β of at least 3,0 would be achieved. This corresponds to a probability of failure p_f less than 0,00135; where $p_f = \varphi(-\beta)$ and φ is the cumulative normal distribution function. The reliability levels also agreed with ASCE-7 practice.

SABS 0160 served as basis for the respective materials-based design standards that were subsequently published. These standards however applied various reference bases and in many cases maintained the option of using allowable stress design in parallel with the limit states design procedures (see Table 1). No systematic reliability calibration was reported in the literature. Bridge design was specified in TMH 7 (1989), providing for both traffic loads, and structural concrete design based on the CEB-*fip* Model Code. The adoption of Eurocode ENV 1997 was considered for geotechnical design, but differences between the load combination schemes in Eurocode ENV 1991-1 and those in SABS 0160 impeded this option.

Table 1 South African structural design standards based on limit states design

Standard	Title	Comments	Date
SABS 0160	The general procedure and loadings to be adopted in the design of buildings	Locally developed	1989 1993
SANS 10100	The structural use of concrete Part 1 Design Part 2 Materials and execution of work	Based on British code: BS 8110	2000
SANS 10162	The structural use of steel Part 1 LS Design of hot rolled steelwork Part 2 LS Design of cold formed steelwork	Based on Canadian code: CAN3-S16-01	2005
SANS 10163-1	The structural use of timber Part 1: Limit states design	Locally developed	2001
SANS 10164	The structural use of masonry Part 1: Unreinforced masonry walling Part 2: Structural design and requirements for reinforced and prestressed masonry	Locally developed	 1987 2003
TMH-7	Code of practice for the design of highway bridges and culverts in South Africa Part 1 General Part 2 Specification of loads Part 3 Structural concrete	Specifications for structural concrete are based on the CEB-*fip* Model Code	1989

3.1.2 South African National Conference on Loading (1998)

The South African National Conference on Loading (SANCL) *Towards the development of a unified approach to design loading on buildings and industrial structures for South Africa* was held in 1998, as reported by Day & Kemp [1999] and introduced by Kemp [1998]. The SANCL was arranged by the Geotechnical and Joint Structural Divisions of SAICE and the SAISC.

In addition to considering the state of SABS 0160 after a decade of use, interest was expressed by various materials groups in the emergence of Eurocode and the possibility of following the Eurocode route in the future. However, the incompatibility between load combination schemes in SABS 0160 and Eurocode as identified by the geotechnical fraternity first needed to be resolved.

The SANCL reviewed the status of SABS 0160:1989 in terms of its general requirements and its specification of actions, together with its relation to the various materials-based design standards. It considered international developments, particularly in North America, Europe and the Eastern Pacific region from an Australian perspective. No preferred reference base for a South African Loading Code emerged, and it was concluded that ISO standards should be used as a primary reference.

The main conclusions drawn at the National Conference on Loading were that a South African Loading Code should form a common basis for all the local materials-based design standards and that these standards should refer to the loading code as their bases of design. Deficiencies in the loading code and inconsistencies with the various materials-based standards, particularly those derived from different reference bases, should be resolved. Provision also needed to be made for geotechnical design. Harmonisation with international structural design practice should be improved. The only feasible method of achieving this was considered to be basing South African standards directly on the relevant ISO standards, due to the differences in the leading international standards.

The SANCL concluded that the SAICE Working Group on the Loading Code needed to be reactivated to revise SABS 0160. It gave some general guidance on how to go about resolving the matter of international harmonisation and the technology base for a future SA Loading Code. It was emphasised that such a revised code should apply to all local materials-based design codes.

3.1.3 SAICE Working Group on the Loading Code

The SAICE Working Group on the Loading Code (WG) was reactivated in 1999 with comprehensive representation of the Joint Structural and Geotechnical Divisions and the various structural materials committees. This group serves as a working group to the SABS Subcommittee SC 5120.61M Construction Standards – *Bases for the design of structures*. The guidelines followed by the WG and its scope of work were formulated from the outcome of the SANCL as summarised in Table 2.

Table 2 Guidelines for revision of the SA Loading Code

	Topic	Guideline
1	Compatibility with international materials code selected as reference for SA standard	Revised standard should be suitable for use in conjunction with any internationally acceptable materials code selected as reference, provided the probabilistic resistance model is compatible with the loading code
2	International compatibility	It should remain a "best practice" practice code and should be compatible with an established international code such as ASCE-7
3	Allowing alternative loading code	It could allow for the alternative use of Eurocode EN 1990 or EN 1991 in conjunction with the materials Eurocodes
4	Relation to SA materials standards	Due to the consistent levels of reliability achieved over the practical range of load ratios, material types and failure mechanisms by the present code, the emphasis in revision should be on achieving consistency with the materials codes.

The following stages in the development of SANS 10160 can be identified:

- **Initial phase:** During the initial phase of activities (1999 – 2003) the WG re-evaluated the SABS 0160 load combination schemes; made an extensive comparison for imposed loads as specified in SABS 0160, ASCE 7-95 and Eurocode ENV 1991-2-1; assessed various alternative standards as reference for wind loads; and reassessed the SABS 0160 specification for crane induced loads. Slow progress was made due to the open ended nature of the investigations and limited resources.
- **Access to Eurocode:** During 2003, the WG obtained access to the work of the Eurocode Subcommittee CEN TC250/SC1 on actions on structures. At this time, Eurocode was in the process of conversion from a voluntary European Standard (ENV) to a normative European Standard (EN). Strict rules were laid down for the implementation of the EN Standard including its adoption by Member States as a national standard. One of the options for load combination given in the revised Eurocode EN 1990 *Basis of structural design* was similar to that of SABS 0160. This assisted in resolving the incompatibility between Eurocode and South African structural design practice.
- **Trial Eurocode procedures:** Fast-tracked access to Eurocode developments made it possible to use the appropriate Eurocode Parts as reference for the topics of load combination/basis of design; imposed loads; wind actions and crane induced actions. Trial formulations of requirements and procedures for a South African standard, which complied with the general terms of reference for the revision of SABS 0160, were compiled during 2004. The results were sufficiently positive to lead to the decision by the end of that year to use Eurocode as reference for the revised SANS 10160.
- **SANS 10160 formulation:** Full reference of SANS 10160 to the relevant Eurocode Parts required a careful assessment of the basis for such a process. The scope of structures to which SABS 0160 applied (buildings and similar industrial structures) was maintained. The scope of actions was revised and extended as per Eurocode, including a formal Part on the

basis of structural design. Development of the revised and extended SANS 10160 was started in 2005.

- **Geotechnical design for buildings:** The special case of providing for the geotechnical design within the scope of application of SANS 10160 was started in 2006. In the absence of a South African standard on geotechnical design, it was necessary to provide not only for geotechnical actions for buildings, but also the basis of geotechnical design.
- **Final compilation:** The structure of SANS 10160 as a single, multi-part standard with an acceptable SABS format was resolved in 2007. Under this scheme, each of the eight parts, could have an independent numbering system and its own set of appendices.

3.1.4 Structural Eurocode summit

The Structural Eurocode Summit was convened in February 2008 by the Joint Structural Division to clarify the relationship between future South African structural design standards and the Eurocode structural standards (Watermeyer 2008). The route taken by the working group to use Eurocode as reference was endorsed. Consensus was also reached that future materials-based standards would be modelled on Eurocode, with clear preference at this stage for adoption of the relevant standards and parts. There was generally no urgency attached to such a process, with the exception of the concrete code where revision based on Eurocode had already started. In most cases, progress with Eurocode implementation in the United Kingdom was identified as a process to be carefully observed.

3.2 Eurocode Program

Many arguments can be presented to substantiate the claim that Eurocode is the most advanced set of structural design standards presently available. These include its advanced nature, its comprehensive scope and its high degrees of harmonisation and unification. These arguments are the primary reasons why relevant Eurocode Parts were selected as reference for SANS 10160. However it is also necessary to critically assess not only the technical contents of Eurocode, but also its context, objectives, stages of development and implementation in relation to South African conditions and structural practice.

The development of Eurocode was initiated by the Commission of the European Community in 1975 in order to remove technical obstacles to trade. *"Within this action program, the commission took the initiative to establish a set of harmonised technical rules for the design of construction works which, in a first stage, would serve as an alternative to the national rules in force in the Member States and, ultimately would replace them"* (Eurocode Foreword).

Eurocode went through a number of stages, of which the last European Union stage started in 1998 was to convert the voluntary ENV Eurocode into the normative EN Eurocode. This process is not yet fully completed for all the 58 Parts of Eurocode. This stage should clearly be distinguished from the next stage of implementation during which each of the 28 Member States is required to compile National Annexes which convert Eurocode into a set of operational national standards. This stage requires extensive efforts by Member State National Mirror Committees.

3.2.1 Main characteristics of Eurocode

The political initiation and sponsorship of Eurocode had an important influence on the nature of the resulting set of standards, particularly in terms of their scope of application. Since the primary objective of Eurocode was to provide access to the common European structural and construction market, it was required to cover a comprehensive scope of structures. The full complement of building, bridge and industrial structures are provided for, as well as all conventional structural materials (concrete, steel, etc). Provision is also made for geotechnical design.

Ten Standards EN 1990 – EN 1999, consisting of fifty-eight Parts (Table 3) are required to cover the extensive scope of the Eurocode (Table 4). A sparse formulation is used to minimise duplication, resulting in the presentation of related material in disjointed sections of the various parts and standards.

Table 3 List of Eurocode Standards, indicating the number of separate Parts

STANDARD	TITLE	Parts
EN 1990	*Basis of structural design*	1
EN 1991	Eurocode 1 : *Actions on structures*	10
EN 1992	Eurocode 2 : *Design of concrete structures*	4
EN 1993	Eurocode 3 : *Design of steel structures*	20
EN 1994	Eurocode 4 : *Design of composite steel and concrete structures*	3
EN 1995	Eurocode 5 : *Design of timber structures*	3
EN 1996	Eurocode 6 : *Design of masonry structures*	4
EN 1997	Eurocode 7 : *Geotechnical design*	2
EN 1998	Eurocode 8 : *Design provisions for earthquake resistance of structures*	6
EN 1999	Eurocode 9 : *Design of aluminium alloy structures*	5
	TOTAL	**58**

Table 4 Scope of Eurocode Standards for structural design

<table>
<tr><th colspan="5">COMMON INPUT INTO STRUCTURAL DESIGN</th></tr>
<tr><td>EN 1990 Basis of structural design</td><td colspan="4">Requirements, limit states design, basic variables, design verification</td></tr>
<tr><td rowspan="2">EN 1991 Actions on structures</td><td colspan="4"><u>General actions:</u> Self-weight, imposed, fire, snow, wind, thermal, execution, accidental</td></tr>
<tr><td colspan="4"><u>Specific structures:</u> Traffic, cranes, machinery, silos, tanks</td></tr>
<tr><th>STRUCTURAL MATERIALS</th><th colspan="4">STRUCTURES</th></tr>
<tr><td>EN 1992 Design of concrete structures</td><td rowspan="6">General & Buildings</td><td rowspan="4">Bridges</td><td>Containment</td><td></td></tr>
<tr><td>EN 1993 Design of steel structures</td><td>Silos, tanks, pipelines</td><td>Towers, masts, piling, crane</td></tr>
<tr><td>EN 1994 Design of composite steel/concrete structures</td><td colspan="2" rowspan="4"></td></tr>
<tr><td>EN 1995 Design of timber structures</td></tr>
<tr><td>EN 1996 Design of masonry structures</td><td rowspan="2"></td></tr>
<tr><td>EN 1999 Design of aluminium structures</td></tr>
<tr><th colspan="5">COMMON FOUDATION DESIGN CONSIDERATIONS</th></tr>
<tr><td>EN 1997 Geotechnical design</td><td>General</td><td colspan="3">Spread footings, pile foundations, retaining structures, embankments</td></tr>
<tr><td>EN 1998 Design provisions for earthquake resistance of structures</td><td>General, (Buildings)</td><td>Bridges</td><td>Silos, tanks</td><td>Towers, masts, chimneys, pipelines</td></tr>
</table>

3.2.2 Eurocode technology base

Due to the strong political and technological support for the development of Eurocode, it is based on a progressive and up-to-date technology base. This is confirmed by the extensive body of literature reporting on the various stages of development. Examples include JCSS (1996) on the basis of design, IABSE (1996) on actions on structures and Implementation of Eurocodes (2005) on the transfer of information on Eurocode. The drafting committees had to consider and reconcile diverse sources of structural engineering technology from various Member States, a process that required maximum use of diplomacy and rational procedures to reach consensus. Where needed, complementary research was commissioned, including scrutiny by specialists from Member States. A more elaborate discussion of the Eurocode technology base is presented in Chapter 1-3.

3.2.3 Member State responsibilities and options

Another consideration which has an influence not only on the Eurocode layout and format, but also on its very nature is the "*recognition (of) the responsibility of regulatory authorities in each Member State ... to determine values related to regulatory safety matters at national level where these continue to vary from State to State.*" Adjustments were also required to deal with the differences in natural and environmental conditions of each Member State.

Where agreement on critical issues could not be reached, options were provided particularly with regard to design approaches and methods of analysis. Many of the values used in the code have been classed as Nationally Determined Parameters (NDPs) to be specified by the individual Member States. The net result is that, in spite of the advanced nature of Eurocode and the high degree of harmonisation amongst Member States and unification between the various components of structural design, it can at best be considered as merely a reference standard. Only when the preferred options to be used in each Member State have been identified and the values assigned to the NDPs will Eurocode become an operational design standard.

4 THE RELATIONSHIP BETWEEN SANS 10160 AND EUROCODE

When compared to SABS 0160, which is the point of departure for the revision of SANS 10160, there are large differences with the related Eurocode Standards EN 1990 *Basis of structural design* and EN 1991 *Actions on structures*. These include the scope of application, actions and procedures, composition and layout. Even the level of specialisation and associated advanced nature of specifications and procedures are substantially different. The way in which Eurocode was to be used in the revision process therefore required careful consideration.

4.1 Derived Basis for SANS 10160

Reference to Eurocode in the development and formulation of SANS 10160 was approached firstly by considering how the information from Eurocode may impact on the function and scope of application of SABS 0160. Related topics included the scope of structures and actions to be covered and setting standards for structural performance within the local regulatory environment. General guidelines were developed for the way in which SANS 10160 was to refer to Eurocode in order to select appropriate Eurocode Parts and procedures.

4.1.1 Scope of application

The decision not to alter the scope of structures covered by SABS 0160, which is buildings and similar industrial structures, was a critical step in resolving the way in which reference was made to Eurocode. This decision also implied that the level of application (i.e. providing only for general structures and design practices) would be maintained thereby avoiding situations where specialist input is required. This approach narrowed down the Eurocode Parts and even sections and specific procedures to be considered.

Although this guideline was not followed strictly, for example when crane induced actions or geotechnical design were considered, there were specific justifications for such cases.

4.1.2 Scope of actions

The next step was to include from Eurocode all additional topics (and actions) that were relevant to the scope of structures and application of SANS 10160. Since provision for geotechnical design played such an important role in initiating the revision of SABS 0160, this topic was identified for special attention, requiring consideration of EN 1997 *Geotechnical design*, in addition to EN 1990. Furthermore, seismic design was inherited from SABS 0160 and required consideration of Eurocode EN 1998 *Design of structures for earthquake resistance*.

Other actions to be considered were thermal actions, actions during execution and construction, accidental actions and actions induced by machinery. *Actions on structures exposed to fire* is one of a set of Eurocode Parts on structural fire design which includes related materials-based standards. This topic was judged to go beyond even an extended scope of SANS 10160, but this might be reconsidered in future.

4.1.3 Regulatory function

Although Eurocode accommodates the varied legal and regulatory conditions of its Member States, it consistently provides for incorporation into the National Standards published by the respective Standards Generating Bodies. The situation in South Africa is distinctly different. Here, the responsibility for structural design standards is taken on by the engineering profession, with support from the relevant materials industries. Publication as a South African National Standard by SABS is used to achieve a certain status, rather than being prescribed by regulatory procedures. In South Africa, codes of practice and the majority of national standards are regarded as statements of good practice and are neither mandatory nor legally enforceable.

SANS 10160, together with the related materials-based design standards, provides the South African design engineer with an effective and convenient way to discharge his or her professional responsibilities in a manner that can be readily demonstrated and justified. This function should be distinguished from a standard that is published as part of a system of regulatory requirements as is the case with Eurocode.

4.1.4 Structural performance and reliability

As indicated above, the level of reliability which applies to SABS 0160 can be expressed in terms of the reliability index β = 3,0 for the reference situation. This requirement was extensively reassessed as reported in Chapter 1-2. In the absence of any indications that this was insufficient, it was decided that the reliability index used in SABS 0160 should be left unchanged. Eurocode, on the other hand, uses a default reference value of β = 3,8 and is therefore more conservative than South African practice. However, Eurocode accepts structural performance to be a matter for determination by Member States. Thus, deviation from the default Eurocode reliability index does not represent any conflict with its requirements. The South African value for this index is also in reasonable agreement with the guidelines from ISO 2394 and is similar to reliability levels applied in ASCE 7.

There were some clear indications that the relatively high level of reliability implied in Eurocode was furthermore applied in a conservative manner during the formulation of design procedures. In the revision process for SANS 10160, it was decided to differentiate between changes in the level of requirements deriving from improved models and changes that were the result of a more conservative approach to structural performance and design. The former were accepted whereas the latter were rejected or adjusted.

The need to adjust for Eurocode's general conservatism applies particularly to standard design situations where the efficiency of design is of paramount importance. This does not apply to the same extent in the case of special situations requiring expert input where the emphasis is more on obtaining suitable solutions for acceptable structural performance than on the economy of the design.

4.2 Reference to Eurocode Parts

Operational reference to Eurocode required the consecutively more detailed steps of selecting the relevant Standards, Parts, sections and procedures, often including selecting among optional alternatives. However the process was not strictly limited to Eurocode material. It was regularly subjected to critical review, particularly to provide for local conditions and existing practice. The detailed processes that were followed are presented in subsequent chapters, with just the general approach taken which is summarised here.

4.2.1 Relevant Eurocode Parts

In addition to EN 1990 *Basis of structural design* and Parts of EN 1991 *Actions on structures*, the general Parts of EN 1997 *Geotechnical design* and EN 1998 *Design of structures for earthquake resistance* are relevant to SANS 10160. The relevant Eurocode Parts are listed in Table 5, with some comments made on how they are referred to in SANS 10160.

4.2.2 Advanced Eurocode procedures

The availability of advanced design procedures for special structures and situations in Eurocode was fully exploited by consciously limiting the level of procedures in SANS 10160 generally to standard structures while still ensuring sufficient compatibility with Eurocode to allow for the use of the advanced Eurocode procedures under South African conditions. Specialist input will however be required for such applications.

4.2.3 Unified Eurocode Standards

An important property of the comprehensive Eurocode that can be exploited in the future is the high degree of consistency and unification that has been achieved between the various Standards and Parts.

The reference of this standard to the Eurocodes also implies a recommendation that the future revision of structural materials-based design standards or the introduction of new standards which are not presently available should do likewise. Such development will not only improve the consistency between this standard and all other South African structural design standards; but by sharing a common basis of design, consistency among the various materials-based standards will also improve.

Table 5 Selected Eurocode Parts

#	Title of Eurocode Part	Comments on Application
EN 1990	Basis of structural design (Annex A1 Buildings)	Compile independent SANS 10160 Part. Serves as reference also to (future) materials-based design standards.
EN 1991-1-1	Actions on structures – Part 1-1: General actions – Densities, self-weight and imposed loads	Apply occupancy classification. Introduce additional imposed loads. Independent selection of load values.
EN 1991-1-2	Actions on structures – General actions – Actions on structures exposed to fire	Considered to be outside present scope of SANS 10160.
EN 1991-1-3	Eurocode 1: Actions on structures – General actions – Snow loads	Not applicable
EN 1991-1-4	Actions on structures – General actions – Wind actions	Procedures for buildings only. Exclude dynamic effects, considered to be specialist topic.
EN 1991-1-5	Actions on structures –General actions – Thermal actions	Newly introduced action. Procedures for buildings only.
EN 1991-1-6	Actions on structures –General actions – Actions during execution	Newly introduced action. Procedures for buildings only.
EN 1991-1-7	Actions on structures –General actions – Accidental actions	Incorporate general procedures for accidental actions into basis of design. Impact & internal explosions considered to be specialist topics.
EN 1991-3	Actions on structures –Actions induced by cranes & machinery	Consider procedures for crane actions. Introduce machinery actions with substantial reformulation.
EN 1997-1	Geotechnical design –General rules	Introduce basis of geotechnical design and actions for buildings.
EN 1998-1	Design of structures for earthquake resistance –General rules seismic actions and rules for buildings	Follow Eurocode format. Use as one of several references. Limit to SA conditions of seismicity. Provide for non-specialist design only.

5 MAIN FEATURES OF REVISION AND CONTENTS OF SANS 10160 PARTS

An outline of the most important features of the eight Parts of SANS 10160 and a summary of their contents are given below. An indication is also given of changes from SABS 0160:1989 where relevant. Additional information on the considerations and motivations for changes and the introduction of new procedures are presented in subsequent chapters.

5.1 Part 1 Basis of Structural Design

ABSTRACT: SANS 10160-1 serves as a general standard to specify procedures for determining actions on structures and structural resistance in accordance with the partial factor limit states design approach. The requirements and procedures are formulated to achieve acceptable levels of safety, serviceability and durability of structures within the scope of application of SANS 10160. Procedures for the basis of structural design include requirements for the specified minimum values for actions on structures presented in SANS 10160-2 to SANS 10160-8; the determination of design values for the effects of combined actions on the structure under a sufficiently severe and varied set of limit states; general requirements for sufficient structural resistance reliability to which the related materials-based design standards should comply.

Limit states design: The following aspects of partial factor limit states basis of structural design are presented in SANS 10160-1:

- a) The general principles for reliability-based structural design are presented in terms of requirements, limit states, basic variables and design verification procedures;
- b) The basis for the determination and combination of actions is presented in SANS 10160-2, SANS 10160-3, SANS 10160-4, SANS 10160-5, SANS 10160-6, SANS 10160-7 and SANS 10160-8;
- c) The reliability basis for structural resistance is provided for implementation in the materials-based design standards.

General topics:

1) The scope, assumptions and exclusions for this standard; normative references; terms, definitions and symbols common to all Parts of this standard.;
2) Requirements for structural design and reliability management, the principles for limit states design and rules for design verification, including accidental design situations;
3) Specifications for the use of testing procedures to be applied in the design of structures;

Informative annexes:

- i) Management of structural reliability, including reliability differentiation;
- ii) Design for consequences of localised failure due to unspecified causes; including the provision of structural robustness;
- iii) Deformation serviceability criteria and the deformation of buildings;
- iv) Guidelines on the planning and application of testing procedures for structural design.

Main features of revision:

The main features of the extensive basis of design provided in SANS 10160-1 include:

- i) An extended reliability framework is presented to provide sufficient reliability levels and to improve consistency of reliability across the range of design situations;
- ii) Provisions are introduced for taking situations and associated actions into account which are not expected during design life, but with such severe consequences that the risks need to be considered; including a proper basis for improved specifications of robustness requirements;
- iii) Improved harmonization with international practice is achieved through reference to ISO 2394 & EN 1990; future revision of materials-based design standards which are referenced to Eurocode will also enhance the unified South African design practice.
- iv) Improved specification of procedures for design assisted by testing is obtained by requiring an equivalent level of reliability to that achieved by the procedures of this Standard; guidance is given on testing procedures and the statistical treatment of the results required for compliance.

5.2 Part 2 Self-Weight and Imposed Loads

<table>
<tr><td colspan="2">ABSTRACT: SANS 10160-2 presents procedures for the treatment of self-weight and imposed loads on buildings. Procedures for determining self-weight of structural and non-structural materials as permanent loads are given, including recommended values of material densities. Minimum characteristic values for imposed loads as variable actions are given for loads on floors as a function of the occupancy and use of the building, and for imposed roof loads and horizontal loads on balustrades and partitions.</td></tr>
<tr><td>Main topics covered:
a) Classification of self-weight actions and imposed loads; and the identification of design situations, in accordance with SANS 10160-1;
b) Stipulations for characteristic values of densities and self-weight of the constructed works;
c) Characteristic values for imposed loads, consisting of
1) Imposed floor loads for residential, social, commercial and administrative occupancies;
2) Imposed floor loads for occupancies with industrial activities, including storage, the use of forklifts, transport vehicles, garages and vehicle traffic areas (excluding bridges);
3) Imposed loads on various classes of roofs; including helicopter landings on building rooftops;
4) Imposed horizontal loads on parapets, partitions walls and guardrails acting as barriers
d) An informative annex providing tables for nominal mass density of construction materials, and nominal mass density and angles of repose for stored materials.</td><td>Main features of revision:
a) Reformulation of the system of occupancy classification;
b) Revision of the stipulated values for imposed loads, in accordance with international practice, with reference to ASCE-7 & EN 1991-1-1;
c) Extension of the occupancies provided for, particularly for the case of industrial activities; including provision for imposed loads due to forklifts and helicopter landings on building roofs;
d) Extension of the tables providing information on densities of materials provided in informative annexes.</td></tr>
</table>

5.3 Part 3 Wind Actions

<table>
<tr><td colspan="2">ABSTRACT: SANS 10160-3 covers procedures for the determination of actions on land-based structures due to natural winds. The scope of application is limited to the general type of buildings and industrial structures (in line with SANS 10160) and is restricted to structures in which wind actions can be treated as quasi-static</td></tr>
<tr><td>Main topics covered:
1. Scope of structures and design situations provided for and structures which are excluded from the scope;
2. Modelling and representation of wind actions, the determination of characteristic values and models for the response of the structure exposed to natural winds;
3. Basis for the calculation of wind speed and wind pressure, in terms of basic regional wind speed, variation with height and the effects of terrain roughness, topography, altitude, neighbouring structures and obstacles;
4. Calculation procedures for wind pressures on surfaces as well as forces on the structure;
5. Coefficients are provided for pressure distribution as a function of area, friction and forces for a wide range of building configurations and geometries;
6. Treatment of internal pressures on roofs and walls with more than one skin;
7. Wind actions on free-standing walls, parapets, fences and signboards;

Annexes (Informative):
1) Effects of the terrain on the wind speed, including transitions between roughness categories, provision for topography effects, the effects of neighbouring structures;
2) Considerations for the design of buildings and structures which fall outside the scope of the current standard;
3) Requirements for the use of wind tunnel testing to assist in the design of structures.</td>
<td>Main features of revision:
1) The description of the wind climate is effectively maintained as given in SABS 0160:1989, but its presentation is modified as follows:
a) basic regional wind speeds are based on equivalent 10 minute average values;
b) the map presenting basic regional wind speeds is modified accordingly, with some updating also of its geographical extent;
c) effective procedures are provided to accommodate the action of inland winds of convective origin, which are difficult to represent by the 10 minute average value;
d) the present terrain categories are modified to provide a more even distribution of wind exposure conditions;
e) procedures for determining shielding effects are included;
f) exponential pressure profiles are maintained.
2) The wind profile terrain category for urban centres is omitted, whilst procedures are provided for shielding effects of neighbouring structures;
3) The wide-ranging additional information on pressure and force coefficients represents a substantial update of the procedures for wind actions on structures;
4) Guidance is given on the use of South African conditions and requirements to provide for design beyond the scope of SANS 10160-3, although specialist input would be required for such applications.</td></tr>
</table>

5.4 Part 4 Seismic Actions and General Requirements for Buildings

ABSTRACT: SANS 10160-4 covers earthquake actions on buildings and provides strategies and rules for the design of buildings subject to earthquake actions. Provisions for actions on structures exposed to earthquakes are revised and updated. The specification of seismic design of standard structures is extended, but procedures are restricted to situations where principles of proper layout and detailing are complied with.

The following aspects of seismic design are covered in SANS 10160-4: a) ground conditions and seismic actions; basic representation of seismic actions; b) considerations in the design of buildings for resistance against earthquakes; c) load effects; structural behaviour; displacements; structural and non-structural component load effects; materials; d) normative annexes on: i) detailing of concrete structures ii) special rules for plain masonry structures.	**Main features of revision:** a) provisions for actions on structures exposed to earthquakes are revised and updated; b) the specification of seismic design of standard structures is extended; c) procedures are restricted to situations where principles of proper layout and detailing are complied with; d) requires the application of advanced procedures, beyond the scope of SANS 10160-4, where these requirements are not met.

5.5 Part 5 Basis of Geotechnical Design and Actions

ABSTRACT: SANS 10160-5 represents an extension of the scope of SANS 10160 to set out the basis for geotechnical design and gives guidance on the determination of geotechnical actions on buildings and industrial structures, including vertical earth loading, earth pressure, ground water and free water pressure, and actions caused by ground movement. Procedures are given for determining representative values for geotechnical actions. The design of geotechnical structures such as slopes, embankments or free-standing retaining structures is not covered by the standard.

The following aspects are covered in SANS 10160-5: a) vertical earth loading; earth pressure; ground water and free water pressure; downdrag or uplift caused by ground movements; deformations caused by ground movement; b) the treatment of geotechnical and geometrical data; the specification of characteristic and design data; c) verification of ultimate and serviceability limit states, including partial factors; d) determination of geotechnical actions e) Informative annexes on: 1) influence of geotechnical categories on engineering requirements 2) guidance on partial material and resistance factors 3) Basic guidance on the design of foundations and evaluation of earth pressures.	**New Part** The main features of SANS 10160-5 include: a) it strives for compatibility with Eurocode to facilitate the use of EN1997-1 for geotechnical design situations beyond the scope of SANS 10160 b) it provides a consistent basis for dealing with geotechnical actions on structures and structural actions on the ground c) it provides information that allows the application of limit states geotechnical design within the scope of the code in the absence of a South African geotechnical design code.

5.6 Part 6 Actions Induced by Cranes and Machinery

ABSTRACT: SANS 10160-6 specifies imposed loads associated with overhead travelling bridge cranes on runway beams at the same level; and also actions induced by a limited range of stationary machinery causing harmonic loading. The standard includes improved provisions for crane induced actions by the introduction of new models and proper specification of the combination of actions.	
The following aspects are covered in SANS 10160-6: a) crane actions include dynamic effects and braking, acceleration and accidental forces; b) actions on structures supporting rotating machines which induce harmonic dynamic effects in one or more planes; c) Informative annexes on: 1) guidance for crane classification; 2) estimating the value of dynamic factor for a crane travelling on rails; 3) serviceability criteria for crane support structures.	**Main features of revision:** a) improved provisions for crane induced actions by the introduction of new models and proper specification of the combination of actions; b) introduction of new provisions for rotating machines which induce harmonic dynamic effects.

5.7 Part 7 Thermal Actions

ABSTACT: SANS 10160-7 introduces new procedures that cover principles and rules for calculating thermal actions on buildings, as well as their structural elements. Its main features are to introduce provisions for thermal actions based on the South African climate, including the classification and representation of actions, the determination of temperatures and temperature gradients in buildings.	
Its main features are the introduction of new provisions for thermal actions based on the South African climate, using Eurocode procedures, including: a) characteristic maximum and minimum shade air temperatures; b) the representation of thermal actions in buildings and structural elements; and c) informative annexes on the determination of thermal parameters and coefficients of linear thermal expansion.	**New Part**

5.8 Part 8 Actions during Execution

ABSTRACT: SANS 10160-8 introduces new procedures that cover principles and general rules for the determination of actions which should be taken into account during the execution of buildings. Its main features are to introduce provisions for actions on structures during execution of the construction works, including actions on the partially completed works and temporary structures. It consists of procedures for the identification of design situations and representation of actions and their effects on the incomplete structure, considering all activities carried out for the physical completion of the work, including construction, fabrication and erection.	
Its main features are the introduction of new provisions for actions on structures during execution of the construction works, including actions on the partially completed works and temporary structures. It consists of: a) all activities carried out for the physical completion of the work (including construction, fabrication, erection): b) design situations and representation of actions; effects of the incomplete structure; construction actions; c) an informative annex on actions during alterations, reconstruction and demolition.	**New Part**

6 OUTLINE OF THE BACKGROUND REPORT

In terms of the general motivation for the compilation of a background report to the revision of SANS 10160 presented in the *Introduction*, the purpose of this report is to record the investigations and assessments made in the revision process. This will promote in-depth consideration of the Draft South African Standard SANS 10160 when it is presented to the profession for approval. It will set out the information on which SANS 10160 is based, and enable an appreciation of the implications of its eventual application. The Background Report should also serve as an important reference to a Commentary, which is intended to be compiled for the use of SANS 10160 in design practice.

The relation between SANS 10160 and the various Eurocode Parts which served as reference is also clarified here. This consideration is important in terms of an assessment of the degree of harmonisation between South African structural design standards and Eurocode that was achieved. In the *Structural Eurocodes Summit* held in February 2008 the trend towards Eurocode in the future in South Africa was a clear conclusion (Watermeyer 2008). The Background Report therefore also provides a useful reference for Standards Committees considering the adoption and/or adaptation of Eurocode for other structural design standards.

Proper understanding of the relationship with Eurocode is important also in practical terms when SANS 10160 serves the function of providing the local input, to be used together with advanced Eurocode procedures and models, for situations beyond the scope of this standard.

Finally the Background Report should serve as reference to the inevitable future revision of this version of SANS 10160. Such revision will most certainly consider further improvement in the harmonisation with Eurocode.

Abstracts of the subsequent chapters of the Background Report are given below:

6.1 Part 1 Basis of Structural Design

6.1.1 Basis for the revision of SANS 10160

A separate journal paper is being compiled where the basis for the revision of SANS 10160 is formulated, with specific reference to the way in which the relation with Eurocode was established and formulated. A draft abstract consists of the following:

Optimal Development of South African Structural Design Standards Referenced to Eurocode as Applied to SANS 10160-1

Retief JV, Dunaiski PE

ABSTRACT

This paper summarises the South African and Eurocode premise for SANS 10160. The strategy that was followed in its development and formulation is then presented. The strategy consists largely of the optimal provision for local requirements, conditions and practice, including the presentation of an effective standard. At the same time, maximum use is made of Eurocode advances, its technology base and as a means of using Eurocode in achieving improved international harmonisation. Illustrations are then given of the implementation of the process in formulating the various design requirements and procedures.

The main objective of the paper is to provide an appreciation for the technical basis of SANS 10160 and its relation to its Eurocode counterpart. It also captures the experience gained in formulating a South African structural standard referenced to Eurocode. This experience may serve as a model to the application of Eurocode to future South African structural standards.

6.1.2 Chapter 1-2 The limit states basis of structural design for SANS 10160-1

Retief JV & Dunaiski PE

ABSTRACT

The rationale for the basis of structural design in terms of reliability based limit states design as implemented in SANS 10160-1 is presented in this Chapter. The objective of the presentation is to provide the background required to judge the acceptability and sufficiency of the stipulations. Such judgement also applies to the reliability features of the actions stipulated in Parts 2 – 8 of SANS 10160. Since Part 1 will also apply to the materials-based design standards according to which structural resistance has to be verified, consistency with these standards needs to taken into account.

The basis of structural design provides the underlying principles on which harmonisation with international practice is achieved. The stipulations of Part 1 are therefore also assessed from the perspective of its contribution to international harmonisation of structural design practice.

Topics which are discussed in more detail include reference reliability levels and reliability differentiation; the performance of various action combination schemes together with the appropriate partial factors; the way in which action combination schemes are applied for the various limit states in SANS 10160-1.

6.1.3 Chapter 1-3 Eurocode from the perspective of the revision of SANS 10160

Retief JV, Dunaiski PE & Holický M

ABSTRACT

The implementation of the Eurocode set of standards for the design of building and civil engineering structures is one of the major recent activities in the development of structural design standards on the international scene. Following an introduction on the development and theory of structural reliability on which Eurocode is based, the terms of reference and objectives of the development of Eurocode are briefly discussed. An overview is given of the set of the ten Eurocode Standards, EN 1990 to EN 1999. The important features of EN 1990 *Basis of structural design* and EN 1991 *Actions on structures*, are described, with emphasis on the actions relevant to the revision of the South African Standard SANS 10160-1989 (Amended).

An analysis is provided of the technological base of Eurocode, including structural mechanics and reliability performance; advances made in standardised structural design practice in terms of harmonisation of structural practice and unification of the design process; and finally its relevance to the South African context.

6.1.4 Chapter 1-4 The reliability basis of Eurocode

Holický M, Retief JV & Dunaiski PE

ABSTRACT

This chapter reviews the reliability basis of Eurocode as developed in EN 1990-2002 *Basis of structural design* as a head Standard to EN 1991 – EN 1999 by capturing and stipulating the general requirements for partial factor Limit States Design verification procedures for actions and material related structural resistance. Taking the International Standard ISO 2394-1998 *General principles on the reliability of structures* as point of departure, it is shown here that EN 1990 develops these principles into operational procedures for design verification across a wide range of structures, but with specific reference to building structures. Such extension consists of providing procedures for treatment of the design situations identified in ISO 2394.

It is concluded that Eurocode EN 1990-2002 represents a significant advance in presenting rational structural design procedures, forming the basis for increased international harmonisation and internal unification of the various components of structural design for a wide range of structures, design situations, a justifiably extended range of actions to be considered, and the range of structural materials, including geotechnical design. The options of procedures and parameters clearly identified as Nationally Determined Parameters allow for harmonised reference to EN 1990 in the revision of SANS 10160.

6.2 Part 2, 7 & 8 General Actions: Self-weight and Imposed Loads; Thermal Actions; Actions during Execution

6.2.1 Chapter 2-1 Review of provisions for general actions in SANS 10160

Retief JV & Dunaiski PE

ABSTRACT

The general actions which are reviewed in this chapter include the provisions for self-weight and imposed loads, representing a basic set of actions to be considered for the design of buildings and similar structures. The revision of the provisions from SABS 1060:1989 is based on Eurocode EN 1991-1-1:2002, and includes additional classes of imposed loads from that standard. Additional general actions which are introduced from Eurocode include provisions for thermal actions from EN 1991-1-5:2004 and actions during the execution of the structure which is based on EN 1991-1-6:2005. Although accidental actions should also be considered to be a general action, only the basis of design for the accidental design situations were taken from EN 1991-1-7:2006 and incorporated into SANS 10160-1, as considered in Chapter 1-2.

The revision of SABS 0160:1989 afforded the opportunity to assess the provisions for imposed loads on building structures in terms of the scope of loads which are provided, together with the classification of occupancies; and finally the specified minimum values to be applied. The assessment of minimum imposed loads is based on a comparison with other standards, with the emphasis on generally accepted international standards.

The standards selected for this purpose are firstly Eurocode ENV 1991-2-1:1995 Part 2-1 General Actions: Densities, self-weight and imposed loads; ASCE-7:95 Minimum Design Loads for Buildings and Other Structures; BS 6399 Part 1:1996 Dead and Imposed Loads; Part 3:1988 Imposed Roof Loads; AS 1170.1-1989 Part 1: Dead and live loads and load combinations . The formulation of the provisions is based on EN 1991-1-1, in accordance with the general use of Eurocode as reference standard for the revised standard SANS 10160. The assessment of imposed loads clearly indicated that the SABS 0160 values are systematically lower than that those from the comparative standards. The specified values were accordingly adjusted by referring to the values from ASCE-7 and EN 1991-1-1 recommended values. The general layout and formulation of EN 1991-1-1 was followed in compiling SANS 10160-2.

The classification system for occupancies was applied in particular, but a more compact and convenient presentation was used. The more detailed and refined classification used in BS EN 1991-1-1 National Annex was also considered. The scope of imposed loads was extended to include imposed loads due to forklifts in industrial buildings and helicopter landing pads on building roofs. Specification of imposed roof loads was modified by differentiating between persistent conditions and the execution stage.

Thermal actions: Extracted only provisions from EN 1991-1-5 for buildings (omitted lengthy section on bridges) and based characteristic maximum and minimum environmental temperatures on TMH7 maps in SANS 10160-7.

Actions during execution: This is an important stage in the life of a structure, apparently representing the conditions where most of structural failures occur. Although the inclusion of provisions for this transient situation does not imply that the design engineer is responsible for a stage over which he may not have any control, the responsibility for structural performance needs to be properly assigned contractually. Only the provisions from EN 1991-1-6 relevant to buildings were retained in SANS 10160-8.

6.3 Part 3 Wind Actions

6.3.1 Chapter 3-1 Review of codification of wind loading for structural design

Goliger AM, Retief JV & Dunaiski PE

ABSTRACT

This chapter presents the results of a review of developments in standardised procedures for the determination of wind actions in the design of structures in order to establish the general basis for the formulation of new provisions for wind actions in SANS 10160. A review is made of options for the development of revised provisions with reference to alternative international standards, with particular attention given to Eurocode EN 1991 *Actions on structures* Part EN 1991-1-4:2005 *General actions – Wind actions*.

An assessment of the implications of applying EN 1991-1-4:2005 procedures to South African conditions and practice is then presented, through a comparison with existing SABS 0160-1989 procedures. The main issues were the representation of the South African thunderstorm climate of strong winds and a general increase in wind actions which would result from the application of Eurocode procedures.

Guidelines are presented for the development of revised SANS 10160 procedures for wind actions derived from EN 1991-1-4:2005 stipulations; but adapted to the South African strong wind climate and present levels of performance and reliability.

6.3.2 Chapter 3-2 Revised wind-loading design procedures

Goliger AM, Retief JV, Dunaiski PE & Kruger AC

ABSTRACT

The assessment of the status of wind load provisions forms an important part of the review of the present South African Loading Code SABS 0160-1989. This chapter presents the considerations taken into account in the formulation of the proposed procedures for determining wind actions on buildings and similar structures, for incorporation into SANS 10160 Part 3.

The main features of the revised procedures for determining wind actions for structural design consist of the following: The scope of structures is limited to consider quasi-static structural response. Sufficient compatibility with the Eurocode procedures is maintained in order to apply them for situations beyond the scope of Part 3, together with the stipulated South African strong wind climate representation. The incorporation of updated and modern procedures for the conversion of the free-stream wind speed into wind loads of Part 3 based on EN 1991-1-4 is considered to be a major improvement.

6.4 Part 4 Seismic Actions and General Requirements for Buildings

6.4.1 Chapter 4-1 Background to the development of procedures for seismic design

Wium JA

ABSTRACT

This chapter provides the background to the clauses in Part 4 of the revised Code. It describes the procedures which were followed during the revision of the sections on seismic loading and gives the motivation for specific issues considered during the revision.

The Code distinguishes between areas of natural and mining-induced seismicity. The choice of an appropriate nominal peak ground acceleration value for areas of natural seismicity is described as well as the motivation for the choice of seismic zones.

The most relevant changes from the existing Code are the adoption of revised response spectra for South Africa, the revision of load factors in the ultimate limit state calculations, and provision of detailing requirements for reinforced concrete elements.

It is believed that a correct structural concept and correct detailing will go a long way in providing earthquake resistant structures in South Africa. The provisions in the revised code aim to achieve such a target.

The revised Code now provides the limitations of the equivalent lateral static force method used in the Code and refers the designer to specialist literature for more advanced methods.

In the revision of the code reference was made to other existing codes. South African construction practice and materials however require that certain items be investigated for local conditions. A list of items to be investigated and further researched is included.

The South African industry has had limited exposure to the design of building structures for seismic loading. It is important that the revised code and the correct design and construction procedures be adopted for all designs in regions of defined seismicity. To this effect, a program of awareness and training is required for the local profession.

6.5 Part 5 Basis for Geotechnical Design and Actions

6.5.1 Chapter 5-1 The development of procedures for geotechnical design

Day PW & Retief JV

ABSTRACT

No provision is made in SABS 0160 for geotechnical actions, let alone geotechnical basis of design. Current practice is for geotechnical design to be carried out using working load design methods even where the structural aspects of the project are designed using limit states design. Part 5 of the code essentially provides the first basis for geotechnical design using limit states design methods in South Africa.

In the absence of a South African geotechnical design code, this chapter sets out the basis of geotechnical design using the partial factor limit states design approach. It provides for the classification of geotechnical actions and the determination of geotechnical and geometric data. Design verification requirements for the ultimate and serviceability requirements are given. The GEO limit state is introduced to handle situations where failure of the ground is likely to govern the design. Factors to be considered in the determination of geotechnical actions are provided.

Informative annexes are provided dealing with engineering requirements for various categories of engineering projects, guidance on the selection of partial material and resistance factors and basic information on the design of foundations and determination of earth pressures. These annexes will probably be withdrawn once a South African geotechnical design code is produced.

One of the main objectives of this Part of the code is to achieve sufficient compatibility with Eurocode to enable the application of EN1997-1 in conjunction with SANS 10160.

6.5 Part 6 Actions Induced by Cranes and Machinery

6.5.1 Chapter 6-1 Investigations into crane load models for codified design

Dymond JS & Dunaiski PE

ABSTRACT

The provision for crane induced actions is one of the most important provisions for industrial structures in a code for loadings on buildings. A comparison between the crane load models in the South African code of practice for loadings on buildings (SABS 0160:1989) and various international codes showed that the SABS 0160:1989 crane load models are over-simplified. For this reason, the revision of the crane load models in the South African loading code was initiated.

The crane load models in the Eurocode EN 1991-3 were identified as the most appropriate models to replace the present provisions. These load models have been investigated in order to determine the load situations allowed for as well as the underlying crane mechanics.

The basis for updating the crane imposed loading provisions for inclusion into SANS 10160 has been developed. Issues to be considered are the selection of the load models based on their exhaustiveness, validity, accuracy and applicability to South African conditions, crane load combinations, provisions for crane support structure fatigue, development of simplified load models for a limited range of simple crane installations and an appropriate, reliability based, crane partial load factor.

6.5.2 Chapter 6-2 Revised provisions for crane induced action

Dunaiski PE & Dymond JS

ABSTRACT

The crane load models in SABS 0160:1989 have been shown to be over-simplistic compared to crane load models in international codes of practice and are therefore being updated for inclusion into the forthcoming revision of the loading code SANS 10160. The crane load models in the Eurocode EN 1991-3 have been identified as the most appropriate models to replace the present provisions.

The verification of the EN 1991-3 load models for actions induced by overhead travelling cranes for inclusion in SANS 10160 has been carried out on the basis of their validity, applicability and exhaustiveness. With the exception of fatigue, the load models from EN 1991-3 were incorporated into SANS 10160 with only minor modifications to adapt them for South African conditions and practices. Two additional load models were included into SANS 10160, viz. misalignment of the crane wheels or rails, and storm wind on open gantries.

The implications of the proposed load models for the designer of crane support structures have been assessed with reference to the additional crane information required as well as the calculation effort involved in the determination of the crane induced actions and subsequent design of the support structure.

6.5.3 Chapter 6-3 Reliability assessment of crane induced actions

Dymond JS, Retief JV & Dunaiski PE

ABSTRACT

The partial load factor applied to loads imposed by overhead travelling cranes in the South African loading code (SABS 0160:1989) is the imposed load factor of 1,6 which has been calibrated for floor loads in office buildings. In contrast, the Eurocode specifies a factor equal to the permanent load factor of 1,35. There is no reliability basis for these crane partial load factors.

A reliability based code calibration has been performed on crane imposed loads for the revision of the South African loading code (SANS 10160). The objective of the calibration was to determine partial load factors for crane loads which result in exceeding a specified target reliability and improve its consistency over a range of choices of materials, loading conditions and structural configurations.

The stochastic modelling of loads lifted by cranes as well as modelling uncertainties for crane imposed loads was developed. Different code formats were investigated to take into account the difference in statistical properties of variables influencing crane loads. Optimal crane partial load factors were determined for the ultimate limit state across a representative range of crane and support structure configurations and classes. The results show that more elaborate code formats are required to improve the consistency of reliability for design crane induced loads.

7 CONCLUSIONS

The revision of SABS 0160:1989 into SANS 10160 took much longer than initially intended. However, advances made also went beyond initial intentions, consisting mainly of the following:

- **SABS 0160 as SA Loading Code:** The essential properties of SABS 0160 as the South African Loading Code for the design of buildings and similar industrial structures have been maintained. This includes maintaining the present levels of structural performance, although allowance is made to improve its consistency. An important consequence is that consistency of SANS 10160 with the present materials-based design standards has been maintained.
- **Specified procedures:** Provision for actions that were specified in SABS 0160 were substantially overhauled and updated. Important additions that have been made to SANS 10160 include a number of design situations and actions, the formal treatment of the basis of structural design, and a unified treatment of geotechnical design, and its interface with structural design.
- **Eurocode and international harmonisation:** Eurocode was used as reference to SANS 10160, and thereby provided access to an extensive source of information, structural engineering technology and experience. Most important however, is the extensive scope of structures, conditions, materials and practices across which Eurocode succeeded in achieving consistency and unification. Much effort was spent to marry the comprehensive nature of Eurocode with the specific needs of a South African Loading Code. From a South African perspective, a degree of international harmonisation has been achieved that went well beyond initial expectations. From a Eurocode perspective, SANS 10160 represents one of the first applications of Eurocode principles beyond its group of Member States. This opens the door to future cooperation and the use of Parts of Eurocode in situations where there is no equivalent South African code. It also contributes concretely to the extended application of Eurocode as reference to other South African structural design standards.

8 REFERENCES

CE (2002). Commision Européenne: *Guidance Paper L : Application and use of Eurocodes*. http://europa.eu.int/comm/enterprise/construction/index.htm .

Day PW & Kemp AR (1999). South African loading code : Past, present and future. Civil Engineering, April 1999, pp 11-12.

Ellingwood B, Galambos TV, MacGregor JG and Cornell CA (1980). *Development of a probability-based load criterion for American National Standard A58*. NBS SP 577, US Department of Commerce, Washington.

IABSE (1996). Colloquium: *Background and application of Eurocode 1*. Delft.

Implementation of Eurocodes (2005). Leonardo da Vinci Pilot Project: *Development of skills facilitating implementation of Eurocodes*. Handbook 1 *Basis of structural design*; Handbook 2 *Reliability backgrounds*; Handbook 3 *Action effects for buildings*; Handbook 4 *Guide to the basis of bridge design related to Eurocode*; Handbook 5 *Design of buildings for the fire situation*.

JCSS (1996). Joint Committee on Structural Safety. *Background documentation: Eurocode 1 (ENV 1991) Part 1 Basis of design*. Report N° 94, ECCS, Brussels. ISBN 92-9147-0022.

Kemp AR (1998) *Loading codes: Background and introduction*, SA National Conference on Loading. SAICE & SAISC, 9 – 10 September 1998, Midrand, South Africa.

Kemp AR, Milford RV, Laurie JAP (1987). *Proposals for a comprehensive limit states formulation for South African structural codes*. The Civil Engineer in South Africa, September 1987, V 29, No 9, pp 351-360.

Milford, RV (1988). *Target safety and SABS 0160 load factors*. The Civil Engineer in South Africa. 30(10) 475 – 481.

SANS 1-1:2003. SOUTH AFRICAN NATIONAL STANDARD: Standard for standards Part 1: The development of national standards and other normative documents. SABS, Pretoria.

SANS 2394 :2004, ISO 2394:1989. SOUTH AFRICAN NATIONAL STANDARD *General principles on reliability for structures* SABS, Pretoria.

Watermeyer R (2008). *Report on Structural Eurocodes Summit: 8 February 2008*. SAICE – IstructE Joint Structural Division.

1-2 The Limit States Basis of Structural Design for SANS 10160-1

Retief JV & Dunaiski PE

1 INTRODUCTION

The concept of standardising *a common basis for defining design rules relevant to the construction and use of the wide majority of buildings and civil engineering works, whatever the nature or combination of the materials used* is formalised with the presentation of the International Standard ISO 2394 *General principles on the reliability for structures* (SANS 2004). Such general presentation of the principles of structural design is taken to the level of operational limit states design procedures as a head Eurocode Standard EN 1990 *Basis of structural design* (2002). This head standard formulates and specifies the procedures to be followed and applied in the standards specifying the actions (EN 1991) and materials-based resistance of structures (EN 1992 – EN 1999).

This Eurocode scheme is followed in the layout of the revised South African Loading Code, the Standard SANS 10160 *Basis of structural design and actions for buildings and industrial structures*. The general principles of structural reliability and their application through limit states design is specified in a separate head Part as SANS 10160 Part 1 *Basis of structural design* (abbreviated as SANS 10160-1). A similar arrangement is followed in the present SA Loading Code SABS 0160-1989 *The general procedures and loadings to be adopted in the design of buildings* (renumbered as SANS 10160-1989; referred to here as SABS 0160 for brevity) through the initial chapters on general design considerations and guidance on limit states design loads. The treatment of the basis of structural design as a separate Part of SANS 10160 is done to emphasise its applicability not only to actions, but also to structural design in general and therefore to the related materials-based design standards.

Justification for treating the general principles of limit states design in a separate Part of SANS 10160 is to allow for its extension in scope as compared to SABS 0160. In SANS 10160-1 provision is made for a more elaborate reliability framework in terms of design situations, differentiated limit states for both the generic ultimate and serviceability limit states, the wider diversity of the extended range of actions which are stipulated, and extension of materials provided for, with specific reference to geotechnical design.

The comprehensive range of structures provided for in Eurocode to include *buildings and civil engineering* structures necessitated the separate treatment of the basis of design in EN 1990 *Basis of structural design*. Such explicit treatment of provisions for the reliability performance of structures was also required to provide a rational basis for harmonisation of the design practice amongst Member States and unification of provisions for actions on structures and resistance for the various structural materials. Alignment of SANS 10160-1 with EN 1990 will therefore facilitate the future extension of the South African standard. The effective way in which alignment with Eurocode can be maintained is demonstrated by such reference to EN 1990 in the formulation of Part 1.

The rationale for the basis of structural design in terms of reliability based limit states design as implemented in SANS 10160-1 is presented in this Chapter. The objective with the presentation is to provide the necessary background in order to judge the acceptability and sufficiency of the stipulations. Such judgement also applies to the reliability features of the actions stipulated in Parts 2 – 8 of SANS 10160. Since Part 1 will also apply to the materials-based design standards according to which structural resistance has to be verified, consistency with these standards also needs to taken into account.

The basis of structural design also provides the underlying principles on which harmonisation with international practice is achieved. The stipulations of Part 1 are therefore also assessed from the perspective of its contribution to international harmonisation of structural design practice.

2 REFERENCE FOR REVISED BASIS OF STRUCTURAL DESIGN

In the formulation of the basis for structural design as presented in SANS 10160-1, Eurocode EN 1990 served as the primary reference. The way in which the present practice, as stipulated in SABS 0160, is taken into account as well as reference to the high level ISO Standard ISO 2394:1998, requires the establishment of certain principles and the setting of objectives for such a process.

2.1 General Objectives with Formulation of Basis of Design

The general objectives with the revision of the SABS 0160-1989 procedures in terms of partial factor based Limit States Design (pfLSD) that have been considered include the following:

- **Updating reliability based procedures:** Considering the more extensive development of reliability based limit states design as standardised in ISO 2394-1989 and applied in EN 1990-2002
- **Improved performance:** Utilise the potential of pfLSD in improving the performance of structures in terms of safety (achieving sufficient reliability) and economy (removing unnecessary/inefficient conservatism)
- **Reliability framework:** Taking account of the extended reliability framework presented in the basis of design, as derived from EN 1990
- **Consistency of reliability:** Therefore achieving the general objective of improving the consistency of reliability across the range of design situations within the scope of the revised standard
- **Array of actions:** Providing for the extended array of actions that can be specified due to the clear formulation of design situations within the reliability framework; Considering the particulars of the respective actions for which stipulations are provided
- **Structural resistance:** Strengthening the provisions and requirements for structural resistance in terms of the application of action combination schemes for the array of structural materials and their failure modes
- **Geotechnical design:** Considering the specific and unique requirements of geotechnical design and the treatment of geotechnical actions, to be consistent with the pfLSD procedures used in the draft standard

2.1.1 Common basis of design

Structural design standards primarily apply a reductionist approach to the design process by treating it in terms of its constituent parts, consisting of the structural elements, the respective loads or actions, selected levels of structural performance, considering different failure modes, which are specified in separate standards for the respective structural materials. Although the constituent parts are integrated by the processes of the conceptual design of the structural system and the determination of load effects through structural analysis, systems considerations can only be specified at a general level.

The various constituent parts of the structural design process represent a diverse range of influences on the performance of the resulting structure. This diversity derives from the differences in variability and uncertainty in design methods and parameters such as the mechanical models, values of loads and material properties, and how the contribution of these parameters are influenced by the geometry of the structure. The sensitivity of the structure to the variability of the parameters represents another class of influence.

Representation of all the conditions that the structure may be exposed to during its life in standardised procedures is a major challenge in the formulation of structural design standards. The implementation of a clearly formulated and specified common basis of design to provide for the diversity of requirements, conditions and elements of the design process is an important instrument to achieve effective design standards.

2.1.2 Unification of steps of design process

The specification of the basis of structural design provides an overarching scheme for achieving the objective for the efficient design of safe, functional and economic structures. The basis of structural design consequently unifies the different steps in the design of a structure with their diverse influences and situations on structural performance as follows:

- **Design steps:** Unification applies to individual steps of the design process such as the different loads resulting from the self-weight, occupancy (imposed loads) or exposure to the environment (wind) or even extreme conditions such as the effects of earthquakes. Similar unification applies to the various failure modes such as axial, flexural, torsion and shear resistance of structural elements, the stability effects of slender elements or subsystems. Combined effects of loads or section forces and failure modes extend the level of complexity, and consequently the need for a consistent basis for their treatment.
- **Design process:** The basis of design also provides unification between consecutive steps of the design process, particularly between the treatment of actions on the structure and resistance, which are conventionally specified in separate standards. Although the separation of actions and structural resistance is made for practical reasons, it is clear that the different steps are integrally related, and are treated as such in the design process. An explicit and rational formulation of the basis of design not only improves the consistency between the specification of actions and resistance in the standards, but also facilitates efficient designs.
- **Structures from different materials:** Unification of the provision for actions and for a number of structural materials standards then ensures consistency in the case where different materials are applied in a single structure. The obvious example is the need for consistency in the treatment of the constructed structure and its foundations. The present fragmented treatment of structural loads on foundations and geotechnical actions on structures leads arguably to significant inefficiencies, and certainly to complications in achieving optimised designs. Harmonisation between the respective materials standards is achieved through their sharing of a common basis of design.

2.2 Reference Sources

The background and motivation for selecting Eurocode as reference to the revision of SANS 10160 is presented in Chapter 1-1, as based on a general assessment of Eurocode (Chapter 1-3) and specifically its reliability basis (Chapter 1-4).

Reference to EN 1990 played an important role in the decision to apply Eurocode as reference to the revision of SABS 0160-1989. Incompatibility of SABS 0160 action combination schemes with those of Eurocode ENV 1991-1:1994 procedures were identified as one of the critical stumbling blocks to applying Eurocode in the development of structural design standards for South Africa (Day 1996, 1997). The realisation that the SABS 0160-1989 action combination scheme can be interpreted as one of the allowable options as Nationally Determined Parameter (NDP) of EN 1990:2002 (SAKO 1999) removed this obstacle. Further investigation confirmed the wide-ranging advantages of such a process (see for example Holický and Retief 2005).

Although EN 1990 served as the primary Eurocode reference, other Standards and Parts also played a role. The general basis of design procedures for accidental actions as presented in EN 1990 is developed further in EN 1991-1-7 *Genera actions – Accidental actions*. Similarly the procedures for design for earthquake resistance as a special case of accidental actions as presented in EN 1998-1 *Design of structures for earthquake resistance – General rules, seismic actions and rules for buildings* were taken into account. The accommodation of geotechnical design from EN 1997-1 *Geotechnical design – General rules* into the basis of design required specific reflection.

Useful background information on the Eurocode *basis of structural design* is given by the JCSS (1996) for the previous version ENV 1991-1:1994 and for EN 1990:2002 by Gulvanessian *et al* (2002).

2.2.1 *General guidelines for the use of principal references*

The (default) use of SABS 0160-1989 as point of departure and the selection of EN 1990-2002 as reference source for its updating is done in the following manner:

- **SABS 0160-1989:** The existing standard is used as basis for the reference level of reliability. The main features of existing practice that are maintained are the following:
  - **Scope of application:** The limited scope of application to buildings and similar industrial structures limits the range of conditions and the complexity of situations to be provided for within the specified reliability procedures;
  - **Reliability level:** Target levels of reliability for the reference situations are generally applied as point of departure for reliability modelling and calibration and even where judgement based decisions have to be made;
  - **Combined reliability effects:** The general approach of basing the combination of the effects of uncertainty on the Turkstra rule [$S_{max} = \max\{S(Q_{1,max}; Q_2); S(Q_1; Q_{2,max})\}$ (see ISO 2394:1998)], implying not only point-in-time values for combined variable actions, but also for the dual scheme for the combination of permanent and variable actions;
  - **Simplified procedures:** Although the reliability framework of the draft standard is extended in accordance with EN 1990-2002 as indicated below, it is simplified as much as possible, only to provide for the limited scope of structures for buildings and similar industrial structures.

- **EN 1990-2002:** This Eurocode standard provides the basis for achieving consistency of reliability across an extensive range of design situations. It also serves the purpose of achieving the following objectives in which reliability requirements play a central role:
  - **International harmonisation:** Since EN 1990-2002 is consistent with ISO 2394-1998 which is accepted internationally as providing the basis for the reliability specification for structural performance, similar harmonisation will be achieved by SANS 10160 Part 1 if it is kept consistent with EN 1990-2002.
  - **Harmonised practice:** In addition to the harmonisation of structural design practice amongst the large number of Member States, which was the primary objective of the development of Eurocode, it also provides significant momentum to the process of international harmonisation.
  - **Unified practice:** Procedures for the treatment of actions on structures and structural resistance for the various structural materials and their combinations result in unified standards with an extended degree of consistency for the respective components of the design process, not only in terms of actions and structural resistance, but at least in principle also among the alternative structural materials as diverse as steel and earth!
  - **Reliability framework:** This could only be achieved in terms of a common reliability framework according to which diverse situations of structural configurations, materials, limit states, design situations, required performance levels and other considerations can be treated at clearly defined and consistent levels of reliability.

2.2.2 *The SABS 0160-1989 limit states procedures and reliability basis*

A concise review of the development and main features of SABS 0160-1989 was given by Kemp (1998) in the introductory lecture of the 1998 South African National Conference on Loading. An important requirement for the development was that the standard would *apply uniformly to all materials*. The basic philosophy applied in the formulation of the standard is elaborated by Milford (1998). Whereas the previous version SABS 0160-1976 was based to a large extent on British codes, substantial calibration of partial factors were applied to derive the 1989 version.

The good reliability performance of SABS 0160-1989 was confirmed by a comparison with Eurocode and ANSI/ASCE-7 by Kemp *et al* (1998). The general agreement between SABS 0160-1989 and ANSI/ASCE-7 was also shown. Eurocode, which was at the ENV stage at that time, had a reliability performance similar to the inefficient procedures used previously. As indicated above, it was the development of alterative combination schemes for Eurocode (SAKO 1999), and their acceptance as Nationally Determined Parameters (EN 1990-2002) that removed the critical incompatibility between South African and Eurocode procedures.

2.2.3 The EN 1990-2002 limit states procedures and reliability basis

The general properties of the Eurocode reliability based partial factor limit states design procedures, as stipulated in EN 1990-2002, are discussed extensively in the accompanying Chapter 1-4 and explained by Gulvanessian *et al* (2001). The most important innovations of EN 1990-2002 relevant to the revision of SANS 10160 include:

- **Reliability levels as national responsibility:** The provision of Nationally Determined Parameters (NDP) allow for setting levels of reliability nationally.
  - The procedures are however clearly developed around a default reference level of reliability expressed in terms of the target reliability index value of β_{target} = 3,8!
- **Limit States extended in terms of Design Situations:** The identification of different conditions applying to the general Ultimate and Serviceability Limit States in terms of Design Situations resulted in a number of developments related to actions to be considered, how they are treated and provisions for structural resistance:
  - **Transient Design Situation:** Provision is formally made for conditions during the execution (construction, fabrication, erection) or maintenance of a structure, resulting in the Eurocode Part EN 1991-1-6 *General actions – Actions during execution.*
  - **Accidental Design Situation:** The special case of accidental situations, which are not expected during the design life of the structure and which can not be provided for economically, is treated as an accidental design situation. A number of cases can be identified as indicated below, with an extensive array of associated Eurocode Parts:
- **Accidental actions and design situations:**
  - **Structural integrity and robustness:** Guidance is given on treatment of general requirements of structural integrity and robustness in terms of 'unidentified' accidental situations
  - **Special accidental actions:** Provision is made for accidental actions related to the specific function or conditions of the structure, such as actions due to impact or internal explosions (but not from explosives).
  - **General accidental actions:** The general principles for treatment of accidental design situations apply to and are elaborated for the effects of:
    - Structural fire design: Provide for actions and materials-based resistance.
    - Seismic design: Although seismic effects are of an accidental nature, their treatment can be selected by Member States subjected to high seismicity to be represented by the sustained design situation.
- **Serviceability Design Situations:** Differentiated provisions for the general Serviceability Limit State are applied, consisting of irreversible, reversible and long-term static effects, in addition to dynamic effects.
- **Geotechnical design:** In addition to the advancement achieved with the development of EN 1997 *Geotechnical Design*, the unification between structural and geotechnical design practice through a common basis of design represents a significant advancement.
- **Combination schemes for actions and other reliability related processes:** In addition to the respective action combination schemes related to the differentiated Limit States and Design Situations, alternative options are provided to be selected as NDP. These include:
  - **Structural actions:** Single or dual expressions for ULS action combinations are allowed to be selected as NDP, including the selection of partial factors. No option is recommended. This results in a dual expression similar to that of SABS 0160-1989 (Holický & Retief 2005).

- **Geotechnical/structural actions:** Three alternative Design Approaches to provide for the combined application of geotechnical and structural actions are presented, to be selected as NDP, again without a recommended option.

An important feature of EN 1990 is that several alternative schemes are presented, giving widely different reliability performance results. These NDP options are then provided without any recommendation being made. The implication is that EN 1990 is not an operational standard, but requires options and values to be selected by Member States before it can be implemented in practical design.

Although the extensive body of information on the conversion of Eurocode into operational national standards by Member States which is presently emerging was not taken into account systematically in the formulation of SANS 10160-1 and other Parts, it is interesting to notice some trends. Various reference levels of reliability are selected, in accordance with existing national practice; where alternative NDP options have been provided in Eurocode, all options are adopted, sometimes indicating under which conditions specific options should be applied, and sometimes leaving such decision to the designer.

3 PRINCIPLES OF LIMIT STATES DESIGN

Since partial factor limit states design is derived from the theory of structural reliability (see Chapter 1-4), the relationship between reliability requirements and the limit states procedures need to be established. This relationship includes the degree of simplification or complexity allowed in the (approximate) limit states procedures. The complexity of the design scheme includes both the differentiated reliability framework of situations and the specification of basic variables and partial factors for each situation.

It is also recognised that structural reliability models cannot capture the effects of gross (human) error, which is an important source of structural failure, if not the dominant one. The application of Quality Management programs provides an important measure against gross error. The interrelationship between Quality Management and structural reliability, and the consequent effect on limit states procedures therefore also needs to be considered.

3.1 Function of Design Standard

A fundamental difference between EN 1990 and SANS 10160-1 derives from the statutory and regulatory function of the respective standards: Eurocode provides for the regulatory requirements of the Member States, whereas South African structural design standards represent a code of acceptable practice as set by the profession and the related industry in order to assist designers in the discharge of their professional responsibilities.

EN 1990 consequently takes the approach of firstly setting high level Principle requirements *for which there is no alternative* (Clause 1.4(2)) requiring design and execution such that the structure *will, during its intended life, with appropriate degrees of reliability and in an economical way: sustain all actions and influences likely to occur during execution and use, and remain fit for the use for which it is required* (Clause 2.1(1)P) and *that it will not be damaged by events such as: explosion, impact, and the consequences of human errors, to an extent disproportionate to the original cause* (Clause 2.1(4)P). Although the bulk of EN 1990 (and Eurocode) then provides *Application Rules* which *are generally recognised rules which comply with the Principles and satisfy their requirements* (Clause 1.4(4)), there is still some concern that the designer is legally required to prove compliance with the Principle rather than the Application Rule in the case of any dispute.

In order to ensure that SANS 10160 is legally interpreted in terms of the design rules, rather than in terms of such high level principles and requirements, these clauses from EN 1990 were excluded from the normative sections of SANS 10160-1. However in order to maintain consistency with EN 1990 the principles for structural performance and robustness against abnormal events for which the normative procedures are deemed to provide compliance are provided in the informative (non-normative) Introduction.

3.2 Provision for Accidental Design Situation and Structural Robustness

Provision for various classes of accidental design situations in a consistent and unified manner represents one of the major advances achieved by Eurocode. The provisions are presented in general terms in EN 1990, with provisions for accidental actions provided in EN 1991-1-7 *General actions – Accidental actions*. Structural fire design is covered in the series of Parts EN 199X-1-2 and seismic design in EN 1998 as important classes of accidental actions. Provisions for specific accidental actions are also included in Parts of EN 1991.

The general strategies to follow and the treatment of accidental design situations and actions are presented in EN 1991-1-7 (Vrouwenvelder *et al* 2005). Provisions are also given for design against impact against structures, mainly from vehicles, and internal gas explosions. It was decided that the accidental actions due to impact and internal explosions are of a rather specialised nature that does not warrant inclusion in SANS 10160. However, the general treatment of the accidental design situation was considered to logically form part of the basis of structural design and was consequently incorporated into SANS 10160-1.

3.2.1 General requirements for accidental actions

Accidental design situations and the associated actions represent conditions which are not expected during the design life of the structure, but with such large consequences that some provision for their occurrence and effects needs to be made. It is therefore required that although damage to the structure is accepted, it will not be damaged to an extent disproportionate to the original cause of the abnormal event.

The level of reliability for accidental design situations is set by the specification of the level of the accidental action applied in the design. The other actions are taken at their nominally unfactored expected values. Resistance is also intended to be taken at a nominal value, provided that sufficient ductility is maintained.

3.2.2 Requirements for structural integrity and robustness

The requirements for structural integrity and robustness are defined in EN 1991-1-7 as a special case of an accidental design situation, sharing the common requirement of preventing disproportionate failure in the case of local failure in the structure. Robustness requirements are consequently formulated in terms of accidental design situations caused by unidentified accidental actions. Although SABS 0160 nominally identified the requirement to provide for structural integrity, the EN 1991-1-7 procedures afforded the opportunity to include the more elaborate and explicit procedures in SANS 10160-1.

The procedures consist essentially of the identification of key structural elements which could cause unacceptable levels of failure if they are lost by some unspecified event. Design measures include the application of notional accidental loads, design to a higher level of reliability, sufficient provision of tying strength to activate and utilise alternative load paths.

The general strategies are presented in the normative sections of SANS 10160-1, whilst quantitative procedures are provided in the informative Annex B. This annex also includes a system of categorisation of consequence classes for buildings. Four levels of consequence classes are used, as opposed to the practice followed in EN 1991-1-7 which uses three levels, with a lower and upper sub-class for the mid level. The four level classification system is used consistently in SANS 10160.

3.2.3 Identified accidental actions

In addition to the provision for earthquake resistance and seismic actions in Part 4 which can be classified as an identified accidental action, other Parts of SANS 10160 may also stipulate provision for specific actions, such as certain classes of impact on crane support structures which are given in Part 6. A separate class of identified accidental action is that which is selected for specific individual structures and facilities due to special conditions which may result from either hazardous conditions or (and) the importance of the structure and consequences of failure. For these cases the general principles should be applied as basis for the design requirements.

The specified value of the accidental action needs to be established in terms of risk optimisation which considers the risk resulting from the probability of the occurrence of the accidental action and the resulting consequences. This risk should then be compared to accepted risk criteria. It is obvious that the risk should not exceed that which is implicit in the safety design of the structure, with probabilities associated with the reliability levels and the (undefined) consequences of failure. These values have been established for specified procedures for accidental actions, such as in SANS 10160-4 for seismic actions. EN 1991-1-7 Annex B (Informative) *Information on risk assessment* provides a general procedure for risk assessment for cases where the level of accidental action needs to be specified. An ISO Standard for risk assessment is also presently under development as ISO/DIS 13824 *General principles on risk assessment of systems involving structures* (ISO TC98 2008)

3.3 Fundamental Reliability Considerations

The provision for reliability performance of limit states design schemes at the basic level is related to the selection (calibration) of values for partial factors. Fundamental reliability considerations are however also used to establish the scheme of partial factors (design scheme) and specific design situations (limit states). The following reliability based considerations therefore need to be taken into account in applying the limit states design procedures of EN 1990 to the revised provisions of SANS 10160-1:

- **Reliability framework:** In order to provide for the extensive scope of application of Eurocode, an extensive reliability framework of design situations with associated differentiated limit states are identified in EN 1990. In addition provision is made to differentiate in terms of reliability classes for structures with associated adjustments in the design procedures. On the one hand this extended framework represents an advance in reliability based limit states design, to be incorporated in SANS 10160-1. On the other hand the scheme needs to be restricted to delineated scope of application of the SA standard.
- **Complexity of design scheme:** The consistency of reliability is improved by the application of a more elaborate and complex design scheme in terms of the formulation of the design functions, sets of partial factors and specification of the basic variables. Again it is necessary to establish how appropriate the elaborate schemes employed in EN 1990 are to SANS 10160-1.
- **Level of reliability:** There is a clear difference in the explicit and implicit levels of reliability as applied in the two respective standards. Differences between economic conditions and technological advances should be accounted for.

In principle appropriate levels of reliability and design schemes need to be established for the complete reliability framework of design situations and the associated differentiated limit states, and the design schemes formulated accordingly, including the calibration of partial factors. The pragmatic process however is to base the scheme of limit states and design schemes on experience based judgement, and to perform calibration for representative situations. A similar approach was taken on adjusting the procedures of EN 1990 to the scope and reliability practice for SANS 10160-1, as carried over from SABS 0160.

3.4 Reliability Requirements and Quality Management

Although only nominal specification is provided for the application of quality management (QM) through the various stages of the process, this matter is closely related to the reliability basis of the design process. Quality management addresses the issue of gross error or human error, which is a major weak point in the application of reliability theory to design procedures. The interrelation between structural reliability and quality management should therefore be considered in terms of both the directions of their relation:

- **QM control of gross error to endorse reliability measures:** QM procedures should be applied to control the (spurious) occurrence of gross or human error, since such a process can not be modelled sufficiently. Provision for sufficient robustness of the structure should be made to limit the extent of failure in the case where the effects of gross error lead to (local) failure.
- **Identification of critical reliability performance issues for QM:** The critical issues needing the attention of QM procedures should be derived and identified from the reliability assessment of the design process. QM should then be applied accordingly in the subsequent execution of the structure.

3.5 Reference Level of Reliability

The selection of appropriate reference levels of reliability has evolved significantly, particularly through ISO 2394 (SANS 2004), since the selection made for SABS 0160-1989 as reported by Milford (1998). This issue is of significant importance due to the deviation of the target reliability for SABS 0160 expressed in terms of the reliability index $\beta_{\text{SABS target}} = 3{,}0$ from the value used in EN 1990-2002 as reference of $\beta_{\text{Eurocode target}} = 3{,}8$. The difference implies a notional failure probability of about an order of magnitude higher for the SABS 0160 as compared to Eurocode! It should however be noted that Eurocode safety levels are considered to be a national responsibility, implicitly allowing the application of the effective level of reliability as an NDP by appropriate selection of values by Member States. There are indications that a diverse range of values is indeed being used in calibrations for the respective National Annexes to the Eurocode Standards.

A comparison between the guidelines for target levels of reliability is presented by Holicky *et al* (2007) and is reproduced here for convenience as Table 1. ISO 2394 applies a more elaborate scheme: it assigns a value of 3,1 for situations of moderate consequences of failure and cost of safety measures, increasing to a value of 3,8 if either the consequences become more severe or safety measures become less costly; with a value of 4,3 if both these conditions apply. This is as compared to the Eurocode value of 3,8 for the reference class of structures RC2.

The motivation for the application of the SABS 0160-1989 target reliability is largely based on calibration to existing acceptable practice. This is similar to ANSI-ASCE7 procedures as derived from the investigation by Ellingwood *et al* (1980), which provides the basis for that standard. The South African practice could also arguably be interpreted to conform to the ISO 2394 guidelines.

The decision to maintain the present reference level of reliability of $\beta_{\text{SABS Limit}} = 3{,}0$ is motivated on the following considerations:

- **Reasonable alignment with international practice:** Agreement with ASCE-7 procedures provides an indication that the present value is not entirely out of line with practice elsewhere. ISO 2394 also provide some mild support for this position.
- **Reasonable basis for difference with Eurocode:** The difference between European and South African practice is therefore similar to that with the USA. It could derive from differences between 'old countries' and regions where more 'new' development takes place, more than differences between advanced and developing economies. It possibly relates effectively to longer use of the structures and facilities, even if this is not explicitly reflected in the specified design life.
- **Tendency towards increased values of actions:** The general tendency of actions as specified not only in Eurocode EN 1991, but also in other standards, is to increase in magnitude. It is not clear whether this results from improved rationality of the models or from creeping conservatism. Upwards adjustment of the reference level of reliability would therefore systematically compound the effect of the creep in action values.
- **Interpretation of target value:** Whereas the Eurocode value is applied as a target to be achieved on average (although it is generally exceeded) (SAKO 1999; the South African application is considered as a constraint ($\beta_{\text{SABS Limit}} = 3{,}0$).

- **Acceptable practice:** The present value is based on acceptable practice. There is no direct evidence that the levels of reliability since the implementation of SABS 0160-1989 are insufficient.
- **Absence of motivation to increase reliability:** Present levels of reliability are therefore maintained, in the absence of motivation to change them. A change would only be justified on strong motivation, either from experience or on theoretical grounds, to do so.

Table 1 Target reliability levels (β) according to ISO 2394 and EN 1990

Relative cost of safety measures	**ISO 2394 Minimum values for β**			
	Consequences of failure			
	Small	Some	Moderate	Great
High	0	1,5 (A)	2,3	3,1 (B)
Moderate	1,3	2,3	3,1(C)	3,8 (D)
Low	2,3	3,1	3,8(D)	4,3(E)
A	for serviceability limit states $\beta = 0$ for reversible and $\beta = 1,5$ for irreversible states			
B	for fatigue limit states $\beta = 2,3$ to 3,1 depending on the possibility of inspection			
For ultimate limit states the safety classes:		C $\beta = 3,1$	D $\beta = 3,8$	E $\beta = 4,3$

Reliability Class	**EN 1990 Minimum values for β**					
	Ultimate LS		Fatigue		Serviceability LS	
Reference period	1 year	50 years	1 year	50 years	1 year	50 years
RC1	4,2	3,3				
RC2	4,7	3,8(F)		1,5 to 3,8	2,9	1,5
RC3	5,2	4,3(G)				
F	With ISO 2394 clause 4.2(b) *moderate safety costs & RC2 consequences*, but EN 1990 is more conservative; EN1990 value agrees with ISO 2394 for either *low safety cost or great consequences*					
G	The EN1990 value for RC3 agrees with ISO 2394 for *low safety cost and great consequences*					
ISO: 2,3 – 3,1	EN: 1,5 – 3,8	Fatigue:	ISO 2394 – restricted range; EN1990 – range from *serviceability LS equivalent* to *ultimate LS*			

The final argument for maintaining the SABS 0160 reference target level of reliability is that to deviate from the present accepted practice would require investigations and motivation for such a step that would be beyond the present mandate and resources of the review group. There is also no apparent indication of the need for such an investigation.

A compilation of the differentiated reliability levels derived from the various sources and relevant to SANS 10160 is presented in Table 2. Differentiation is applied in terms of the limit states, characteristics of the various failure mechanisms and the reliability classes which consider the consequences of failure.

3.6 Limit States and Design Situations

The generic ultimate and serviceability limit states are differentiated in terms of design situations to provide for appropriate reliability requirements in terms of action combination schemes; the specification of partial factors γ, including the action combination factors ψ; and criteria in the case of serviceability.

The general scheme of limit states in terms of design situations is shown schematically in Figure 1. A summary of the characteristics of the limit states as they are applied in SANS 10160-1 is given in Table 3, including related target reliability levels as they are inferred from the compilation of Table 2.

Table 2 Differentiated target reliability levels (β) derived from various sources

PERFORMANCE CLASS	β	SOURCE
Ductile, gradual modes of failure (Reference)	**3,0**	Milford (1998) SABS 0160-1989
Brittle, sudden modes of failure	4,0	
Connection details between components	4,5	
Safety class (SC) → Reference Class	3,1	ISO 2349 EN 1990
SC – *Consequences Great* or *Cost Moderate*	3,8	
SC – *Consequences Great* and *Cost Moderate*	4,3	
Fatigue – *Inspection possible*	2,3/1,5	ISO 2349 (EN 1990 – 1,5)
Fatigue – *Inspection not possible*	3,1	ISO 2349 (EN 1990 – 3,8)
Serviceability - *Irreversible*	1,5	ISO 2349, EN 1990
Serviceability - *Reversible*	0	ISO 2349

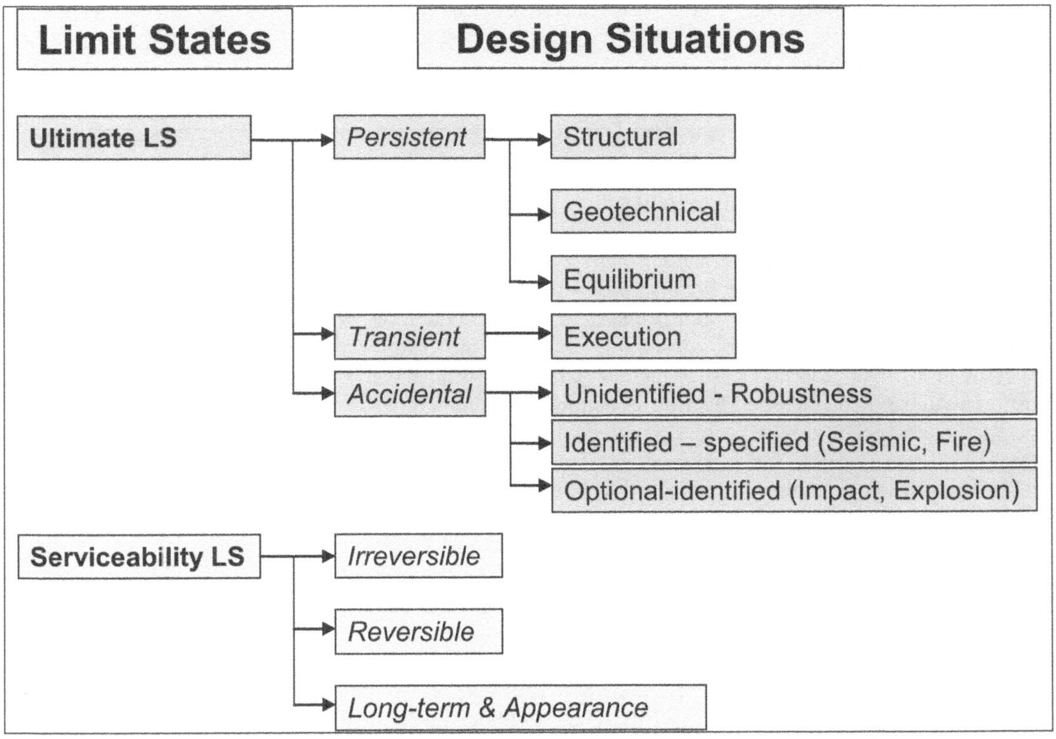

Figure 1 Limit states in terms of design situations

Table 3 Limit states and design situations and related reliability

General Limit State	Design Situation	Specific Limit State or Situation	Reliability Specification	β
Ultimate	Persistent	- Structural (STR) - Equilibrium (EQU) - Geotechnical (GEO)	Reliability Class: {RC1; **RC2**; RC3; RC4}	{2,3; **3,0**; 3,8; 4,3}
	Transient		Adjusted for reduced reference period	
	Fatigue	Inspection - Possible - Not possible	Treated as: - SLS – Irreversible - As ULS	 {1,5} {3,0}
	Accidental	General: Derived from - Function, occupation - Exposure	Based on risk assessment - Reliability implied by specification of accidental action A_d - Event not expected during design life of structure	Selected values
		Actions specified, e.g. - Imposed vehicle impact; - Internal wind pressure		
		Unidentified - Consequence Class	Robustness strategies - {CC1; **CC2**; CC3;CC4}	
		Seismic design	- No collapse limit - Damage limit	
Serviceability	Irreversible	Effect only remedied by maintenance	Increased reliability	1,5
	Reversible	- Static - Vibration	Reduced reliability	0
	Long-term	Creep; Consolidation		

3.7 Specification of basic variables

Basic variables are defined as *part of a specified set of variables representing physical quantities which characterize actions and environmental influences, geometrical quantities, and material properties including soil properties*. The specification of characteristic values of basic variables introduces a degree of conservative bias into the design process. It therefore influences the specification of appropriate partial factors (Chapter 1-4 §5.4). It should however be noted that a differentiated scheme of specification is applied, as summarized in Table 4. Generally actions and geometry are specified at average (expected, nominal) values and material or product properties at fractile values of 5/95%. It should be noted that such a differentiation cannot be applied to the values of soil properties since they are involved in both geotechnical actions on structures and geotechnical resistance.

4 DESIGN FUNCTION FOR COMBINATION OF ACTIONS

As indicated previously, the design function for the combination of permanent and variable actions represented a significant difference between the ENV version of Eurocode and SABS 0160. The reconciliation of the fundamental differences in the treatment of the combined effect of permanent and variable actions on structures is therefore important.

Although the guideline given by the SA National Conference on Loading (Kemp 1998; Kemp *et al* 1998) for maintaining the SABS 0160 combination scheme was taken as point of departure for SANS 10160-1, the need for providing for the extended scope of application, particularly with regard to geotechnical actions, had to be considered.

Table 4 Summary of defined characteristic values of basic variables

Basic Variable	General Specification	Comments
	ACTIONS	
Permanent	Nominal dimensions and mean unit masses	Expected (average) value
- large variability	- not specified	Principle not applied for limited scope of structures
Variable	Probability of 0,02 per annum	Expected (average, mean) *maximum* value over 50 years
- combination (ψQ_k)	- arbitrary-point-in-time value	Expected value at *any* time, based on Turkstra rule
Accidental	Specified for individual situations, projects	*Not expected* during design life
Fatigue	According to materials standards	Generally average fatigue actions
Geotechnical	According to EN 1997-1 *Geotechnical design – General*	*Cautious* estimate of expected values
	MATERIAL & PRODUCT PROPERTIES, GEOMETRICAL DATA	
Material (X_k) Product (R_k)	Prescribed probability, generally 5/95% fractile	
Geotechnical	According to EN 1997-1 *Geotechnical design – General*	*Cautious* estimate of expected/fractile values
Geometry (a_k)	Specified dimensions on drawings	Average (nominal) value
- imperfections	According to materials standards	Not treated in geometry terms

This requires efficient procedures for situations where permanent actions dominate, for which the SABS 0160 procedure is evidently not performing well. The possibility of improving consistency with the various options for action combination schemes provided in EN 1990 also needed to be considered.

A revision of the scheme for the combination of permanent and variable actions was consequently incorporated in the formulation of SANS 10160-1. The assessment of alternative schemes is based on structural reliability modelling.

4.1 General Methodology for Reliability Assessment of Design Function

An assessment of the reliability performance of the SABS 0160 and EN 1990 design functions is reported by Holický & Retief (2005). A different methodology as outlined by Ter Haar & Retief (2001) was used to develop and calibrate the revised scheme implemented in SANS 10160-1. The method consists firstly of the determination of a single graph which represents the global safety factor (resistance/load effect) required to achieve a target level of reliability. Compliance of alternative design functions can then be assessed without the need of recalculation of the reliability performance. This methodology, which was found to be more effective and clear, is used below to reassess the SABS 0160 and EN 1990 design functions and to present and motivate the SANS 10160-1 function. A similar method was used by Brozzetti & Sedlacek (2000) for calibration of the EN 1990 alternatives.

4.2 Reliability Requirements for Design Functions

The general procedures followed for the derivation of partial factors for Eurocode as set out in EN 1990 Annex C (Informative) *Basis for partial factor design and reliability analysis* are applied in

Chapter 1-4 of this report to derive typical load factors. A similar treatment of structural resistance is given by Holický *et al* (2007). Only a brief outline of reliability modelling is given here. The reliability performance function $g(X)$ of the limit state is expressed as a function of the basic variables of resistance (R); permanent actions (G) and variable actions (Q) as follows:

$$g(X) \;=\; R-(G+Q) \;=\; 0 \qquad (1)$$

A summary of probabilistic models for the basic variables $\{R, G, Q\}$ is presented in Table 5. The models consist of the type of distribution; the mean value as ratio to the characteristic value (μ_X/X_k) and the coefficient of variation (V_X) or standard deviation (σ_X). The models applied in the calibration of SABS 0160 as indicated are used subsequently in representative reliability analysis. Values for Q are included at various parametric characteristic levels, although the specified values are (close to) the expected maximum values according to Table 4. Additional information is presented from Holický (2002). Resistance is treated parametrically with a lognormal distributed and a representative range of values for its coefficient of variation V_R.

Table 5 Probability models for representative basic variables

Variable X	Application		Distribution	Mean k_X (μ_X/X_k)	p_k (%)	V_X	σ_X
Permanent load	SABS 0160		Normal	1,05	-	0,10	(0,10)
Imposed office floor	SABS 0160		Gumbel (Type I)	0,96	36	0,25	(0,24)
	Characteristic	95%	Gumbel (Type I)	0,71	7		0,24
		98%		0,64	3		
Wind lifetime maximum	SABS 0160		Gumbel (Type I)	0,41	0,03	0,52	(0,21)
	Holicky		Gumbel (Type I)	0,70	6	0,35	(0,25)
Variable action: Q	Parametric		Gumbel (Type I)	1	**43**	-	(0,25)
				0,75	**10**		
				0,68	**5**		
				0,60	**2**		
				0,56	**1**		
Resistance	Parametric [5% characteristic]		Lognormal	1 [1,28]	- [5]	0,10;	
						0,15	
						0,20	
						0,25	

The reliability requirement, against which design functions can be assessed, is obtained by the solution of Equation (1) using the First Order Second Moment (FOSM) method (Ang & Tang 1984). The SABS 0160 models for permanent load (G), variable load (Q) represented by imposed floor load, and a parametric resistance (R) with a coefficient of variation (V_R) of 0,15, are used as a representative case.

An inverse FOSM solution is used to obtain the target reliability of $\beta_T = 3{,}0$. From the results the dimensionless *global safety factor* (*GSF*) can be obtained as the ratio of the mean values of resistance and total actions as given in Equation (2). The results are obtained for the parametric range of the dimensionless ratio of variable to total actions as given by Equation (3) for the range of values $\chi\{0-1\}$.

$$GSF_{mean} = \left(\frac{\mu_R}{\mu_G + \mu_Q} \right) \tag{2}$$

$$\chi_{mean} = \left(\frac{\mu_Q}{\mu_G + \mu_Q} \right) \tag{3}$$

The function $f\{GSF_{mean}; \chi_{mean}\}$ represents the *basic reliability requirement* for the specific case. Since it is expressed in terms of the mean values of the basic variables, it can also be referred to as the *unbiased reliability function* or $f_{R,\,u}$ against which design functions can be measured. The result for the reference case is shown in Figure 2.

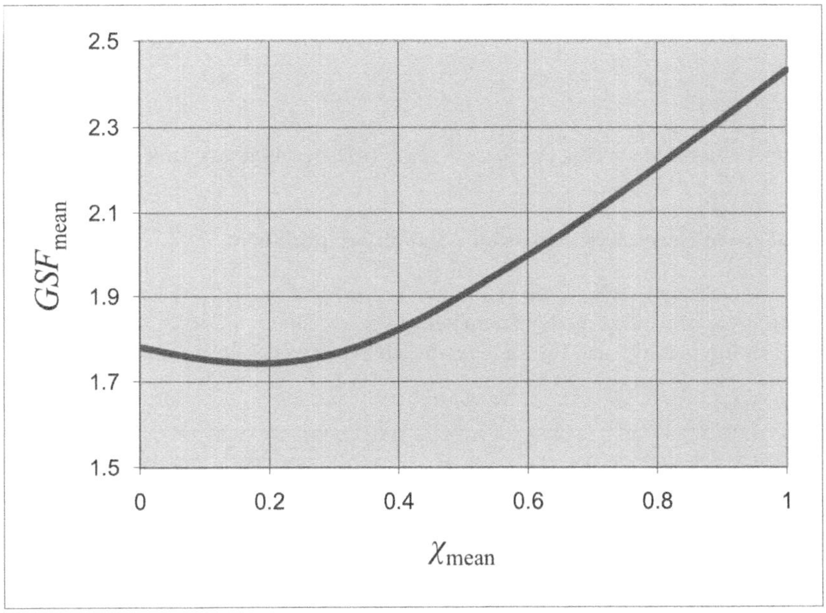

Figure 2 Reliability requirement $f_{R,\,u}$ for β_T = 3,0, in terms of mean values of basic variables

A prominent feature of $f_{R,\,u}$ is the non-linear nature of the graph, particularly across the range of load ratios which is typical of practical situations with χ ranging from values of 0,3 to 0,6, with the value increasing significantly for situations where Q dominates ($\chi \rightarrow 1$), and to a lesser extent where G dominates ($\chi \rightarrow 0$).

The design functions are expressed in terms of the characteristic values of the basic variables $\{R_k; G_k; Q_k\}$. The unbiased reliability requirement $f_{R,\,u}$ is therefore converted to the characteristic requirement $f_{R,\,k}$ by expressing Equations (2) & (3) in terms of $\{R_k; G_k; Q_k\}$. This can be achieved by a direct transformation, without the need for a separate FOSM solution. In this manner the reliability implications of the specification of the characteristic values of the basic variables can easily be assessed.

The effect of expressing the reliability requirement in terms of characteristic design variables is demonstrated in Figure 3. The nominal values of k_X as specified in Table 5 are used in Figure 3(a); for Figure 3(b) a value of k_Q = 0,9 is used. The reduced *GSF* value required to achieve the target level of reliability due to the increased conservatism provided by the *characteristic bias* in the specification of the design variables is evident. It is also clear that the required *GSF* is rather sensitive to such characteristic bias.

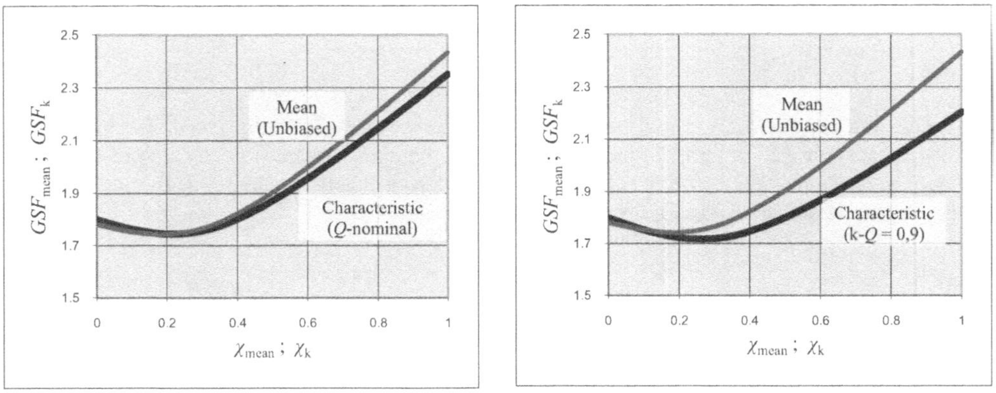

(a) Nominal characteristic bias for Q_k (k_Q =0,96)

(b) Conservative characteristic bias for Q_k (k_Q = 0,90)

Figure 3 Influence of characteristic bias on reliability requirement

The influence of various model parameters on $f_{R,k}$ is shown in Figure 4 (a) for different values of the target level of reliability, and (b) for a parametric range of resistance variability for the reference reliability level.

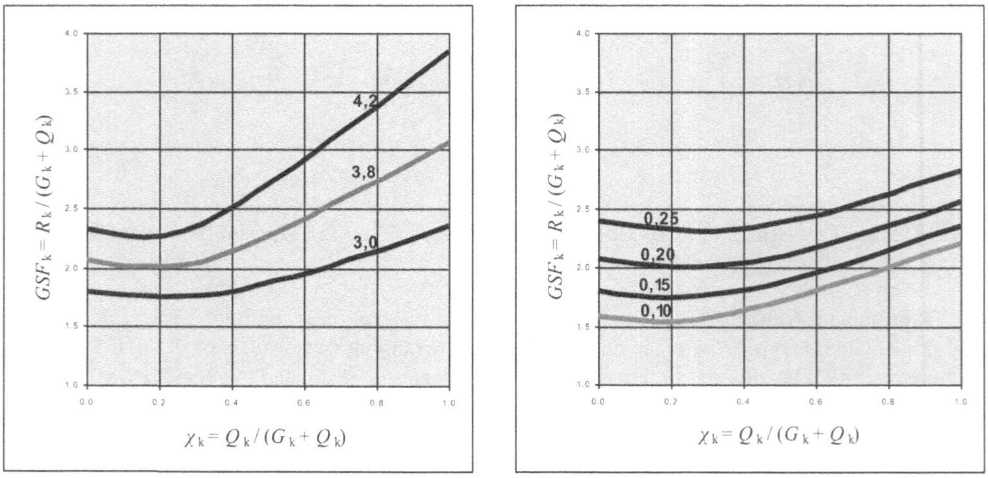

(a) Reliability requirement as function of β_T {3,0; 3,8; 4,2}

(b) Reliability requirement as function of V_R {0,10; 0,15; 0,20; 0,25}

Figure 4 Parameter influence on reliability requirement

4.3 Action Combination Schemes

A summary of the various expressions for the combination of permanent and variable actions used in SABS 0160:1989 and EN 1990:2002 respectively is provided in Table 6. The scheme used in ASCE-7 is also shown due to its similarity to that of SABS 0160. Since the present assessment considers the combination of permanent and variable actions, the tabulated expressions are simplified to provide for only a single variable action and recommended or typical values for the partial factors (γ_G; γ_Q; ξ; ψ_0).

Table 6 Expressions for combination of permanent (G_k) and variable (Q_k) actions (simplified)

Standard	#	Expression for Single Variable Action
SABS 0160	4 (e)	$1,5D_n$
	4 (f)	$1,2D_n + 1,6Q_k$ (Imposed) $1,2D_n + 1,3Q_k$ (Wind)
EN 1990	6.10	$1,35G_k + 1,5Q_k$ $(\gamma_G G_k + \gamma_Q Q_k)$
	6.10 (a)	$1,35G_k + 1,5\psi_0 Q_k$ = $1,35G_k + 1,05Q_k$ ($\psi_0 = 0,7$ typically)
	6.10 (b)	$1,35\xi G_k + 1,5Q_k$ = $1,15G_k + 1,5Q_k$ ($\xi = 0,85$)
	6.10 (b UK)	$1,35\xi G_k + 1,5Q_k$ = $1,25G_k + 1,5Q_k$ ($\xi = 0,925$)
	6.10 (a-mod)	$\gamma_G G_k$
ASCE-7	Clause 2.3.2-1	$1,4D_n$
	Clause 2.3.2-2	$1,2D_n + 1,6Q_k$

The SABS 0160 scheme is quite similar to that used in ASCE-7 for both the general design function given as Equation 4 (f) as well as the case where permanent actions dominate, as given in Equation 4 (e).

In EN 1990, allowance is made for the use of a dual function, given as Expression 6.10 (a) & (b), as alternative to the initial Expression 6.10, to be selected by Member States as an NDP option. The Expression 6.10 (a) & (b) is sometimes interpreted in terms of a Turkstra rule application, taking the variable and permanent actions alternatively as the *leading action* with reduced values for the *accompanying action*.

In EN 1990 Annex A1 for buildings, a special interpretation of Expression 6.10 (a) is identified, which only considers permanent actions for the situation where the action is dominated by this class of action (Expression 6.10 (a-mod), also known as the Finnish version. This is similar to SABS 0160 Expression 4 (e). No recommendation is made however to the value for γ_G in the case of Expression 6.10 (a-mod).

The three widely different schemes allowed by EN 1990 as shown in Table 6 demonstrate the accommodating nature of Eurocode. In addition, the partial factors are also classified as NDP for which values are to be selected nationally. As an example of the effective tolerance allowed by EN 1990, the UK National Annex allows *both options* of Expression 6.10 or 6.10 (a) & (b), to be decided by the designer; and modifies ξ (0,85 → 0,925) to change γ_G (1,15 → 1,25) for 6.10 (b)!

Such wide tolerances allowed in EN 1990 can therefore clearly accommodate the SABS 0160 scheme for the combination of permanent and variable actions. However it also presents an opportunity to consider a revised scheme for SANS 10160-1! The internal motivation for a revised scheme would be to improve the reliability performance of the design function; an external motivation would be to improve harmonisation with (mainstream) Eurocode practice.

4.4 Reliability Performance of Alternative Design Functions

When the design functions listed in Table 6 are plotted on the $\{\chi_k;\ GSF_k\}$-graph, a linear graph results. As indicated above, the alternative design functions represent different approaches to approximating the non-linear requirement function with (simple) linear design functions.

The primary objective of the design function is to achieve a GSF_k which exceeds the function $f_{R,\,k}$ across the range of practical values for χ_k; such exceedance for exceptional structures, where χ_k is respectively close to 0 or 1, would ensure reliability requirements under these conditions. Large exceedance of GSF_k implies unnecessary conservatism, which has economic implications. This could result in implicit bias for example between concrete structures which tend to cover the region of lower values of χ_k, and steel structures for which higher values generally apply. Consistent reliability is achieved when the design function(s) closely approximates the reliability requirement.

4.4.1 Reliability performance of SABS 0160 function

The reliability performance of the SABS 0160 Expressions 4 (e) & (f) is shown in Figure 5. In Figure 5(a) the results are shown when a resistance factor $\gamma_R = 1/\varphi_R = 1{,}42$ is used to achieve a probability of resistance failure $p_{f,\,R} = 10^{-2}$ as specified in SABS 0160.

The reliability requirements are generally satisfied by the primary Expression 4 (f) over a wide range of practical χ_k values. At G_k dominating conditions Expression 4 (e) takes over albeit in an inefficient manner, with a sharp transition and large conservatism for G_k only. The transition occurs at $\chi_k = 0{,}158$ or $Q_k = 0{,}1875G_k$.

Insufficient reliability is achieved at the other extreme, where $\chi_k > 0{,}8$ or $Q_k > 4G_k$. A possible solution would be to increase the resistance factor γ_R, with results as illustrated in Figure 5(b). This is however a hypothetical solution since it would not be proper to solve insufficiencies in action functions by adjusting resistance. Nevertheless it provides an illustration of the vertical shift of the design function due to an increase in the resistance factor.

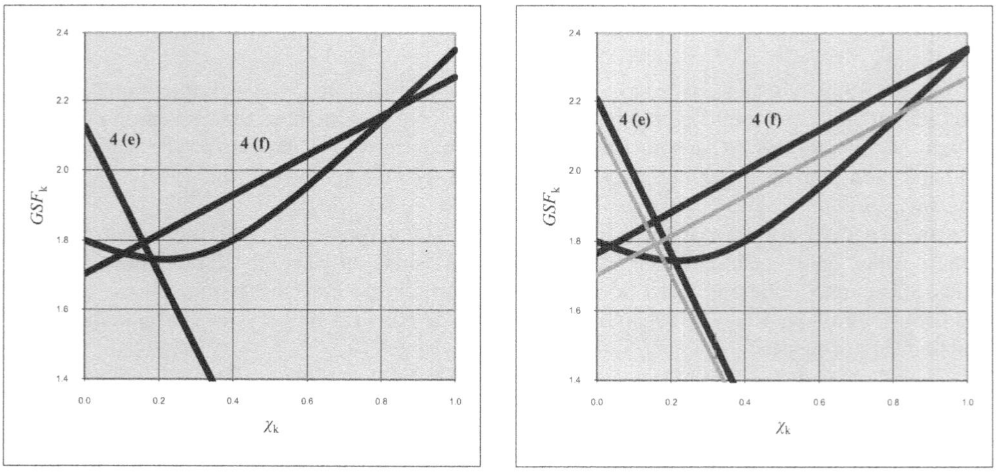

(a) Resistance at $\gamma_R = 1/\varphi_R = 1{,}42$ ($p_{f,\,R} = 10^{-2}$)

(b) Resistance at $\gamma_R = 1/\varphi_R = 1{,}47$ ($p_{f,\,R} = 0{,}5*10^{-2}$)

Figure 5 Reliability compliance of SABS 0160 design functions

4.4.2 Reliability performance of EN 1990 combination functions

The reliability performance of EN 1990 Expression 6.10 is given in Figure 6 for the case under investigation. The resistance factor is set in accordance with the EN 1990 procedure of setting $\beta_{T,R} = \alpha_R\beta_T = 0,8 \times 3,0 = 2,4$ or $p_{f,R} = 0,8*10^{-2}$ which requires $\gamma_R = 1,43$.

As far as satisfying the reliability requirements, Expression 6.10 generally achieves this objective across the practical range of design conditions. As shown in Figure 6(a), a large degree of conservatism is however achieved at the lower practical ranges of χ_k and decreasing as χ_k increases towards the crossover value at $\chi_k > 0,75$ ($Q_k > 3G_k$) where the reliability is insufficient.

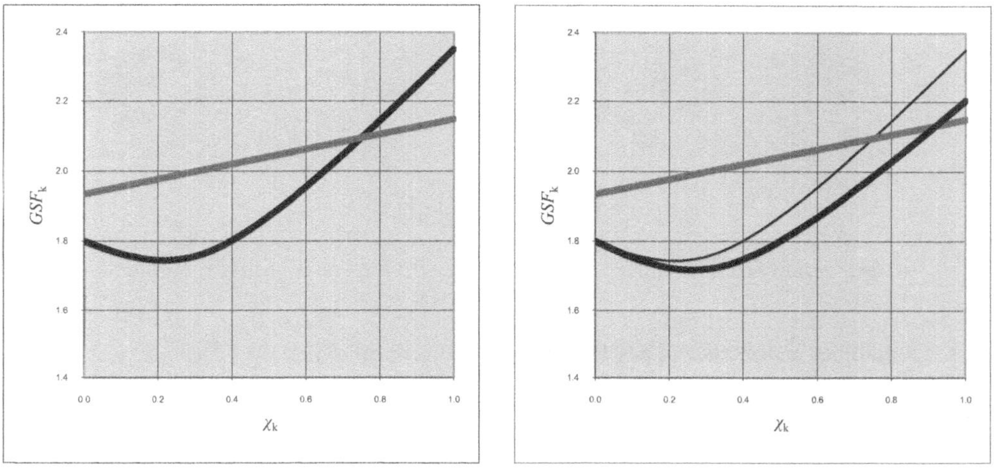

(a) Nominal characteristic bias for Q_k ($k_Q = 0,96$)

(b) Increased characteristic bias for Q_k ($k_Q = 0,90$)

Figure 6 Reliability compliance of EN 1990 Expression 6.10

The effect of a more conservative characteristic specification of Q_k, as illustrated in Figure 3(b), is shown in Figure 6(b). Although the formal specification of Q_k in EN 1990 is that it is nominally the expected lifetime maximum value (see Table 5), the general conservatism applied in Eurocode will likely be closer to that presented in Figure 6(b) than the case used in this assessment as shown in Figure 6(a).

The comparison of EN 1990 Expressions 6.10 (a) & (b) are shown in Figure 7(a), with the UK modification shown in Figure 7(b). The large and variable over-conservatism of Expression 6.10 in the mid ranges of χ_k is improved by the dual Expressions 6.10 (a) & (b). The consistency of reliability is also improved. The lack of reliability for $\chi_k > 0,6$ ($Q_k > 1,5G_k$) is more acute, although the same argument about larger conservatism for Q_k as discussed for Expression 6.10 applies here. The UK modification of Expression 6.10 (b) improves the consistency of reliability up to $\chi_k = 0,5$ but with some over-conservatism; the lack of reliability for large χ_k is moderated somewhat.

4.5 Revised SANS 10160 Scheme for Combination of Permanent and Variable Actions

Whilst maintaining the basic tenet of the SABS 0160 scheme for the combination of G_k & Q_k, its modification was investigated. The technical motivation for the reassessment was to improve the reliability performance of the design functions. From a harmonisation point of view the possibility of improving consistency with one of the main EN 1990 options was also a consideration. The reliability performance of reference functions as assessed above was used to devise a scheme that fits into the relationship with Eurocode, provides for the revised scope of application and could be applied effectively and economically.

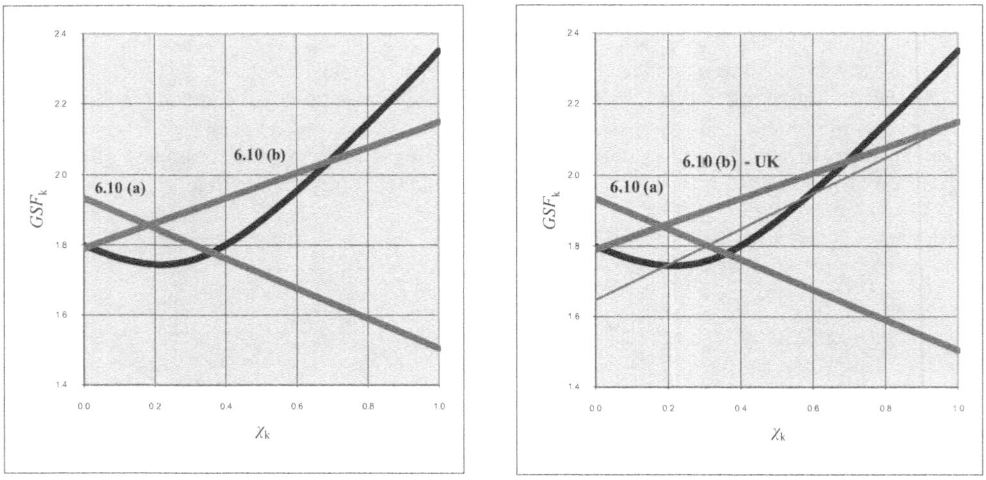

(a) Expression 6.10 (a) & (b) | (b) Expression 6.10 (a) & (b UK)

Figure 7 Reliability compliance of EN 1990 Expression 6.10 (a) & (b)

4.5.1 Outline of scheme

The basic scheme of the design function for the combination of G_k & Q_k from SABS 0160 Expression 4 (f) has been maintained in SANS 10160-1 without any modification. Equation (5) below is based on SANS 10160-1 Clause 7.3.2.1 Expression (6), assigned as the STR combination case which applies to consideration of internal failure, as related to the strength of the structure.

The partial factor for permanent action γ_G = 1,2 is maintained; although γ_Q should be calibrated for each variable action, the value of γ_W = 1,3 for wind being the only exception to the general value of γ_Q = 1,6 and it is stipulated as before. The combination scheme for multiple variable actions is also maintained. This scheme, which is based on the Turkstra rule, will be discussed below.

$$E_{d,STR} \quad = \quad 1{,}2G_k \quad + \quad 1{,}6Q_{k,1} \quad + \quad \sum_{i>1} \psi_i 1{,}6Q_{k,i} \qquad (5)$$

The STR expression should apply predominantly in the design of structures. The design should also be checked for the dominant G_k case by, testing for compliance to the STR-P Expression (7) (Clause 7.3.2.2), shown here as Equation (6). Through a calibration process a value of γ_G = 1,35 was determined; with Q_k factored at γ_Q = 1,0 added to Expression 6 (e) of SABS 0160. The STR-P expression can therefore be seen as a simplified version of Expression 6.10 (a) from EN 1990. The factoring of the Q_k component of STR-P can be compared to typical values of are $\psi_0\gamma_Q$ = 0,9 – 1,05 for Expression 6.10 (a).

$$E_{d,STR-P} = \quad 1{,}35G_k \ + \ 1{,}0Q_k \qquad (6)$$

Only a single variable action is considered in STR-P; obviously the most severe variable action should be used for this design check. The motivation for such simplification is that since the case for dominant G_k is considered, the contribution from accompanying Q_k should be minimal.

The STR-P will only result in a more unfavourable action effect in situations where the permanent action is *large* in comparison to the variable action. The intersection between STR and

STR-P takes place at χ_k = 0,20, therefore STR-P only controls the design when $G_k > 4Q_k$! In SABS 0160 the transition point is at χ_k = 0,16 resulting in Expression 6 (e) to control if $G_k > 5{,}3Q_k$.

4.5.2 Reliability performance of STR & STR-P functions

The reliability performance of the STR & STR-P scheme for the combination of permanent and variable actions employed in SANS 10160-1 is shown in Figure 8 for the reference case introduced in §5.2. The STR function is exactly the same as that of Expression 6 (f) of SABS 0160 as shown in Figure 5; the STR-P function should be compared to that for expression 6 (e) as shown in Figure 8.

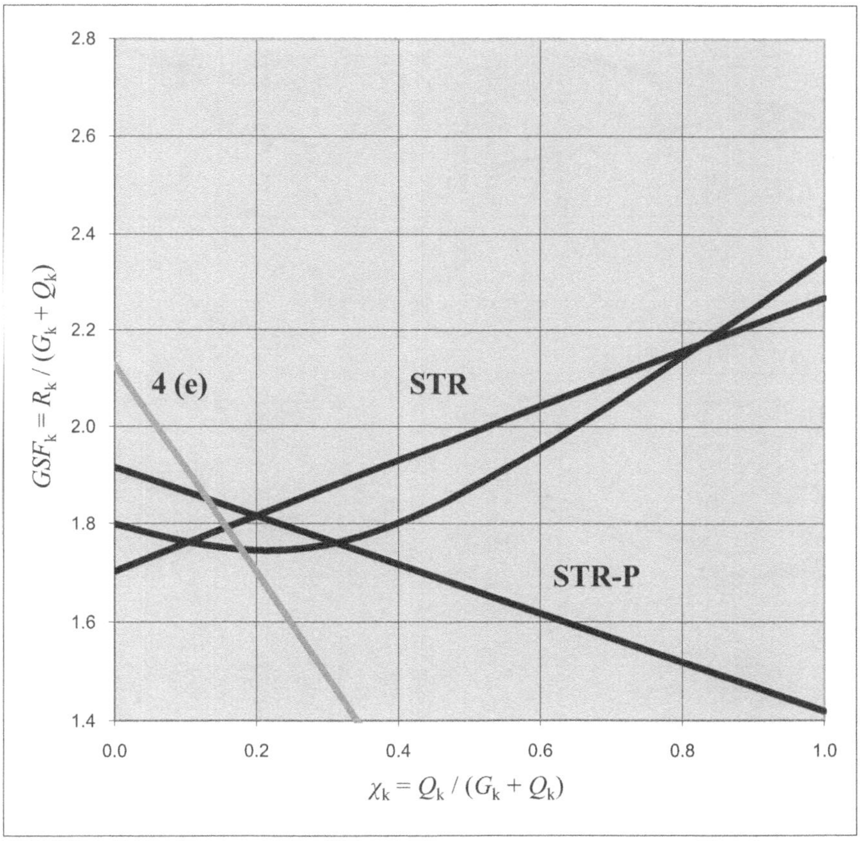

Figure 8 Reliability performance of SANS 10160-1 design functions STR & STR-P (β_T = 3,0 ; γ_R = 1,42 ($p_{f,R}$, = 10^{-2}))

The sufficiency and consistency of reliability of the SANS 10160-1 design scheme is evident from Figure 8. Although the intersection between the two functions STR & STR-P has shifted towards the practical range of χ_k-values, as compared to SABS 0160 Expressions (e) & (f), this marginal shift is fully compensated for by the less abrupt transition, the omission of over-conservatism for G_k-only case (χ_k = 0) and the close approximation of the reliability function across the range of χ_k {0 – 0,6}.

4.5.3 Parametric assessment of effect of resistance variability

The reliability performance of the modified SANS 10160-1 design functions is presented in Figure 9. The effect of the variability of the resistance, as given by V_R {0,10; 0,15; 0,20; 0,25} is shown. The resistance factor γ_R is determined to comply with the limit of $p_{f,R} = 10^{-2}$ for the probability of the resistance below the design (factored) value.

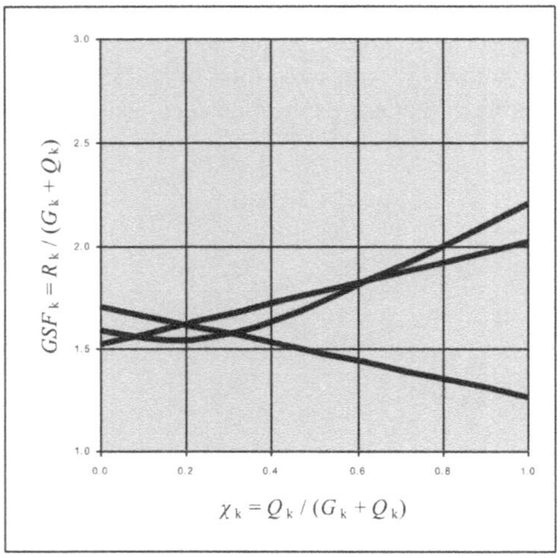

(a) $V_R = 0,10$
$\gamma_R = 1,26$ ($p_{f,R}$, $= 10^{-2}$)

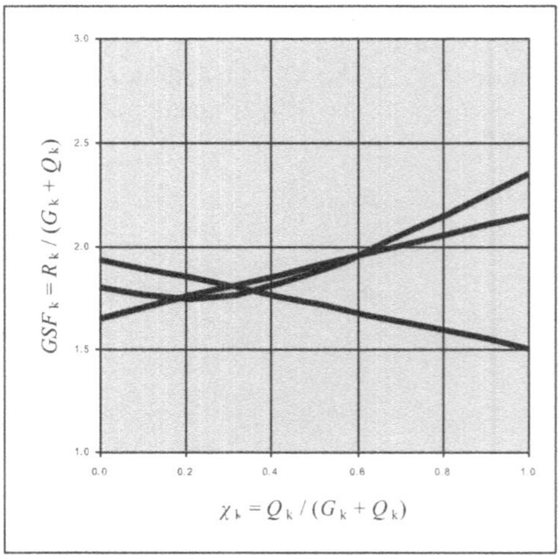

(a) $V_R = 0,15$
$\gamma_R = 1,42$ ($p_{f,R}$, $= 10^{-2}$)

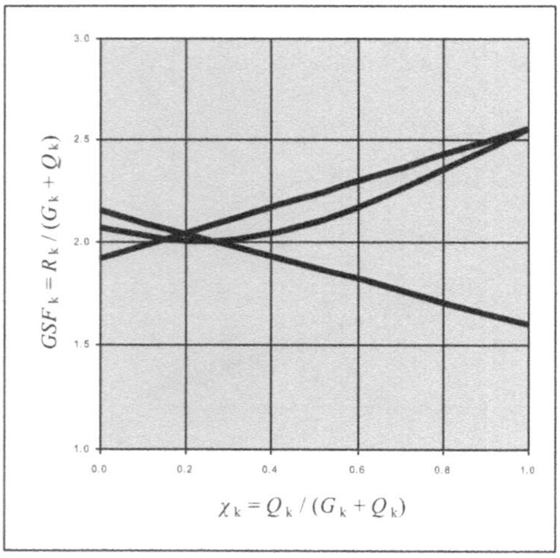

(a) $V_R = 0,20$
$\gamma_R = 1,59$ ($p_{f,R}$, $= 10^{-2}$)

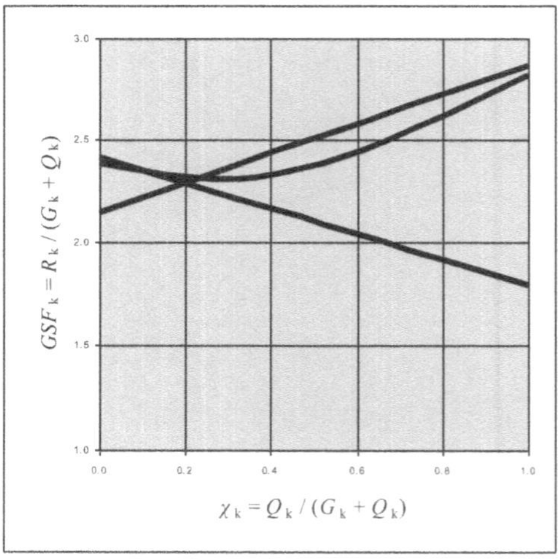

(a) $V_R = 0,25$
$\gamma_R = 1,79$ ($p_{f,R}$, $= 10^{-2}$)

Figure 9 Reliability performance of SANS 10160 design functions across parametric range of resistances

The parametric analysis indicates insufficient reliability for Q_k-dominated situations in the case of low variability for resistance ($V_R = 0,10$) and marginally insufficient reliability for G_k - dominated situations in the case of high variability ($V_R = 0,25$). Since the insufficiency is directly related to the resistance models, this effect needs to be taken into account in the calibration of resistance requirements as stipulated in the materials design standards.

4.6 Reliability differentiation and classification

Reliability differentiation is formally introduced in SANS 10160-1 by providing for reliability levels which differ from the reference level which is used in the assessment above. Such differentiation is based on a reliability classification of structures. It is consistent with the EN 1990 procedures. It is however presented informatively, as opposed to the normative stipulations of EN 1990.

4.6.1 Reliability classes

The system of reliability classification presented in SANS 10160-1 is a modified version from that of EN 1990 Annex B (Informative). Provision for differentiated reliability classes are provided for in Annex A (Informative) *Management of structural reliability for construction works* as allowed for in Clause 4.4. A four level classification system is applied, as compared to the Eurocode scheme of three levels, where the mid-class is divided in two sub-classes. A four level classification system is applied consistently in SANS 10160, including the provisions for accidental design situations, seismic design (Part 4), geotechnical design (Part 5) and crane classification (Part 6).

The pertinent features of the reliability classification is summarised in Table 7. The normative provisions of SANS 10160 are formulated and calibrated for Reliability Class 2 (RC2) at the reference target level of reliability $\beta_T = 3{,}0$ as indicated. The reference reliabilities for the reliability classes are derived from SABS 0160. The values for the multiplication factor K_F which may be applied to the partial factors for actions $[\gamma_F]$ to be used in fundamental combinations for persistent design situations, according to Clause A3.2.2 are also given in Table 7, as derived from EN 1990 Annex B. In accordance with EN 1990 it is stated that reliability differentiation may also be applied through partial factors on resistance γ_M but this is not normally used (Clause A3.2.4).

Table 7 Reliability classification of structures

Reliability class	**Consequence of failure**		Reference reliability β_T	Multiplication factor K_F
	Life loss	**Consequences**		
RC1	**Low**	**Low:** economic, social, environmental	2,5	0,9
RC2	**Medium**	**Medium:** economic, social **Considerable:** environmental	3,0	1,0
RC3	**High**	**Very great:** economic, social, environmental	3,5	1,1
RC4		**Post-disaster** function; **Beyond boundary**	4,0	1,2

4.6.2 Design performance for different reliability classes

Figure 10 shows the application of the multiplication factors of Table 7 which are linearly interpolated to the reliability levels shown in Figure 2(a), for the cases of $\beta_T = 3{,}8$ ($K_F = 1{,}16$) and $\beta_T = 4{,}2$ ($K_F = 1{,}22$), with the resistance factor kept at $\varphi_R = 0{,}705$ ($p_{f,R}, = 10^{-2}$).

The adjustment achieved through the multiplication factor K_F is increasingly insufficient to adjust for higher reliability classes. The multiplication factors for higher reliability classes should therefore be considered to be a nominal adjustment of the reliability requirements.

The inclusion of an adjustment also for resistance would provide some remedy. However, a resistance adjustment merely shifts the design functions vertically in the diagram, and this will not properly provide for insufficient reliability for situations where variable actions are dominating. Proper calibration would be needed to derive optimal adjustment factors.

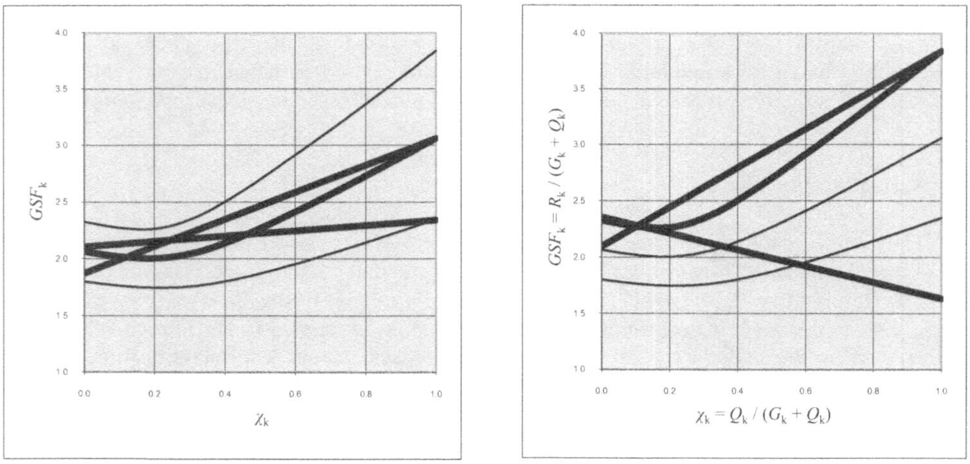

(a) Target reliability $\beta_{\text{target}} = 3{,}8$
$K_{F,G} = K_{F,Q} = 1{,}16$; $K_{F,R} = 1{,}0$

(a) Target reliability $\beta_{\text{target}} = 4{,}2$
$K_{F,G} = K_{F,Q} = 1{,}22$; $K_{F,R} = 1{,}0$

Figure 10 Performance of SANS 10160 design functions adjusted for target reliability

5 PARTIAL FACTORS FOR VARIABLE ACTIONS (γ_Q)

EN 1990 introduces a single partial factor for all variable actions γ_Q with a recommended value of 1,5 as indicated above. In SANS 10160 the practice followed in SABS 0160 to allow the determination of partial factors for each individual variable action Q_i separately, is maintained. The general procedure of applying imposed office floor load (Q_{Ofl}) as a representative variable load, as described above, should then be adapted to provide for specific variable actions.

In this process the other classes of imposed loads, wind and thermal actions, geotechnical actions on structures and crane induced actions fall within the scope of SANS 10160 and therefore need to be considered. Although the specification of the requirements for individual variable actions is properly part of Parts 2 – 8, the reliability basis of design is still treated as part of Part 1, and is therefore briefly reviewed here.

In the case of EN1990 the partial factor for variable actions is generically set at the (recommended) value of 1,5 for *all* variable actions, although its value is an NDP, allowing different values to be selected by Member States. In contrast, sets of combination factors $\{\psi_0;\ \psi_1;\ \psi_2\}$ are set individually for the variable actions (see §6). This illustrates the importance given in Eurocode to the combination of variable actions. In contrast, the SABS 0160 practice of setting γ_Q individually for each variable action is maintained in SANS 10160-1, similar to the practice followed by ASCE-7. However generally the default value of 1,6 is used in these cases, using different values only with sufficient justification.

The partial action factors for the respective variable actions provided for in SANS 10160 are subsequently evaluated. The combination of multiple variable actions and the related combination factors are treated in a subsequent section.

5.1 Imposed Floor Loads

The imposed floor loads are specified in SANS 10160-2 in terms of occupancy classification of the building area. An important consideration is the possibility of crowding, generally in the case of public areas. This is one of the three main probability models for imposed floor loading presented in the JCSS Model Code (2002), as extracted from CIB compilations on actions on structures. The other two mechanisms are the sustained imposed floor load associated with one period of occupancy, typically of eight year duration, and intermittent loads with much shorter duration which are related to the occupancy.

Nevertheless the imposed floor loads stipulated in EN 1991-1-1 are still based on non-probabilistic considerations (Hemmert-Halswick *et al* 1988). Development of indicative models for imposed floor loads are reported by Holický (2005), but this process has not been applied systematically in EN 1991-1-1 *Actions on structures – General actions: Densities, Self-weight and imposed loads*. Similarly the international practice to which SABS 0160 imposed load values were compared (Retief *et al* 2001) were based on values ultimately selected by judgement. As reported in Chapter 2-1 the values specified in SANS 10160-2 are primarily based on values applied in international standards, mainly that of EN 1991-1-1 and ASCE-7.

The partial factor for imposed floor loads of γ_I = 1,6 is therefore based on the parametric calibration reported above, using the probability model listed in Table 5. The same partial factor is then applied generically to all imposed loads specified in SANS 10160-2, including industrial floor areas, fork lifts and traffic loads for parking garages; roofs and helicopter roof landing areas; horizontal loads on partitions and balustrades. It should be noted that a probability model for imposed roof loads was used to adjust the specified values for such loads, but the partial load factor of 1,6 is maintained.

5.2 Wind Actions

The partial factor for wind actions used in SABS 0160 of γ_W = 1,3 derives from the (anomalously) low value for the ratio $k_{W\cdot}$ of mean to characteristic (specified) value of the probability model as shown in Table 5, as compared to models generally used, e.g. given by Holický (2002). The effects of the two alternative wind load probability models are shown in Figure 11.

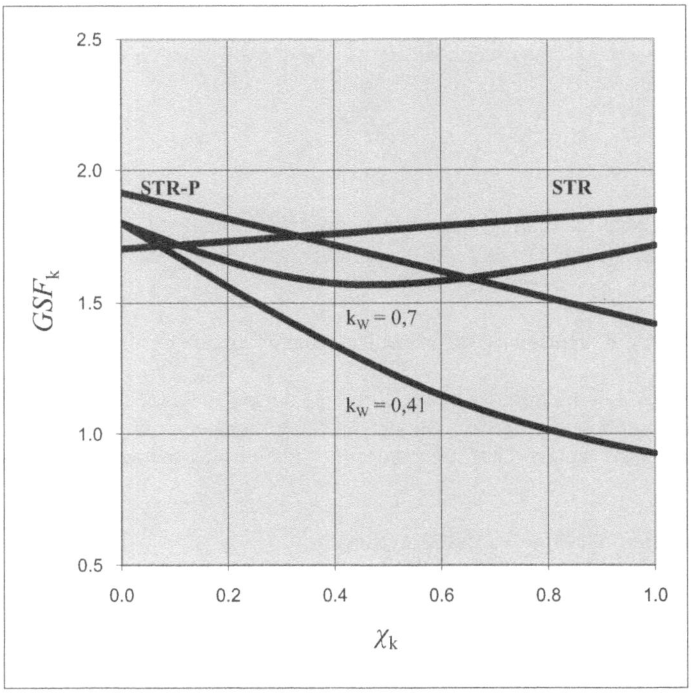

Figure 11 Wind design functions compared to alternative probability models ($k_{W\cdot} = Q_{W,k}/\mu_{W\cdot}$)

When the existing SABS 0160 wind partial factor of 1,3 is applied to the SANS 10160-1 dual design function, the respective sets of partial factors are {1,2; 1,3} for the STR combination and {1,35; 1,0} for the STR-P combination. These functions are compared in Figure 11 to the two alternative probability models for wind actions as listed in Table 5, with mean values of 0,42 (SABS 0160) and 0,7 (Holický) of the characteristic value. It is clear that the design functions are over-conservative if a mean of 0,42 applies, but more reasonably provide for the more realistic value of 0,7.

5.3 Seismic Actions

Since seismic actions are classified as an accidental action, which is not expected during the design life of the structure, and for which a degree of damage determined (in principle) through risk assessment is accepted, the design value of the action (basic specified ground motion) is directly specified and no partial factor needs to be established. This approach represents a substantial deviation from SABS 0160 practice, as derived from Eurocode EN 1990 & EN 1998-1.

5.4 Geotechnical Actions on Buildings

The practice of separating actions and resistance followed in structural design cannot be applied in geotechnical design, where an integral treatment is required. Although provision for geotechnical design for buildings is nominally provided for in SANS 10160-1, such integral treatment is provided in SANS 10160-5 *Basis of geotechnical design and actions*. The partial factors for geotechnical design are largely determined from experience based judgement, as reported in Chapter 5-1.

5.5 Crane-induced Actions

The partial factor for crane induced actions was derived from an extensive reliability analysis as reported in Chapter 6-3.

5.6 Thermal Actions

The partial factor for thermal actions was taken generically at the value of 1,6 without any calibration or even consideration of extreme temperature probability models.

5.7 Actions during Execution or Different Reference Periods

Adjustment of actions applied to the structure during the execution of the building, or for different design life reference periods, is done by determining appropriate values for the actions, whilst the values of partial factors for sustained design situations are maintained.

5.8 Combination of Multiple Variable Actions

The concept of the Turkstra rule (Equation 7) to combine multiple variable actions is well established as a reasonable approximation, albeit non-conservative according to ISO 2394. It should be noted that the partial action factor is applied to an average or *(arbitrary) point-in-time* value, as indicated in Equation (5) above. According to the Turkstra rule each variable action should be taken *in turn* as the leading action, and the maximum action effect taken, which could vary throughout the section or element in the structure.

$$\max Q_{combined} = \max[(Q_{1,\max} + Q_{2,expected}); (Q_{1,expected} + Q_{2,\max})] \quad (7)$$

An additional factor ψ is introduced to convert the characteristic value of $Q_{k,i}$ into its accompanying value $\psi_i Q_{k,i}$. In SABS 0160 ψ is selected to obtain a point-in-time estimate of Q. A more elaborate three-tiered scheme of combination factors $\{\psi_0; \psi_1; \psi_2\}$ is introduced in ISO 2394 and applied in EN 1990. See Chapter 1-4 §4.1 for the definition and recommended values and §5.5 – 5.6 for theoretical analysis of the values.

The elaborate scheme of combination factors may be warranted for the extensive scope of structures, design situations, actions and their combined occurrence, structural materials and their respective failure modes, provided for in EN 1990. Furthermore the recommended values for the combination factors appear to be relative conservative, as is indeed indicated by the assessment presented in Chapter 1-4.

It was therefore decided to maintain the single value scheme for the combination factor from SABS 0160 and to interpret it at its Turkstra rule based point-in-time value. Values for ψ were updated to provide for the extended range of variable actions, the revised values for imposed loads (see Chapter 2-1) and probability models for imposed loads provided in the JCSS Model Code (2001) that became available from CEB (1989).

The ψ-values selected for SANS 10160 are compared in Table 8 to the equivalent values from SABS 0160, the CIB-derived values and the EN 1990 combination (ψ_0) and quasi-permanent (ψ_2) values, the latter considered to be equivalent to the point-in-time values on which the selection is based.

5.9 Serviceability Limit State

Provisions for the serviceability limit state (SLS) were substantially reviewed for SANS 10160-1, in accordance with the requirements of EN 1990. The most important development is the introduction of differentiated serviceability performance levels, which then required a review of the associated serviceability criteria.

Although the criteria are provided in an informative annex, their normative status is enhanced by specifying that deviation is only allowable with proper justification. The normative status of requirements for the SLS is consistent with the status of SANS 10160 as a professional code of practice, as opposed to regulatory standards which are primarily concerned with public safety.

5.9.1 Differentiated serviceability levels

Serviceability performance levels are differentiated as {Irreversible; Reversible; Long-term; Appearance}, as indicated in Figure 1. The Irreversible SLS applies *where some consequences of actions exceeding the specified service requirements will remain when the actions are removed*; and the less severe Reversible SLS *where no consequences of actions exceeding the specified service requirements will remain when the actions are removed.* Long-term and Appearance SLS represent the lowest level of performance requirements. Although no proper reliability assessment is generally applied to the SLS, an indication of target levels of reliability are taken as summarised in Table 3; further refinement of the scheme of target reliabilities may be feasible.

5.9.2 Action combination scheme for the SLS

The action combination scheme presented in EN 1990 is followed in SANS 10160-1, but with the simplification which results in the use of a single combination factor ψ, as opposed to the three level scheme of EN 1990 which is used in the formulation of differentiated SLS action combinations. The single value for ψ results in the same expression which applies to Reversible, Long-term &

Appearance SLS. The SABS 0160 partial factors are also maintained in SANS 10160-1. Examples of the expressions for SLS action combinations are given in Table 9.

Table 8 Comparison of SANS 10160 Combination factors for actions to other references

<table>
<tr><th>Category</th><th>Specific use</th><th>SABS 0160</th><th>Derived from JCSS</th><th>EN 1990 ψ_0/ ψ_2</th><th>SANS 10160 ψ</th></tr>
<tr><td>A</td><td>Domestic and residential</td><td rowspan="4">0,3</td><td>0,25; 0,30; 0,40</td><td rowspan="2">0,7/0,3</td><td>0,3</td></tr>
<tr><td>B</td><td>Public areas (no crowding)</td><td>0,23; 0,28; 0,34</td><td>0,3</td></tr>
<tr><td>C</td><td>Public areas (congregation)</td><td>0,33</td><td rowspan="2">0,7/0,6</td><td>0,3</td></tr>
<tr><td>D</td><td>Shopping areas</td><td>0,26</td><td>0,3</td></tr>
<tr><td>E1:</td><td>Light industrial use</td><td>?</td><td>0,33</td><td rowspan="3">1,0/0,8</td><td>0,5</td></tr>
<tr><td>E2:</td><td>Industrial use</td><td>?</td><td>0,60</td><td>0,6</td></tr>
<tr><td>E3:</td><td>Storage areas</td><td>0,6</td><td>0,70</td><td>0,8</td></tr>
<tr><td>F</td><td>Parking (vehicles ≤ 25 kN)</td><td>0,6</td><td></td><td>0,7/0,6</td><td>0,8</td></tr>
<tr><td>G</td><td>Parking (25 – 160 kN)</td><td>-</td><td></td><td>0,7/0,3</td><td>0,3</td></tr>
<tr><td>FL_1 – FL_6</td><td>Fork lifts</td><td>-</td><td></td><td></td><td>0,6</td></tr>
<tr><td>H</td><td>Inaccessible roof</td><td>0</td><td></td><td>0/0</td><td>0</td></tr>
<tr><td>J</td><td>Accessible flat roof (ex A-D)</td><td rowspan="2">0,3</td><td></td><td></td><td>0,3</td></tr>
<tr><td>K</td><td>Accessible flat roofs (A – D)</td><td></td><td></td><td>As floor</td></tr>
<tr><td>HCL1/2</td><td>Helicopter load</td><td>-</td><td></td><td></td><td>0</td></tr>
<tr><td>Wind</td><td>Accompanying
Serviceability</td><td>0</td><td></td><td>0,6/0</td><td>0
0,3</td></tr>
</table>

Table 9 Example SANS 10160 combination schemes for the serviceability limit state

<table>
<tr><th colspan="4">SERVICEABILITY LIMIT STATE</th></tr>
<tr><td rowspan="2">Irreversible</td><td>$1{,}1G_k + 1{,}0Q_k + \sum_{i>1}\psi_i Q_{k,i}$</td><td>(5)</td><td>Imposed leading (at characteristic value)</td></tr>
<tr><td>$1{,}1G_k + 0{,}6Q_k + \sum_{i>1}\psi_i Q_{k,i}$</td><td>(6)</td><td>Wind leading (at reduced characteristic value)</td></tr>
<tr><td>Reversible; Long-term; Appearance</td><td>$1{,}1G_k + \sum\psi_i Q_{k,i}$</td><td>(7)</td><td>All variable actions at arbitrary-point-in-time value</td></tr>
</table>

5.9.3 Serviceability criteria

The serviceability criteria as set in SABS 0160 are based on the International Standard ISO 4356:1977. These criteria were reclassified to provide separately for the irreversible and the reversible/long-term/appearance SLS respectively. The limiting values were systematically reviewed and assessed. The criteria are presented in SANS 10160-1 Annex C (Informative) *Recommended criteria for deformation of buildings*. Background to assist in the application of the SLS criteria is presented in Annex D (Informative) *Deformation of buildings*, which is based again on ISO 4356:1977.

6 RELIABILITY REQUIREMENTS FOR RESISTANCE

Although reliability requirements for structural resistance are of equal importance as compared to the treatment of actions, their specification is largely left to the relevant materials-based design standards. Furthermore the diversity of resistance is equal to that of the range of actions to be considered. The consistent treatment of structural resistance across the wide range of structural materials and subsequent structural systems and configurations requires careful consideration. Such treatment can only be based on rational reliability models for structural performance.

Inherent diversity of resistance derives from the use of structural materials with a wide range of strength and related properties, different degrees of control and even availability of information on these properties. The range of characteristics is clearly illustrated when considering the design of steel structures versus foundation design, in spite of their respective relevance to the reliability performance of the building. Similarly the various modes of failure and the subsequent consequences vary across the range of structural materials, configurations and conditions.

The specification of the consistent treatment of structural resistance in terms of *the basis of design for structural resistance* really represents the requirements which are set to standards committees for materials-based design standards, as opposed to stipulations to be followed by the designer. This important topic is not developed any further here. The elaborate procedures for reliability based development of standards for structural design provided in EN 1990 serve as an important guide to materials standards committees. Some aspects of the reliability basis for resistance are discussed by Holický *et al* (2007).

7 CONCLUSIONS

The specification of the basis of structural design in SANS 10160-1 is derived from Eurocode EN 1990. Critical elements are however taken and maintained from SABS 0160. The function of SANS 10160-1 is to establish an updated and improved basis for structural design in South Africa, which provides for local environmental and institutional conditions, and will ensure a smooth transition from existing practice.

The most important elements of SABS 0160 which are maintained in SANS 10160-1 include:

- The scope of application,
- The function of the standard to provide for standard (non-specialist) practice;
- The reference reliability level, including the application of simplified procedures such as the application of the Turkstra rule for combinations of actions.

An important outcome of the consistency that has been maintained in SANS 10160-1 with SABS 0160 is that existing materials-based limit states design standards can be directly applied together with SANS 10160!

The scope of provisions for the basis of structural design is however expanded substantially, as derived from EN 1990, including:

- The formal treatment of the basis of design in a separate Part of the standard;
- The extended reliability framework in terms of differentiated limit states for both the ultimate and the serviceability limit states,
- The related action combination schemes;
- The more formal reliability basis for the reliability framework, reliability differentiation, specification of basic variables and partial factors, the basis for design being assisted by testing.

SANS 10160-1 also accommodates the extended range of actions included from Eurocode into Parts 2 to 8. Provision for geotechnical design and actions as treated in SANS 10160-5, as derived from EN 1997-1 *Geotechnical design – General rules*, has a notable influence on the provisions for action combination schemes.

The requirements for the basis for structural resistance are only implied in SANS 10160-1, but explicit specification would really relate to standards committees for materials-based design standards and would therefore lie outside the scope of design procedures. Sufficient consistency with EN 1990 has been maintained so that it could serve as basis for future development of the next generation of materials-based standards.

SANS 10160-1 *Basis of structural design* thus represent an important link between South African structural design practice and Eurocode as a reference and source of information from which structural design procedures are adapted to South African conditions and scaled to local practice.

8 REFERENCES

Ang AH-S and Tang WH (1984). *Probability concepts in engineering planning and design: Volume 2 Decision, risk and reliability.* John Wiley & Sons, New York.

Brozzetti J, Sedlacek G (2000). *Background to the combination rules (6.10) or (6.10a) and (6.10b) in prEN 1990 Basis of design.* Document CEN/TC 250 N419. CEN/TC 250 Secretariat.

CIB (1989). Conseil International du Batiment pour la Recherche, l'Etude et la Documentation. *Actions on structures – Live loads in buildings.* Report 116, Rotterdam.

CEN (2002). European Committee for Standardisation EN 1990:2002 *Eurocode – Basis of structural design.* Brussels.

Day PW (1996). *Implementation of Eurocode 7 in South Africa.* Eurocode – Towards implementation, Institution of Structural Engineers, London.

Day PW (1997). *Limit States Design – a South African perspective.* XIV International Conference on Soil Mechanics and Foundation Engineering, Hamburg.

Day PW & Kemp AR (1999). South African loading code : Past, present and future. Civil Engineering, April 1999, pp 11-12.

Ellingwood B, Galambos TV, MacGregor JG and Cornell CA (1980). Development of a probability-based load criterion for American National Standard A58. NBS Special Publication SP 577, National Bureau of Standards, US Department of Commerce, Washington.

Gulvanessian H, Calgaro J-A & Holický M (2002). *Designer s guide to EN 1990 Eurocode: Basis of structural design.* Thomas Telford. 192 pp.

Hemmert-Halswick A, Sedlacek G, Merzenich G (1988). *Eurocode actions on structures: Background document Part 6: Imposed loads on floors and roofs.*

Holický (2002). *Conventional probabilistic models of basic variables.* Klokner Institute, CTU Prague (Private communication).

Holický (2005). Chapter I *Self-weight and imposed loads on buildings.* Leonardo da Vinci Pilot Project – Implementation of Eurocodes, Handbook 3 *Action effects for buildings.*

Holický M & Retief JV (2005). *Reliability Assessment of Alternative Load Combination Schemes for Structural Design*. Journal SAICE, Vol. 25, No 1, pp 15 – 20, 2005.

Holický M, Retief JV & Dunaiski PE (2007). *The reliability basis of design for structural resistance.* Zingoni A (Ed) Recent Developments in Structural Engineering, Mechanics and Computation. Millpress, Rotterdam.

ISO 4356:1977. International Standards Organisation. *Basis for the design of structures – Deformations of buildings at the serviceability limit states*. International Standard.

ISO 2394:1998. International Standards Organisation. *General Principles on Reliability for Structures*. International Standard. [Also issued as SANS 2394:2004]

ISO TC98 (2008). ISO/DIS 13824 *General principles on risk assessment of systems involving structures* TC98 *Bases for design of structures* SC2 *Reliability of structures* WG11.

JCSS (1996). Joint Committee on Structural Safety. *Background documentation: Eurocode 1 (ENV 1991) Part 1 Basis of design.* Report N° 94, ECCS, Brussels. ISBN 92-9147-0022.

JCSS (2001). Joint Committee on Structural Safety. *Probabilistic model code.* http://www.jcss.ethz.ch/.

Kemp AR (1998) *Loading codes: Background and introduction*, SA National Conference on Loading. SAICE & SAISC, 9 – 10 September 1998, Midrand, South Africa.

Kemp AR, Milford RV and Laurie JAP (1987). Proposals for a comprehensive limit states formulation for South African structural codes. The Civil Engineer in South Africa, 29 (9), 351-360.

Kemp AR, Mahachi J & Milford AR (1998). *Comparisons of international loading codes and options for South Africa.* SA National Conference on Loading. SAICE & SAISC, 9 – 10 September 1998, Midrand, South Africa.

Melchers RE (1987). *Structural reliability analysis and prediction.* Ellis Horwood, Chicester, England.

Milford RV (1998). SABS 0160: Philosophy and load factors. South African National Conference on Loading. SAICE & SAISC, 9 – 10 September 1998, Midrand, South Africa.

SABS (1994). SA Bureau of Standards, Pretoria. *Code of Practice: The general procedures and loadings to be adopted in the design of buildings***.** SABS 0160:1989 (Amended 1990, 1991, 1994)**.**

SAKO (1999). Joint Committee of NKB and INSTA-B. *Basis of Design of Structures. Proposal for Modification of Partial Safety Factors in Eurocodes.* NKB Committee and Work Report 1999:01 E, pp. 55.

Ter Haar TR & Retief JV (2001). *A Methodology for Structural Code Calibration.* International Conference: Safety, Risk and Reliability, Malta, March 21-23 2001.

Vrouwenvelder A (2002). *Developments towards full probabilistic design codes.* Structural Safety 24 (2002) 417-432.

Vrouwenvelder A, Stieffel U & Harding G (2005) *Eurocode 1, Part 1-7, Accidental Actions, Background Document*, First Draft. CEN TC 250 SC1, not published.

1-3

Review of Eurocode From the Perspective of the Revision of SANS 10160

Retief JV, Dunaiski PE & Holický M

1 INTRODUCTION

The development of the Eurocode set of standards for the design of building and civil engineering structures is one of the major recent activities in the development of structural design standards on the international scene. The publication of the EN version EN 1990 to EN 1999 follows from the conversion of the previous voluntary Eurocode ENV 1991 to ENV 1999 into normative European Standards. This process which is presently at an advanced state of completion, presents the realisation of the cumulative progress made with codified structural design over several decades.

The next phase of implementation of Eurocode as National Standards by Member States has already started. This phase consists of the publication of the various Eurocode Parts as National Standards, together with a National Annex for each Part. This step will take Eurocode fully from a set of reference standards which contains many undefined parameters and unspecified options to operational design standards for each Member State, with a high degree of harmonisation across Europe. Implementation will require the withdrawal of all conflicting National Standards by Member States within a limited period of time.

In this Chapter an overview is given of the set of ten Eurocode Standards for structural design EN 1990 to EN 1999, describing the scope of structures, structural materials and design situations provided for. Particular emphasis is placed on the extensive and comprehensive scope of Eurocode and the way in which unification has been achieved through the implementation of a common system of structural reliability. Aspects of relevance to the revision of SANS 10160 are specifically considered and treated in more detail. A more extensive review of the Eurocode *Basis of structural design* is presented Chapter 1-4.

2 OVERVIEW OF EUROCODE STANDARDS FOR STRUCTURAL DESIGN

It is convenient to take a broad view of Eurocode in terms of its development, general basis, layout and structure, and relation to Member State National Standards, before its technical contents are considered in subsequent sections of this chapter.

2.1 Stages of Development

The Eurocode Standards for the design of structures have evolved through a number of clearly different stages, with different terms of reference applying to each stage: The development was initiated in 1975, and *the objective of the programme was the elimination of technical obstacles to trade and the harmonisation of technical specifications* for a common European construction industry (Eurocode Foreword).

The present stage consists of its development into a set of European Standards (CE 2002), each consisting of a number of Parts, which are in the process of being published by the European Committee for Standardisation (CEN).

In the process of developing Eurocodes EN 1990 – EN 1999, provision is made to accommodate the requirements, conditions and to some degree the preferences of Member States by allowing adjustment of the European Standards. In addition to the updating of the previous set of voluntary Eurocodes ENV 1991 – ENV 1999, the provision for adjustment of the European Standards by Member States to provide for their application as Member State National Standards represents one of the main features of this stage of development.

Implementation of Eurocodes as Member State National Standards (e.g. BS EN 1990 in the UK) represents the final stage of the development of Eurocodes. Although this stage has already started, it still requires substantial effort to be completed: Member State Mirror Committees need to draft a National Annex for each part, to be published by the National Standards Generating Body (e.g. BS EN 1990 National Annex), before any Eurocode Part becomes operational nationally. *The National annex may only contain information on those parameters which are left open in the Eurocode for national choice, known as Nationally Determined Parameters, to be used for the design of buildings and civil engineering works to be constructed in the country concerned.* It is also required that *conflicting National Standards shall be withdrawn* (Eurocode Foreword).

The operational use of Eurocodes as National Standards represents the final stage of the process, although it does not form part of its development as such. In accordance with the initial objectives for Eurocode, this stage represents the use of harmonised standards amongst the present twenty-eight Member States (Austria, Belgium, Cyprus, Czech Republic, Denmark, Estonia, Finland, France, Germany, Greece, Hungary, Iceland, Ireland, Italy, Latvia, Lithuania, Luxembourg, Malta, Netherlands, Norway, Poland, Portugal, Slovakia, Slovenia, Spain, Sweden, Switzerland and United Kingdom).

The final formal Eurocode stage consists of maintenance, which provides for the application of the necessary amendments, promotion of the use of Eurocodes amongst its Member States and outside Europe, increased harmonisation, future revisions and further development (CEN/TC 250 2005).

2.2 Objectives for Development of Eurocode

According to Directives of the European Commission (CE 2002) the Eurocodes are intended to *provide common design methods ... to be used as reference documents for Member States to*:

- *Prove the compliance of buildings and civil engineering works ... with Essential Requirement No 1 Mechanical resistance and stability ... and a part of No 2 Safety in the case of fire*
- *Facilitate the exchange of construction services, to improve the functioning of the internal market;*
- *To increase the competitiveness of the European civil engineering firms, contractors, designers and product manufacturers in their world-wide activities.*

In facilitating the internal European market, Eurocode provides the technical and performance basis for harmonised technical European specifications (hEN) and European Technical Approvals (ETA) for construction products, in that it serves *as a framework for drawing up harmonized technical specification for construction products (hENs and ETAs)* (Eurocode Foreword). *Material hENs and ETAs shall take the technical requirements of the EN Eurocodes into account so that the assumptions of design according to the EN Eurocodes are met* (CE 2002).

2.3 Consistency with ISO Standard on Reliability for Structures

The Eurocodes are generally compatible with the ISO Standards on structural design, and in particular with ISO 2394:1998 *General principles on reliability for structures*. The Eurocodes develop the ISO 2394 concepts at higher levels of abstraction into practical and operational design rules. This is primarily achieved through the head Eurocode Standard EN 1990 *Basis of structural design*, which is then applied practically in the procedures of the subsequent Standards EN 1991 – EN 1999.

In ISO 2394 two alternative approaches are presented through which appropriate levels of reliability can be achieved: (i) Full probabilistic modelling of structural reliability, (ii) Limit states design procedures for which partial factors and specification of basic variables are established (calibrated) on reliability principles. Although Eurocode nominally allows the first method, its approach is effectively based on reliability based partial factor limit states design procedures.

The prominent status for the development of the reliability principles of limit states design can however be traced back to the reference of Eurocode to ISO 2394. The implications of the innovation of introducing a separate standard, EN 1990, formulating the general design approach and philosophy are discussed below.

2.4 Scope and Arrangement of the Eurocode Standards

As indicated in the Introduction, the set of Eurocodes consists of the ten Standards EN 1990 – EN 1999 as listed in Table 1. Respective Standards are published as independent Parts, with the number of Parts used to cover individual topics also indicated in Table 1. In the case of EN 1991, the respective Parts cover the various actions on structures, consisting of seven general actions and three specific actions. The number of Parts for concrete design was consolidated to four in the conversion of ENV 1992 to EN 1992, but the large number of Parts (twenty) was maintained in EN 1993 for steel structures. The complete list of Eurocode Standards and Parts is listed in Annex A.

Table 1 List of Eurocode Standards, indicating the number of separate Parts

STANDARD	TITLE	# of Parts
EN 1990	*Basis of structural design*	1
EN 1991	Eurocode 1 : *Actions on structures*	10
EN 1992	Eurocode 2 : *Design of concrete structures*	4
EN 1993	Eurocode 3 : *Design of steel structures*	20
EN 1994	Eurocode 4 : *Design of composite steel and concrete structures*	3
EN 1995	Eurocode 5 : *Design of timber structures*	3
EN 1996	Eurocode 6 : *Design of masonry structures*	4
EN 1997	Eurocode 7 : *Geotechnical design*	2
EN 1998	Eurocode 8 : *Design provisions for earthquake resistance of structures*	6
EN 1999	Eurocode 9 : *Design of aluminium alloy structures*	5
	TOTAL	**58**

The fifty-eight Eurocode Parts can be arranged to indicate their respective groupings and relations as presented in Table 2. In addition to first presenting *general rules* (sometimes the term *common rules* are used) for each Standard, the range of structures includes *buildings, bridges*, with the remaining structures that can broadly be classified into *industrial* and *containment structures*.

The rules for *buildings* are generally combined with the general/common rules in the parts EN 199X-1-1, together with specific provision made in EN 1990 *Basis of structural design* Annex A1 *Buildings* and EN 1991-1-X *Actions on structures - General actions*, providing for *self-weight and imposed loads, fire, snow, wind*, etc.

The rules for *bridges* are given in EN 1991-2 for traffic loads, EN 199X-2 for concrete, steel, composite and timber materials respectively and EN 1998-2 for earthquake resistance of bridges. It should be noted that provision is also made for bridges in EN 1990 *Basis of structural design* Annex A2 and EN 1991-1-X *Actions on structures - General actions*: Part 4 *Wind actions*; Part 5 *Thermal actions*; Part 6 *Actions during execution*; Part 7 *Accidental actions*.

Another prominent theme in Eurocode is that of *structural fire design*: In addition to a separate Part, EN 1991-1-2 *General actions – Actions on structures exposed to fire*, six respective Parts are provided for the various structural materials for concrete, steel, composite, timber, masonry and aluminium alloy in EN 199X-1-2 respectively. Such explicit and comprehensive treatment of structural fire design directly derives from the fact that this topic represents part of Essential Requirement #2 of the European Commission Directive (CE 2002).

Table 2 **Summary and arrangement of the 58 Eurocode Parts**

Standard (Specifications)	Buildings & General	Fire	Bridges	Containment	Industrial
EN 1990 Basis of Design	**Annex A1 (Normative) Buildings**		A2 (N) Bridges		
EN 1991 Actions on structures	**Self-weight; Imposed**; Snow; **Wind**; **Thermal**; **Execution**; **Accidental**	Actions on structures exposed to fire	Traffic loads on bridges	Silos & tanks	**Cranes & machinery**
EN 1992 Concrete structures	General; buildings	Fire design	Concrete Bridges	Liquid retaining	
EN 1993 Steel structures	General ; buildings; Joints	Fire design	Steel Bridges	Shells	Crane support
Plated I; Plated II	Cold formed; Stainless S			Silos; Tanks	Towers & masts
Toughness; Tension	High strength; Fatigue			Pipelines	Chimney; Piling
EN 1994 Composite	General; buildings	Fire design	Composite Bridges		
EN 1995 Timber	General, buildings	Fire design	Timber bridges		
EN 1996 Masonry	General; Execution; Simplified	Fire design			
EN 1997 Geotechnical design	**General rules**				
	Investigation & testing				
EN 1998 Earthquake resistance	**General rules, actions**		Bridges	Silos, tanks, pipes	Tower, mast, chimney
	Strengthening & repair				
	Foundations, retaining				
EN 1999 Aluminium Alloy	General rules; Shells	Fire design			
	Fatigue; Sheeting				

NOTE Shaded cell indicates Parts relevant to SANS 10160

2.5 Level of Application of Eurocode Procedures

In the Eurocode Foreword the field of application is stated as follows: *to provide common structural design rules for everyday use for the design of whole structures and component products of both traditional and innovative nature. Unusual forms of construction or design conditions are not specifically covered and additional expert consideration will be required by the designer in such cases.*

Nevertheless certain rather specialist structures and design conditions are covered within the wide scope of Eurocode! An important example within the scope of the SA loading code is the procedures provided for dynamic effects of structures exposed to wind loads. Structural fire design is another example, which is defined and goes beyond the (present) scope of the SA loading code.

The effective advanced nature of design procedures provided for in Eurocode is a direct implementation of the intention with its development, which is to provide comprehensively for the enhancement of trade across Europe in the building and civil engineering field. However it does raise a concern that the procedures may be applied by the designer without having the specialist knowledge for using the procedures.

2.6 Nature of Formulation of Eurocode Standards and Parts

Another implication of the comprehensive scope of application of Eurocode is that the large bulk of the requirements and design rules needs to be presented in an economic and sparse manner. This required the implementation of two main measures:

- No duplication or redundancy is allowed, requiring that procedures or rules are only presented once. The implication is that related topics are presented in different sections, or even Parts of a Standard.
- The layout of various topics is to compile the general or common design rules in the first Part of a Standard, with subsequent Parts *only* providing the additional design rules that need to be considered for specific applications, such as Part 199X-2 providing the rules for bridges in addition (or modifying) the rules given in Part EN 199X-1-1.

2.7 Provision for National Considerations

The high degree of harmonisation achieved by a common structural design Standard applicable to all Member States could only be achieved through allowance for national choice of Nationally Determined Parameters (NDP). The incorporation of NDP options in the National Annex for each Eurocode Part provide some controlled release of the requirements of the European Standard, to provide for *values to be used where a symbol only is given, ... country specific data (geographic, climatic, etc.), ... procedure to be used where alternative procedures are given, ... decisions on the application of informative annexes, non-contradictory complementary information (NCCI) to assist the user* (CE 2002).

Provisions for national conditions, responsibility for safety and a limited degree of preference in selected practice are made within a transparent framework of allowances and constraints. Early indications of the effects of NDP options show the effectiveness of balancing the need for harmonisation and special conditions and preferences by careful formulation and selection of parameters. The high degree of harmonisation achieved, but with transparent allowance of freedom for national choices, represents a major achievement of Eurocode.

On the other hand there is also clear evidence of the use of NDP options to adjust Eurocode procedures nationally as much as possible to achieve procedures which are as close as possible to existing local practice. In some cases the process reflects proper concerns about the implications of the effects of Eurocode procedures with regard to present requirements, although this may not always be the case.

As a result a concerted effort is made to gather information on the selection of NDP options by Member States in order to improve harmonisation further in the future. The compilation of all the NDP values and options selected by Member States into a database set up by the Ispra Joint Research Centre in Italy forms part of this process.

2.8 Differentiated Normative Requirements

The general distinction between normative stipulations and recommendations or information is refined in Eurocode through the following differentiated process:

- **Principle clauses** (indicated with a P e.g. Clause 2.1(1)P), for which there is no alternative (specifying that *compliance **shall** be proved*) and
- **Application rules** which are generally recognised rules which comply with the principles and satisfy their requirements (specifying that *compliance **should** be proved*).

- **Alternative design rules** are permissible
  - But may not be acceptable for CE Marking in structural product standards.
  - Where an alternative is allowed as an NDP in a clause, this is indicated through a NOTE to the clause. These clauses are listed in the Foreword to each respective Part.

- **Informative guidance** is also provided in the form of
  - A NOTE within the normative text provides information and guidance, which ***may*** be used;
  - An Annex (Informative) provides more extensive guidance, supporting procedures and information;
  - The National Annex may specify whether an informative annex may or may not be used, or *should* be used normatively!

An illustration of clauses representing Principle requirements and Application Rules is presented in Appendix B.

3 EUROCODE TECHNOLOGY BASIS AND CHARACTERISTICS

At face value Eurocode represents an extensive advance in structural design standardisation in terms of its comprehensive scope, which is consistently covered by a large number of Standards and Parts, representing a high degree of harmonisation across a diverse range of conditions and practice across the Member States.

Of equal importance however, is the technology basis which it represents, both in terms of the technology from which Eurocode has been developed, and the associated support for its implementation which is presently under development.

3.1 Technological Basis for Eurocode

The development of Eurocode provided the opportunity to incorporate information from a wide technology base into an integral set of standards. The best procedures and structural mechanics models were selected from Member States or from recent research, often done specifically for implementation in the Standards. The following examples illustrate this process:

- Developed from selected standards from Member States:
  - Procedures for robustness design against unidentified accidental actions are based on British standardised design practice for robustness; various models are used in procedures for impact and internal explosions (Vrouwenvelder *et al* 2005)
  - Crane induced actions are derived from a widely accepted German DIN Standard (Sedlacek & Grotman 1996), which also serves as basis for similar ISO and Australian standards.
- Introduction of new standards and procedures based on research and model standards:
  - **Accidental actions:** Introducing a new class of design situations, actions and resistance stipulations, thereby providing a logical framework for robustness, fire design, and earthquake resistance amongst other situations
  - **Structural fire design:** Developed extensively in concerted research on a Global Fire Safety Concept (Schleich 2005); incorporating structural behaviour – BRE Cardington fire tests
  - **Self-weight and imposed floor load:** Procedures are based on the CIB Model Codes (CIB 1989a; CIB 1989b) and a synthesis of European practice (Hemmert-Halswick *et al* 1988)
  - **Pressure coefficients for wind actions:** Derived from extensive wind tunnel testing (Vrouwenvelder & Steenbergen 2001);
  - **Traffic loads on bridges:** Based on extensive surveys and research (Croce 2001);
  - **Geotechnical design:** incorporated into the process of structural design with the necessary unification (Savidu *et al* 2001)

3.2 Supporting Information

The Eurocode Standards are also supported by a wealth of information in the form of guidebooks, introduction courses, handbooks and software. The supporting information is intended to explain the procedures and assist in their introduction and use, as opposed to the background information which provides the source for the selection and formulation of the requirements and design rules. Examples of supporting information include:

- Introductory courses are presented as part of the activities of the European wide implementation process:
  - Development of skills facilitating implementation of Eurocodes. Handbook 1 – 5. (Leonardo da Vinci Pilot Project 2005);
  - Eurocodes Workshop: *Dissemination of information for training.* Brussels 18 – 20 February 2008. (JRC/DG-ENTR 2008)
- Designers Guides: A series of Designer's Guides is being published, generally by key members of the development program, such as:
  - Gulvanessian *et al* (2002) *Designers guide to EN 1990 Eurocode*,
  - Narayanan & Beeby (2005) *Designers guide to EN 1992 Design of concrete structures*
- National activities to provide information and training, of which the British activities are typical examples, but which are also of relevance from the South African perspective:
  - Eurocodes News, published by Thomas Telford (quarterly)
  - Courses, background reports, software, published by the Institution of Structural Engineering, the UK Concrete Centre, the UK Steel Construction Institute, the British Standards Institute, the UK Department for Communities and Local Government, etc.

4 EUROCODE EN 1990:2002 *BASIS OF STRUCTURAL DESIGN*

The extensive scope of Eurocode is indicated in the outline given above: It provides for a comprehensive range of structural materials and structure classes, providing for a diverse range of natural conditions and structural engineering practice legacies, presented in ten Standards which consist of fifty-eight Parts. Such a comprehensive scope for Eurocode clearly requires a strong central basis in order to achieve the intended harmonisation amongst Member States externally, and to ensure proper unification among the various Parts, rules and procedures internally.

Such central basis is provided by the explicit formulation of the reliability based limit states design procedure, as indicated above. A more elaborate exposition of the reliability basis of EN 1990 and the way in which it is applied in limit states design verification procedures is given in Chapter 1-4.

According to Gulvanessian *et al* (2002) the Eurocode Standard EN 1990:2002 *Basis of Structural Design* plays a key role as *the head document of the Eurocode suite and describes the principles and requirements for safety, serviceability and durability of structures*. EN 1990 therefore plays a central role in the unification of structural design into an integral process, as presented across the various Standards and Parts. The relationship between EN 1990 as head document to the Standards EN 1991 – EN 1999 is shown schematically in Figure 1.

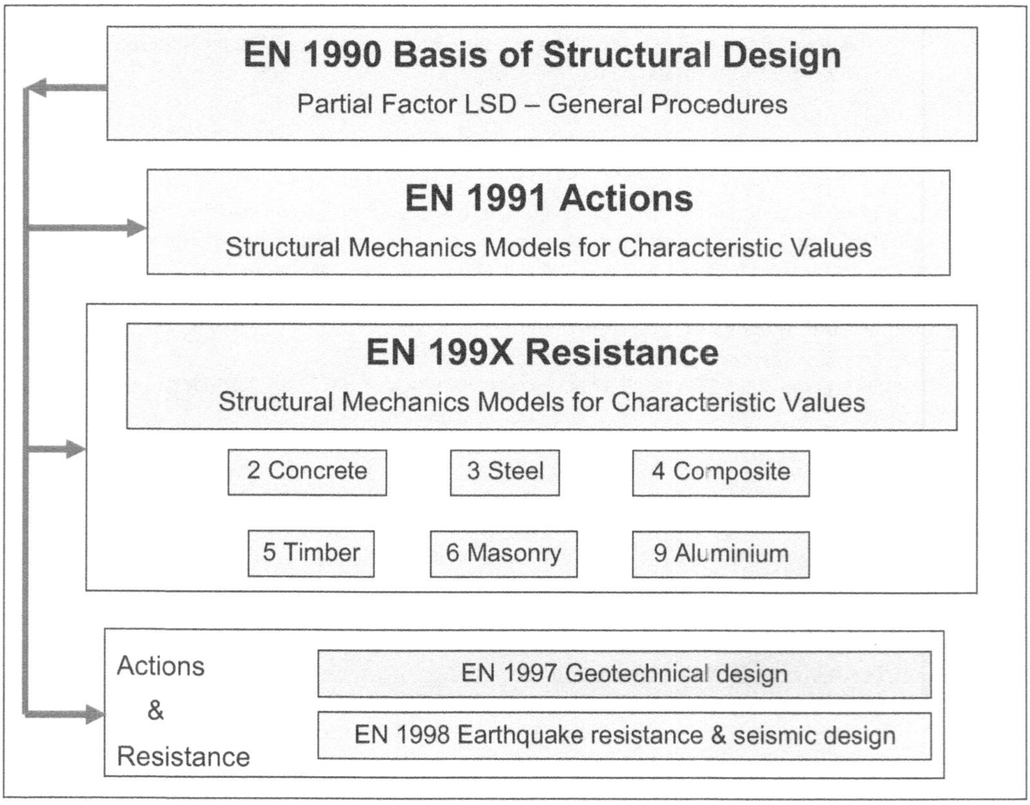

Figure 1 Schematic arrangement of Eurocode Standards

Although the function of the EN 1990 *Basis of structural design* requirements is that they provide important information regarding the specification of design actions on the structure (as determined generally from EN 1991) the requirements are not limited to this function. EN 1990 also sets important requirements regarding structural performance in terms of appropriate (design) resistance against these actions. Therefore the EN 1990 requirements apply to EN 1991 as well as to EN 1992 – EN 1999.

Incidentally the separate arrangement for EN 1997 *Geotechnical design* and EN 1998 *Design of structures for earthquake resistance* in Figure 1 derives from the characteristics of these two design fields of application that make it more effective to treat them integrally, as opposed to having separate procedures for actions and resistance. Such an arrangement is often interpreted in terms of that the two fields are related to the support of the structure.

4.1 Function of Head Standard EN 1990

The function of EN 1990 is often said to provide the *material independent* requirements for structural design. This includes the principles of structural performance in reliability terms, the setting of a related reliability framework and the principles of limit states design procedures through which design could be done and compliance verified. The limit states procedure consists generally of the specification of actions on structures and structural resistance, and their relationship, in terms of basic variables, and an array of partial factors, for an extensive arrangement of limit states.

4.1.1 Unified design of individual structures

The obvious unification achieved in Eurocode by the use of a common *basis of structural design* is the integral and serial design process for actions and resistance, with the unique incorporation of geotechnical and seismic design. In fact the need for a unified and harmonised approach to consider earthquake resistance played a key role in the initiation and technical motivation of the Eurocode development (Bossenmayer 2003).

4.1.2 Unification of design for various structural materials

Another important and unique degree of unification that is achieved in Eurocode through sharing a common *basis of structural design* is the parallel consistency between the design requirements and rules for the various structural materials. This effect is specifically important in the case of EN 1994 *Design of composite steel and concrete structures*, which are fully consistent with EN 1992 *Design of concrete structures* and EN 1993 *Design of steel structures*!

However, such consistency is also achieved for the design rules for such obviously diverse materials as steel and geotechnical materials at the extremes, but including aluminium alloy, timber and masonry. The coherent application of limit states design procedures for the various materials represents an important achievement for Eurocode.

4.2 Principles and Application of the Basis of Structural Design as Presented in EN 1990

The principles of the basis of structural design as presented in EN 1990 essentially consist of formulating the requirements for structural performance in reliability terms. The application rules are then presented in terms of partial factor limit states procedures, first in general terms to define the various limit states, and subsequently in design verification procedures for specific limit states.

It is important to note that the application rules are considered to be sufficient to satisfy the requirements specified by the Principle clauses, always of course within the scope of application of these rules: Clause 1.4(4) *The Application Rules are generally recognised rules which comply with the Principles and satisfy their requirements*.

The basic requirement for structural performance is given in Clause 2.1(1)P *A structure shall be designed and executed in such a way that it will, during its intended life, with appropriate degrees of reliability and in an economical way*

- *sustain all actions and influences likely to occur during execution and use, and*
- *remain fit for the use for which it is required*

These requirements are then met by applying design verification such that *limit states shall be related to design situations* (EN 1990 Clause 3.1(3)P) and *the selected design situations shall be sufficiently severe and varied so as to encompass all conditions that can reasonably be foreseen to occur during the execution and use of the structure* (Clause 3.2(3)P).

4.3 Scope and Contents

Following the development from general requirements to specific procedures and application rules generally applied in the formulation of Eurocode, in EN 1990 a non-normative foreword is followed by six normative Sections and two Annexes (Normative) and three Annexes (Informative):

- **Foreword:** Provides a standard Eurocode foreword, matters relating to EN 1990 and its National Annex:
    - Standard foreword, including background, field of application, relation to national standards and harmonised technical specifications (ENs and ETAs) for CE Marking of structural products;
    - EN 1990 scope of describing the principles and requirements for safety, serviceability and durability of structures; guidelines for design cases beyond the Eurocode scope;
    - NDP clauses, all placed in Annex A1 (Normative) for buildings and Annex A2 (Normative) for bridges.
- **Section 1 General:** The scope of structures and requirements for their safety, serviceability and durability; normative references; assumptions regarding appropriately qualified personnel, supervision, materials, use and maintenance, and terms and definitions are presented in this Section.
- **Section 2 -4 Requirements, principles of LSD & basic variables:**
    - Basic requirements of reliability and economy; reliability management procedures; the design working life; durability requirements; quality management;
    - Limit states and design situations;
    - Characteristic values of basic variables of actions, including their classification; material and product properties; and geometrical data.
- **Section 5 Structural analysis and design assisted by testing:** Structural modelling including static and dynamic actions; structural fire design; design based on a combination of tests and calculations.
- **Section 6 Verification by the partial factor method:** General specification for design values of the effects of actions; material and product properties; geometrical data; design resistance *in symbolic terms*; action combination schemes for limit states and design situations.
- **Annex A1 (Normative) Application for buildings:** Values for combination factors and partial factors for the respective limit states are presented, for equilibrium (EQU), structural/geotechnical (STR/GEO), accidental and seismic and serviceability; recommended numerical values for combination and partial factors.
- **Annex A2 (Normative) Application for bridges:** A similar application of the general specifications of Sections relevant to bridge structures is presented.
- **Informative Annexes:**
    - Annex B: Management of structural reliability for construction works giving guidance regarding reliability differentiation; in terms of consequence classes;
    - Annex C: Basis for partial factor design and reliability analysis giving information and theoretical background to the partial factor method.
    - Annex D: Design assisted by testing, giving guidance in considering types of tests; their planning; derivation of design values; statistical principles.

4.4 Abstract of Introductory Sections

Pertinent features of the introductory sections of EN 1990 are the following:

4.4.1 Scope of Standard:

The scope of the standard is so defined as to establish principles and requirements for the safety, serviceability and durability of structures for a comprehensive range of *buildings and civil engineering works*, with specific reference to the inclusion of provisions for:

- geotechnical aspects;
- structural fire design and situations involving earthquakes;
- execution and temporary structures.

4.4.2 Reliability differentiation:

In addition to specifying the principle of achieving the required levels of reliability, reliability differentiation is allowed by reliability management by:

- adoption of different levels of reliability;
- measures to reduce errors in design and execution, and gross human errors;
- various design matters, such as robustness (structural integrity), investigation of soils and environmental influences, mechanical models used and detailing.

4.4.3 Design situations:

The concept of design situations (sets of physical conditions representing the real conditions during a certain interval for which the design will demonstrate that relevant limit states are not exceeded) is introduced, with design situations classified as

- **Persistent** referring to normal use;
- **Transient** for shorter reference periods;
- **Accidental** for exceptional conditions;
- **Seismic** to provide for earthquake resistance.

4.4.4 Basic variable characteristic values:

The specifications of characteristic values for basic variables are summarised in Table 3.

4.5 Technology Basis for EN 1990

The technology basis for Eurocode discussed above can be illustrated by the wealth of information on the basis of design which is available in support of EN 1990. Pertinent literature related to the background to EN 1990, models and procedures for its calibration by Member States, and guidelines for its introduction to and use by designers are summarised in Table 4.

Table 3 Summary of defined characteristic values of basic variables

BASIC VARIABLES	GENERAL SPECIFICATIONS	COMMENTS
	Actions	
Permanent - large variability - sensitive structure	Nominal dimensions and mean unit masses - use 5/95% (lower/upper) fractile values	Expected (average) value Structure is sensitive to variability of permanent actions
Variable - combination ($\psi_o Q_k$) - frequent ($\psi_1 Q_k$) - quasi-permanent ($\psi_2 Q_k$)	Probability of 0,02 per annum - characteristic equivalent together with leading variable action - exceeded 0,01 of reference period - exceeded 0,5 of reference period	Expected maximum over 50 years Equivalent expected combined maximum Average value
Accidental	Specified for individual projects	Not expected during design life
Fatigue	According to relevant parts of EN 1991	Generally average fatigue actions
Geotechnical	According to EN 1997-1 *Geotechnical design – General*	*Cautious* estimate of expected values
	Material & Product Properties, Geometrical Data	
Material (X_k) Product (R_k)	Prescribed probability, generally 5/95% fractile	
Geometry (a_k)	Specified dimensions on drawings	Nominal (average) value
- imperfections	Prescribed fractile given in EN 1992 to EN 1999	Related to resistance reliability

Table 4 Extract of technology basis for EN 1990

	Stage	Reference
1	Reference	ISO 2394 (1998/1986) *General principles on reliability for structures*
2	Background	JCSS (1996) *Background documentation: Eurocode 1* (ENV 1991) *Part 1 Basis of design*
3	Background	IABSE (1996). *Colloquium Background and application of Eurocode 1.*
4	Calibration	JCSS (2001) *Probabilistic model code*
5	National calibration	SAKO (1999) *Basis of design of structures: Proposal for modification of partial safety factors in Eurocodes*
6	Introduction	Leonardo da Vinci Project (2005). Implementation of Eurocodes: *Development of skills facilitating implementation of Eurocodes.* Handbook 1 *Basis of structural design*
7	Application	Gulvanessian H, Calgaro J-A & Holický M (2002). *Designer's guide to EN 1990 Eurocode: Basis of structural design*

5 OVERVIEW OF EUROCODE EN 1991 *ACTIONS ON STRUCTURES*

The specification of actions in EN 1991 *Actions on structures*, which provides for actions from both environmental and occupational sources, is complemented by actions originating from geotechnical processes in EN 1997-1 *Geotechnical design – General rules* and seismic processes in EN 1998-1 *Design provisions for earthquake resistance of structures* respectively.

The general basis for actions as specified in the voluntary Eurocode version ENV 1991 is presented in an IABSE colloquium (IABSE 1996), with updated information on the normative Eurocode version EN 1991 in a Leonardo da Vinci Project seminar (2005).

The various actions on structures specified in Eurocode EN 1991 are presented in ten separate Parts. The Eurocode Parts specifying the various actions are listed in Table 5, with the actions potentially relevant to the revision of SANS 10160 shown in colour code. The specification of actions on structures in EN 1991 is done in accordance with EN 1990, which serves the function of providing the basis for the determination of actions on structures, including the classification system for actions, defining characteristic values, and other related requirements and design rules for determining their appropriate properties.

Table 5 Actions on structures specified in Eurocode

EN 1991	*Actions on structures*
General actions	
EN 1991-1-1	*General actions - Densities, self-weight, imposed loads for buildings*
EN 1991-1-2	*General actions - Actions on structures exposed to fire*
EN 1991-1-3	*General actions - Snow loads*
EN 1991-1-4	*General actions - Wind actions*
EN 1991-1-5	*General actions - Thermal actions*
EN 1991-1-6	*General actions - Actions during execution*
EN 1991-1-7	*General actions - Accidental actions due to impact and explosions*
Actions related to function of the structure	
EN 1991-2	*Traffic loads on bridges*
EN 1991-3	*Actions induced by cranes and machinery*
EN 1991-4	*Actions in silos and tanks*
Actions through foundations of structures	
EN 1997-1	*Geotechnical design: General rules*
EN 1998-1	*Design provisions for earthquake resistance of structures: General rules - Seismic actions and general requirements for structures*
Colour code:	Relevant to the revision of SA Loading Code
	Outside scope of SA Loading Code

Actions are classified in accordance with the requirements of EN 1990 terms of:

- Variation in time – permanent, variable or accidental
- By their origin – direct or indirect
- By their spatial variation – fixed or free
- By their nature or structural response – static or dynamic

Such classification has a direct bearing on how an action is to be treated in determining action effects. Specific meaning of these terms is provided by their definition as given in Table 6.

Table 6 Classification of actions

Term	Definition
Permanent action (*G*)	Action that is likely to act throughout a given reference period and for which the variation in magnitude with time is negligible, or for which the variation is always in the same direction (monotonic) until the action attaints a certain limit value NOTE If the variability of *G* cannot be ignored, or the structure is sensitive to variations in G, two values $G_{k,sup}$ and $G_{k,inf}$ should be used..
Variable action (*Q*)	Action for which the variation with time is neither negligible nor monotonic
Accidental action (*A*)	Action, usually of short duration but of significant magnitude, that is unlikely to occur on a given structure during its design working life NOTE 1 An accidental action can be expected in many cases to cause severe consequences unless appropriate measures are taken. NOTE 2 Impact, snow, wind and seismic actions may be variable or accidental actions, depending on the available information on statistical distributions
Seismic action (A_E)	Action that arises due to earthquake ground motions
Geotechnical action	Action transmitted to the structure by ground, fill or groundwater
Fixed action	Action that has a fixed distribution and position over the structure or structural member such that the magnitude and direction of the action are determined unambiguously for the whole structure or structural member if this magnitude and direction are determined at one point on the structure or structural member
Free action	Action that may have various spatial distributions over the structure
Dynamic action	Action that causes significant acceleration of the structure of structural members
Static action	Action that does not cause significant acceleration of the structure or structural members
Quasi-static action	Dynamic action represented by an equivalent static action in a static model

5.1 General Actions

5.1.1 Self-weight, densities and imposed loads for buildings (EN 1991-1-1)

Actions due to self-weight and imposed loads are specified in EN 1991-1-1 *General actions – Densities, self-weight, imposed loads for buildings.*

Self-weight of construction works:

The self-weight of the construction works should, in most cases, be represented by a single characteristic value and be calculated on the basis of the nominal dimensions and the characteristic values of the densities (EN 1991-1-1 Clause 5.1(1)).

Self-weight is specified as a permanent action with single characteristic value consisting of structural and non-structural elements and fixed services. EN 1990 specifies that upper and lower values *shall* be used when the variability cannot be considered as small.

According to CIB (1989a) the variability of non-structural self-weight could be high. Where the self-weight can vary with time, it should be taken into account by the use of upper and lower characteristic values. If there is doubt as to the permanency of self-weight it should be treated as a variable imposed load (Holicky 2001).

- **Annex (Informative):** Densities of materials:
  - Building materials
  - Additional materials for bridges and stored materials
  - Angle of repose for specific stored materials.

Imposed loads on buildings:

Imposed floor and roof loads for buildings are those arriving from their occupancy, deriving from normal use, furniture or moveable objects, and anticipated rare events (concentration of people or stacking of objects) (CIB 1989b).

- Within one storey or roof, imposed loads shall be taken into account as a free action to be applied to the most unfavourable part of the influence area of the load effect under consideration.
- Loads on other relevant storeys may be assumed as uniformly distributed fixed actions.

Imposed loads are categorised according to occupancy as

- Residential (A); Office (B); Congregated people (C); Shopping (D);
- Storage and industrial E1 – storage; E2 – industrial;
- Actions induced by forklifts FL1 to FL6
- Traffic and parking in buildings for vehicle gross weight: F < 30 kN; G 30 to 160 kN;
- Roofs: not accessible (H); accessible (I (as A to D));
    - Roof helicopter landing (HC1 < 20 kN; HC2 20 kN to 60 kN)
- Partition walls and parapets: Occupational areas (A to G).

Values for specified imposed loads are based on the comparison of practice from various countries (Hemmert-Halswick *et al* 1988) and against reliability based models (CIB 1989b). Values of imposed loads are specified as NDP. Ranges of values are allowed, with recommended values generally tending to be at the upper limit of the range (55%: versus mid range – 15%; lower range – 30%).

5.1.2 Actions on structures exposed to fire (EN 1991-1-2)

Provision for fire safety is an Essential Requirement (ER2 *Safety in the case of fire*) of the European Commission Construction Products Directive (CPD) (CE 2002, Holicky & Markova 2005). The objective ultimately is to develop performance based procedures, based on the Global Fire Safety Concept (Schleich 2005). Structural fire actions are classified as accidental actions, and are to be combined with other actions specified in EN 1991 according to the action combination rules of EN 1990. Provision is made for design based on compliance with

- prescriptive rules based on thermal actions given by a nominal fire;
- performance-based design using physically based parametric fires.

Specifications for structural fire actions from EN 1991-1-2 *General actions – Actions on structures exposed to fire … are intended to be used in conjunction with the fire design Parts* (Parts 199X-1-2) *of EN 1992 to EN 1996 and EN 1999* (for the respective structural materials of concrete, steel, composite steel/concrete, timber, masonry and aluminium alloy) *which give rules for designing structures for fire resistance.*

5.1.3 Wind actions (EN 1991-1-4)

Specifications of natural wind actions for the design of *buildings and civil engineering works with heights up to 200 m* and *bridges having no span greater than 200 m, provided that they satisfy the criteria for dynamic response,* are provided in EN 1991-1-4 *General actions – Wind actions*. The specifications are *intended to predict characteristic wind actions on land-based structures, their components and appendages.* Pertinent features of the wind action model are (Vrouwenvelder 2001):

- **basic wind velocity:** characteristic 10 minute mean wind velocity; wind map with indicative values generally 20 – 30 m/s;
- **terrain categories:** I – sea/coastal, to V – 15% coverage of buildings > 15 m;
- **variation with height**: logarithmic; starting from 1 – 10 m;

- **pressure coefficients:** elaborate set of tables from latest wind engineering research;
- **classification:** variable; fixed (distribution across structure for given direction); quasi-static, dynamic (causing resonance), aero-elastic (response influencing air flow);
- **non-standard configurations:** – procedures for structural size and dynamic factors;

The following structures or structural response to wind actions are excluded from the scope of the Part on wind actions:

- lattice towers with non-parallel chords, guyed masts and chimneys,
- torsional vibration of buildings,
- bridge deck vibrations from transverse wind turbulence, cable supported bridges,
- vibrations where more than the fundamental mode needs to be considered.

5.1.4 Thermal actions (EN 1991-1-5)

The calculation *of thermal actions on buildings, bridges and other structures including their structural elements* is specified in EN 1991-1-5 *General actions – Thermal actions*. Provision is made for changes in the temperature of structural elements for structures exposed to daily and seasonal climatic changes, and for structures in which thermal actions are mainly a function of their use, such as cooling towers, silos, tanks, warm and cold storage facilities, hot and cold services.

Thermal actions are considered to be variable and indirect. Characteristic values have an annual probability of being exceeded of 2%, equivalent to a return period of 50 years. Thermal actions are based on maximum and minimum shaded air temperatures, adjusted for elevation, with season, colour and orientation taken as parameters to provide for radiation effects.

5.1.5 Actions during execution (EN 1991-1-6)

Requirements for *the determination of actions which should be taken into account during execution of buildings and civil engineering works* are specified in EN 1991-1-6 *General actions – Actions during execution*. The Standard provides the classification of actions, identification of design situations with regard to the respective limit states, the representation of actions, normative annexes providing supplementary rules for buildings and bridges, and an informative annex on actions on structures during alteration, reconstruction or demolition.

5.1.6 Accidental actions due to impact and explosions (EN 1991-1-7)

The general treatment of accidental actions is specified in EN 1990. The scope of the Part EN 1991-1-7 *General actions – Accidental actions* includes together with actions, also *rules for safeguarding buildings and other civil engineering works against accidental actions. For buildings it also provides strategies to limit the consequences of localised failure caused by an unspecified accidental event.* Exceptional natural conditions (snow, wind) not to be expected during the life of the structure can also be accepted as accidental situations (Vrouwenvelder *et al* 2005).

General principles:

Strategies to consider accidental actions and consequences of localised failure consist of:

- avoid disproportionate damage by preventing, reducing the accidental action;
- design to sustain the action;
- strategies against the effects of localised failure through enhanced redundancy;
- key element design on which the structure is particularly reliant;
- prescriptive rules for minimum ductility to provide for integrity.

Failure from an unspecified cause:

The requirements for ensuring sufficient robustness of structures are specified in terms of provisions for failure from an unspecified cause. Requirements to consider the consequences of localised failure are given in a normative section. The requirements are supplemented by procedures given in Annex A (Informative) *Design for consequences of localised failure in a building structure from an unspecified cause*. In this annex rules and methods are provided for designing buildings:

- to sustain localised failure from unspecified cause without disproportionate collapse;
- in terms of consequence classification of buildings.

Impact and internal explosions:

Specifications for the treatment of accidental situations due to impact from road vehicles, fork lift trucks, rail traffic, ship traffic on rivers and canals, and others. Limited provisions for internal explosions are supplemented by an informative annex considering dust explosions in rooms and silos, natural gas explosions, and explosions in road and rail tunnels.

Risk assessment procedures:

Annex B (Informative) *Information on risk assessment* provides procedures for risk assessment. The intention is that where design for accidental design situations is specified for individual projects, the design approach and appropriate accidental actions should be based on risk assessment, for which guidelines are given in this annex.

5.2 Actions Induced by Cranes and Machinery (EN 1991-3)

5.2.1 Actions induced by cranes

The Eurocode provisions for actions induced by cranes on supporting building structures are based on the respective German DIN Standard (Sedlacek & Grotman 1996). The selection of the German Standard represents a case where the most advanced procedures are selected from a Member State for incorporation into Eurocode. The German standard also forms the basis for the equivalent ISO and Australian standards.

Specifications for the ... *imposed loads (models and representative values) associated with cranes on runway beams ... which include, when relevant, dynamic effects and braking, acceleration and accidental forces* ... are given in EN 1991-3 *Actions induced by cranes and machinery*, Section 2 *Actions induced by hoists and cranes on runway beams*. The following are the main topics which are treated in Section 2:

- **Design situations and action combinations:** The Standard provides for the identification of the relevant design situations, including the combination of actions, for the ultimate, fatigue, accidental and the serviceability limit state.

- **Groups of loads:** The different ways in which various actions are to be combined into a group of loads to be considered as a single characteristic crane action is clearly specified. Provision is made for seven ULS groups, a test load group and two accidental load groups.

- **Fatigue:** The effects of fatigue loads are taken into account in terms of the fatigue damage equivalent load which takes the distribution of the hoist loads and the effects of the variation of the crane positions into account

5.2.2 Actions induced by stationary machinery

Section 3 of EN 1991-3 *applies to structures supporting rotating machines which induce dynamic effects in one or more planes. This section presents methods to determine the dynamic behaviour and action effects to verify the safety of the structure.*

- **Design situations** are selected for verifying that the service conditions of the machinery are in compliance with the requirements regarding the machine itself as well as the impact of the machine on the surroundings. The relevant design situations for the ultimate and the fatigue limit states are also identified.

- **Actions:** The actions induced by stationary machinery are classified as permanent, variable and accidental actions, in accordance with EN 1990:
  - **Permanent actions** during service include the self-weight of all fixed and movable parts of the machinery as well as static actions from service such as friction forces, forces caused by thermal expansion and forces caused by flow of fluids and gases.
  - **Variable actions** are related to accelerated masses.
  - **Accidental actions** may occur from accidental magnification of the eccentricity of rotating masses, mis-synchronisation between generator and machines or impact effects (water hammer) from pipes during sudden shutdown.

6 GEOTECHNICAL DESIGN AND ACTIONS ON STRUCTURES: EN 1997 *GEOTECHNICAL DESIGN*

The development of EN 1997 *Geotechnical design* represents one of the major advances achieved by Eurocode. The compilation of a comprehensive standard for limit states geotechnical design represents an important advance in geotechnical practice.

- A noteworthy aspect of this achievement is the degree of harmonisation that could be attained in the process across Europe.
- Another significant achievement is the unification between geotechnical and structural design represented by Eurocode the procedures incorporated in EN 1997 and EN 1990.

EN 1997-1 *Geotechnical design – General rules* provides an integral treatment of the basis of geotechnical design, geotechnical actions and resistance, in a unified manner, with reference to EN 1990 and EN 1991. The provision of general rules for geotechnical design of EN 1997-1 is supplemented by Part EN 1997-2 on the performance and evaluation of field and laboratory testing. EN 1997-1 covers the following topics, of which only the first topic is summarised here:

6.1 Introductory Topics of EN 1997-1

- **Basis of geotechnical design**, including three alternative approaches for the ultimate limit state (Design Approach 1, 2 and 3); which are also given in EN 1990 Annex A1 for buildings. Additional information is given in two annexes:
  - Annex A (Normative) providing recommended values for partial factors;
  - Annex B (Informative) providing background to the three alternative design approaches.

- **Geotechnical practice:** Various sections consider aspects of general geotechnical practice, including:
  - The collection of geotechnical data;
  - Supervision of construction, monitoring and maintenance;
  - Treatment of fill, dewatering and ground improvement.

- **Geotechnical configurations:** Treatment of a number of configurations are considered, such as:
  - Spread and pile foundations, anchorages, retaining structures, embankments;
  - Design situations such as hydraulic failure and overall stability.

- **Informative annexes:** Sample procedures and methods for determining limit values are presented in various annexes for:
  - Earth pressure, bearing resistance or settlement;
  - Checklist for supervision.

6.2 Geotechnical actions

Geotechnical actions are defined as actions transmitted to the structure by the ground, fill or groundwater. *The selection of characteristic values for geotechnical parameters shall be based on results and derived values from laboratory and field tests, complemented by well established experience*, and *shall be selected as a cautious estimate of the value affecting the occurrence of the limit state. A cautious estimate of the mean value is a selection of the mean value of the limited set of geotechnical parameter values, with a confidence level of 95%; where local failure is concerned, a cautious estimate of the low value is the 5% fractile.*

6.3 Geotechnical design

In the procedures presenting provisions for geotechnical design, the way in which partial factors for actions & action effects, geotechnical material properties and resistances are distributed are stipulated. This is given for the ultimate limit state verification for persistent and transient design situations. Important features of the geotechnical design procedures are:

- Combination schemes are given to provide for respective situations where geotechnical (GEO) or structural (STR) actions, or combinations (STR & GEO) are relevant.
- The combination schemes are given as three alternative *Design Approaches* (DA1; DA2; DA3), which are allowed as NDP options. One of these approaches needs to be selected in the National Annex.
  - The alternative design approaches represent widely different performance requirements: This indicates that although Eurocode represents a high degree of harmonisation, the tolerances allowed for national NDP options are substantial.
  - Since the Design Approaches also include the alternative structural action combination schemes allowed in EN 1990, these options represent a wide array of options which need to be selected for the respective Nation Annexes, in addition to the selection of partial factor values!

The characteristics of the alternative Design Approaches and their relevance to South African conditions are discussed more elaborately in Chapter 5-1, and the implications of the various action combination schemes for structural effect are discussed in Chapter 1-2.

7 EN 1998 *DESIGN OF STRUCTURES FOR EARTHQUAKE RESISTANCE*

The Standard EN 1998 *Design of structures for earthquake resistance* is separated into six Parts, with EN 1998-1 *General rules, seismic actions and rules for buildings* mainly being considered here. The other Parts consider general issues such as strengthening and repair, and specific structures such as bridges; silos, tanks and pipelines; towers, masts and chimneys; and a Part considering foundations, retaining structures and geotechnical aspects.

The general rules of EN 1998-1 consider performance requirements, ground conditions, seismic actions, general rules for buildings and specific rules for the various structural materials. Some general requirements include:

- **Basis of design for earthquake resistance:** According to EN 1998-1 the purpose of earthquake resistant design is *to ensure, that in the event of an earthquake: human lives are protected; damage is limited; structures important for civil protection remain operational.* In accordance with the classification of seismic design situations as accidental, it is noted that *the attainment of these goals is only partially possible and only measurable in probabilistic terms*. Fundamental requirements and compliance criteria for earthquake resistant design are:
  - **No-collapse requirement:** *to withstand design seismic action ... without local or global collapse, thus retaining its structural integrity and residual load bearing capacity after seismic events. The design seismic action is expressed in terms of a) the reference action associated with a reference probability of occurrence* (10% recommended) ... *in 50 years, and b) the importance factor γ_I to take account of reliability differentiation.*
    - **The ultimate limit state** is associated with compliance with the no-collapse requirement, failure to comply with which *may endanger the safety of people.* For the ultimate limit state *the structural system shall be verified as having the **resistance** and **energy dissipating capacity** specified.*
  - **Damage limitation requirement:** *to withstand a seismic action, ... without the occurrence of damage and the associated limitations of use,* (with a) *probability of exceedance* (10% recommended) *in 10 years.*
    - **The serviceability limit state** is associated with damage limit compliance *corresponding to states beyond which specified service requirements are no longer met.*

- **Guiding principles:** *The guiding principles governing conceptual design against seismic hazard are:*
  - *structural simplicity; uniformity, symmetry and redundancy;*
  - *bi-directional resistance and stiffness; torsional resistance and stiffness;*
  - *diaphragmatic behaviour at storey level;*
  - *adequate foundation.*

8 MATERIAL-BASED STRUCTURAL RESISTANCE

In this chapter Eurocode is reviewed from the perspective of its relevance to the revised SA loading code SANS 10160. Therefore the relation of the materials-based design standards EN 1992 to EN 1996 & EN 1999 (structural concrete, steel, composite, timber, masonry and alloy aluminium) are considered mainly in terms of their relation to its implementation to the basis of structural design from EN 1990 and actions on structures from EN 1991. The materials-based Eurocode Standards are however also relevant to the future development of South African structural design standards.

An extensive series of calibration studies for structural concrete, steel, steel/concrete composites and timber are summarised by Sedlacek *et al* [1996]. The calibration is generally based on the principle of separating resistance calibration from actions by requiring resistance reliability $\beta_R = \alpha_R \beta_T$ where $\alpha_R = 0,8$ and $\beta_T =$ the target level of reliability of 3,8. These investigations are taken further in calibration exercises for the selection of NDP values for the respective National Annexes, e.g. SAKO (1999)

9 FURTHER DEVELOPMENT OF THE EUROCODES

Planning for the future development of the Eurocodes is addressed in *Report on evolution of EN Eurocodes* (CEN 2005) prepared by CEN TC250 Ad-hoc group *Maintenance and future actions*. The planning includes maintenance, improvement of harmonisation, promotion and improvement of the Standards.

- **Maintenance:** Remove errors, misleading statements or inconsistencies, and collect the technical background including the non-contradictory information from the National Annexes.
- **Harmonisation:** Limit the NDPs to geographical, geological or climatic conditions or to specific levels of protection. Information will be prepared for Member States to enable them to change their NDPs in order to reduce divergence from the recommended values provided by the EN Eurocodes.
- **Implementation and promotion:** With the harmonised EN Eurocodes international collaboration within Member States and various other countries should be possible. Promotion should be aimed at:
  - Member States involved in the preparation of the EN Eurocodes or that have recently joined the EU;
  - Countries outside EU and EFTA.
- **Improvement:** It follows from the available experience and use of the EN Eurocodes that other topics may need to be improved or developed anew. In particular the following items are considered:
  - load combination rules for ultimate limit states
  - geotechnical design
  - serviceability and fatigue limit states,
  - robustness requirements,
  - accidental actions,
  - reliability differentiation,
  - methods of risk assessment.

10 SUMMARY AND CONCLUSIONS

The final ten Eurocode Standards EN 1990 to EN 1999, consisting of fifty-eight Parts, provide unified procedures for structural design for buildings and civil engineering works, and will result in significant harmonisation of structural engineering practice and the industry across Europe. The process will be finalised by the introduction of the National Annex for the respective Parts and the replacement of existing standards by Member Countries. Advances include the scope of structures provided for, level of standardisation of design, its advanced technological base and the harmonisation and unification of the Standards.

10.1 Scope of Structural Design

Eurocode provides *for the structural design of buildings and civil engineering works, including geotechnical aspects, structural fire design, situations involving earthquakes, execution and temporary structures* (EN 1990:2002), providing for:

- **Structures** The comprehensive range of:
  - Buildings; bridges; silos, tanks and containment structures; towers, masts and chimneys; pipelines; piling; crane support structures; geotechnical structures;
  - Principles can also be applied to *the design of special construction works (e.g. nuclear installations, dams) but provisions other than provided by the Eurocodes will probably be necessary* (Gulvanessian *et al* 2002).
- **Design situations** The Limit States are extended to provide for an elaborate set of design situations for which associated limit states with their design rules are stipulated:
  - Transient, persistent and accidental situations,
  - Including fire and earthquakes; durability; fatigue;
  - Differentiated levels of serviceability – irreversible, reversible, long-term, appearance.
- **Actions** Provisions are made for the following:
  - Revised and extended rules for the standard range of actions such as self-weight, imposed floor and roof loads, snow and wind actions, traffic loads on bridges, crane induced actions, actions in silos and tanks

  - Proper provisions are developed for actions due to thermal effects and during execution of the structure
  - Accidental actions due to fire, earthquakes, impact, internal explosions, unidentified sources (robustness requirements)
- **Structural materials:** Provision is made for:
  - The complete set of conventional structural materials of concrete, steel, steel/concrete composite construction, timber, masonry, aluminium alloy
  - Geotechnical design
- **Natural conditions:** Wide ranges of natural conditions across Europe provides for
  - Snow conditions from Scandinavia to the southern Mediterranean regions, with associated ranges for thermal loads
  - Earthquake frequencies from low seismicity to those so high that seismic loads need to be considered as a variable load
  - Extensive range of strong wind conditions (reference wind from 20 to 30 (36) m/s)

10.2 Unified and Harmonised Procedures

The introduction of a separate Standard EN 1990 *The basis for structural design* which specifies all the common requirements for actions and material-independent considerations provides unification of the different Parts of Eurocodes, ensuring:

- **Unified treatment** in the different components of a given structure, such as actions, materials-based structural design, foundation design and earthquake resistance;
- **Consistency** between the alternative structural materials,
  - which could be considered in design alternatives;
  - which are often used together in one structure;
  - which are servicing the whole structural engineering industry;
- **Harmonisation** of the practice of Member States by reaching consensus regarding very diverse approaches taken traditionally;
  - where harmonisation could not be achieved, this is clearly identified through the NDP options; allowing for future reduction of differences;
  - allowing even for using the wider European body of experience and insight in setting national safety standards;
  - improving the functioning of the European market; improving the competitiveness of the structural engineering industry.
- **International harmonisation** The consistency of EN 1990 with ISO 2394 provides the basis for improved international harmonisation of structural engineering practice,
  - thereby allowing the sharing of research and experience in the process of improving the rational basis of structural engineering;

10.3 Technological Basis for Eurocodes

The advances achieved by the Eurocodes were only possible because of the sound and extensive technology base which derived from the varied sources from Member States and strong support through concerted research, development and ultimately standardisation. Furthermore the technology base is substantially captured in background reports and publications, including various model codes. Such information is particularly useful when the Eurocodes are used as reference for standards development elsewhere.

A rich set of literature is being developed to support the implementation and ultimate use of the Eurocodes. The extensive market for the use of the Eurocodes provides sufficient motivation and justification for the development of explanatory handbooks, guidebooks and supporting software and design aids.

10.4 Relevance of the Eurocodes to the South African context

It is imperative that South African practice should keep track with international developments, particularly with its open economy in general, and market opportunities in structural engineering materials, products and services in particular. Europe is an important trade partner of South Africa. Harmonisation of structural engineering practice would be beneficial to such trade relations. Such harmonisation with European practice would even be beneficial to trade relations with the rest of Africa and elsewhere.

The Eurocode Standards have demonstrated the importance of a common basis for structural design through the use of a separate Standard EN 1990 as the basis for unification of the different components of structural design. The present revision of the South African Standard SANS 10160 therefore provides a useful opportunity to benefit from the advances made by the development and implementation of the Eurocodes.

11 REFERENCES

Bossenmayer H (2003). Personal communication.

BRE (2003). UK Building Research Establishment, *Review of SAKO Report.* Gulvanessian H – personal communication.

Brozzetti J, Sedlacek G (2000). *Background to the combination rules (6.10) or (6.10a) and (6.10b) in prEN 1990 Basis of design.* Document CEN/TC 250 N419. CEN/TC 250 Secretariat.

CE (2002). Commision Européenne. Guidance Paper L *Application and use of Eurocodes.* Document CONSTRUCT 01/483 Rev.1. Brussels.

CEN (2002). Committé Européan de Normalisation. EN 1990:2002 Eurocode – Basis of structural design. European Committee for Standardisation, Brussels.

CEN/TC 250 (2005). Ad-hoc group "Maintenance and future actions", Report on evolution of EN Eurocodes.

CIB (1989a). Conseil International du Batiment pour la Recherche l'Etude et la Documentation. *Actions on structures – Self-weight loads.*, Report 115, Rotterdam.

CIB (1989b). Conseil International du Batiment pour la Recherche l'Etude et la Documentation . *Actions on structures – Live loads in buildings.* Report 116, Rotterdam.

Croce P (2001). *Traffic loads on road bridges.* Pisa University, Department of Structural Engineering (2001). *Proceedings of the seminar on EN 1991 – Actions on structures.* Leonardo da Vinci Pilot Project.

Gulvanessian H, Calgaro J-A & Holický M (2002). *Designer's guide to EN 1990 Eurocode: Basis of structural design.* Thomas Telford. 192 pp.

Hemmert-Halswick A, Sedlacek G, Merzenich G (1988). *Eurocode actions on structures: Background document Part 6: Imposed loads on floors and roofs.*

Holický M (2001). *Densities, self-weight and imposed loads on buildings.* Pisa University Department of Structural Engineering (2001). *Proceedings of the seminar on EN 1991 – Actions on structures.* Leonardo da Vinci Pilot Project.

Holický M, Markova J (2005). *General principles of fire safety.* Chapter IV in Implementation of Eurocodes Handbook 1 *Basis of structural design.* Leonardo da Vinci Pilot Project.

Holický M, Retief JV (2005). *Reliability Assessment of Alternative Load Combination Schemes for Structural Design.* Journal SAICE, Vol. 25, No 1, pp 15 – 20, 2005.

Holický M, Vrouwenvelder T (2005). *Basic concepts of structural reliability.* Chapter I in Implementation of Eurocodes Handbook 2 *Basis of structural reliability and risk engineering.* Leonardo da Vinci Project.

IABSE (1996). *Colloquium Background and application of Eurocode 1.* Delft.

ISO (1993). International Standards Organisation. ISO 3534 *Statistics- Vocabulary and Symbols*: Part 1: Probability and general statistical terms, Part 2: Statistical quality control.

ISO (1997). International Standards Organisation. ISO 12491 *Statistical methods for quality control of building materials and components*. 1997.

ISO (1998). International Standards Organisation. ISO 2394:1998 *General Principles on Reliability for Structures*. International Standard. [Also issued as SANS 2394:2004]

JCSS (1996). Joint Committee on Structural Safety. *Background documentation: Eurocode 1 (ENV 1991) Part 1 Basis of design.* Report N° 94, ECCS, Brussels. ISBN 92-9147-0022.

JCSS (2001). Joint Committee on Structural Safety. *Probabilistic model code.* JCSS working materials, http://www.jcss.ethz.ch/.

JRC/DG-ENTR (2008). *Eurocodes: background and applications.* Joint Research Centre and Directorate General – Enterprise and Industry, Brussels 12 – 18 February 2008. http://eurocodes.jrc.ec.europa.eu

Leonardo da Vinci Project (2005). Implementation of Eurocodes: *Development of skills facilitating implementation of Eurocodes.* Handbook 1 *Basis of structural design*; Handbook 2 *Reliability backgrounds*; Handbook 3 *Action effects for buildings*; Handbook 4 *Guide to the basis of bridge design related to Eurocode*; Handbook 5 *Design of buildings for the fire situation.*

Narayanan RS, Beeby AW (2005) *Designers guide to EN 1992-1-1 and EN 1992-1-2* Eurocode 2: *Design of concrete structures. General rules and rules for buildings and structural fire design.* Thomas Telford.

SABS (1989). SA Bureau of Standards, Pretoria. SABS 0160:1989 (Amended 1990)**:** *Code of Practice: The general procedures and loadings to be adopted in the design of buildings***.**

SAKO (1999). Joint Committee of NKB and INSTA-B. *Basis of Design of Structures. Proposal for Modification of Partial Safety Factors in Eurocodes.* NKB Committee and Work Reports 1999:01 E, pp. 55.

Savidu A, Gulvanessian H, Pavlovic M (2001). *Geotechnical limit states.* Pisa University Department of Structural Engineering (2001). *Proceedings of the seminar on EN 1991 – Actions on structures.* Leonardo da Vinci Pilot Project.

Sedlacek G, Grotman D (1996). *Actions induced by cranes.* IABSE Colloquium Background and application of Eurocode 1. Delft.

Sedlacek G, Gulvanessian H (1996). *Eurocode 1, Basis of design and actions on structures, Part 2.1 Densities, self-weight, imposed loads.* IABSE Colloquium Background and application of Eurocode 1. Delft.

Sedlacek G, Brozzetti J, Hanswille G, Litzner HU, Canisius TDG (1996). *Relationship between Eurocode 1 and the "material" oriented Eurocodes.* IABSE Colloquium Background and application of Eurocode 1. Delft.

Schleich J-B (2005). *Fire safe design of buildings.* Implementation of Eurocodes Handbook 5 *Design of buildings for the fire situation,* Chapter I. Leonardo da Vinci Pilot Project.

Thomas Telford (quarterly). *Eurocode News.* Available on Eurocodes Expert website: www.eurocodes.co.uk/newsletters.aspx

Vrouwenvelder T, Steenbergen H (2001). *Wind actions.* Pisa University Department of Structural Engineering (2001). *Proceedings of the seminar on EN 1991 – Actions on structures.* Leonardo da Vinci Pilot Project.

Vrouwenvelder A, Stieffel U, Harding G (2005). *Eurocode 1, Part 1.7 Accidental actions – Background document.* First draft.

Vrouwenvelder T (2005). *Accidental actions on buildings.* Implementation of Eurocodes, Handbook 3, Action effects for buildings, Leonardo da Vinci Project.

APPENDIX A LIST OF EUROCODE STANDARDS AND PARTS

0	**EN 1990**	**Basis of structural design**
Annex A1 (Normative)		Buildings
Annex A2 (Normative)		Bridges
1	**EN 1991**	**Eurocode 1 : Actions on structures**
1.1	EN 1991-1-1	General actions - Densities, self-weight, imposed loads for buildings
1.2	EN 1991-1-2	General actions - Actions on structures exposed to fire
1.3	EN 1991-1-3	General actions - Snow loads
1.4	EN 1991-1-4	General actions - Wind actions
1.5	EN 1991-1-5	General actions - Thermal actions
1.6	EN 1991-1-6	General actions - Actions during execution
1.7	EN 1991-1-7	General actions - Accidental actions due to impact and explosions
1.8	EN 1991-2	Traffic loads on bridges
1.9	EN 1991-3	Actions induced by cranes and machinery
1.10	EN 1991-4	Actions in silos and tanks
2	**EN 1992**	**Eurocode 2 : Design of concrete structures**
2.1	EN 1992-1-1	General rules, and rules for buildings
2.2	EN 1992-1-2	General rules - Structural fire design
2.3	EN 1992-2	Concrete bridges
2.4	EN 1992-3	Liquid retaining and containment structures
3	**EN 1993**	**Eurocode 3 : Design of steel structures**
3.1	EN 1993-1-1	General rules, and rules for buildings
3.2	EN 1993-1-2	General – Structural fire design
3.3	EN 1993-1-3	General – Cold formed thin gauge members and sheeting
3.4	EN 1993-1-4	General – Structures in stainless steel
3.5	EN 1993-1-5	General – Strength and stability of planar plated structures without transverse loading
3.6	EN 1993-1-6	General – Strength and stability of shell structures
3.7	EN 1993-1-7	General – Design values for plated structures subjected to out of plane loading
3.8	EN 1993-1-8	General – Design of joints
3.9	EN 1993-1-9	General – Fatigue strength
3.10	EN 1993-1-10	General – Material toughness and through thickness assessment
3.11	EN 1993-1-11	General – Design of structures with tension components
3.12	EN 1993-1-12	General - Supplementary rules for high strength steels
3.13	EN 1993-2	Bridges
3.14	EN 1993-3-1	Towers, masts and chimneys – Towers and masts
3.15	EN 1993-3-2	Towers, masts and chimneys – Chimneys
3.16	EN 1993-4-1	Silos, tanks and pipelines – Silos
3.17	EN 1993-4-2	Silos, tanks and pipelines – Tanks
3.18	EN 1993-4-3	Silos, tanks and pipelines – Pipelines
3.19	EN 1993-5	Piling
3.20	EN 1993-6	Crane supporting structures

4	**EN 1994**	**Eurocode 4 : Design of composite steel and concrete structures**
4.1	EN 1994-1-1	General rules, and rules for buildings
4.2	EN 1994-1-2	General rules - Structural fire design
4.3	EN 1994-2	Composite bridges
5	**EN 1995**	**Eurocode 5 : Design of timber structures**
5.1	EN 1995-1-1	General rules, and rules for buildings
5.2	EN 1995-1-2	General rules - Structural fire design
5.3	EN 1995-2	Bridges
6	**EN 1996**	**Eurocode 6 : Design of masonry structures**
6.1	EN 1996-1-1	General rules for buildings - Rules for reinforced and unreinforced masonry
6.2	EN 1996-1-2	General rules - Structural fire design
6.3	EN 1996-2	Selection of materials (?)
6.4	EN 1996-3	Simplified calculation methods for masonry structures
7	**EN 1997**	**Eurocode 7 : Geotechnical design**
7.1	EN 1997-1	General rules
7.2	EN 1997-2	Ground investigation and testing
8	**EN 1998**	**Eurocode 8 : Design provisions for earthquake resistance of structures**
8.1	EN 1998-1	General rules, seismic actions and rules for buildings
8.2	EN 1998-2	Bridges
8.3	EN 1998-3	General rules - Strengthening and repair
8.4	EN 1998-4	Silos, tanks and pipelines
8.5	EN 1998-5	Foundations, retaining structures and geotechnical aspects
8.6	EN 1998-6	Towers, masts and chimneys
9	**EN 1999**	**Eurocode 9 : Design of aluminium alloy structures**
9.1	EN 1999-1-1	General rules, and rules for buildings
9.2	EN 1999-1-2	General rules - Structural fire design
9.3	EN 1999-1-3	Additional rules for fatigue
9.4	EN 1999-1-4	Trapezoidal sheeting
9.5	EN 1999-1-5	Shell structures

APPENDIX B CLAUSES FROM EN 1990 ILLUSTRATING PRINCIPLES AND APPLICATION RULES

<table>
<tr><th colspan="2">PRINCIPLES, APPLICATION RULES, DEFINITIONS and NOTES</th></tr>
<tr><td>Principle clause</td><td>2.1(1)P A structure <u>shall</u> be designed and executed in such a way that it will, during its intended life, with appropriate degrees of reliability and in an economic way
- sustain all actions and influences likely to occur during execution and use, and
- remain fit for the use for which it is required.

2.1(4)P A structure <u>shall</u> be designed and executed in such a way that it will not be damaged by events such as explosion, impact and the consequences of human errors, to an extent disproportionate to the original cause.

NOTE 1 The events to be taken into account are those agreed for an individual project with the client and the relevant authority.</td></tr>
<tr><td>Application Rule</td><td>2.1(6) The basic requirements <u>should</u> be met:
- by the choice of suitable materials,
- by appropriate design and detailing, and
- by specifying control procedures for design, production, execution and use relevant to the particulars of the project.

2.2(3) The choice of the levels of reliability for a particular structure <u>should</u> take account of the relevant factors, including :
- the possible cause and /or mode of attaining a limit state ;
- the possible consequences of failure in terms of risk to life, injury, potential economical losses;
- public aversion to failure ;
- the expense and procedures necessary to reduce the risk of failure.</td></tr>
</table>

1-4 The Reliability Basis of Eurocode

Holický M, Retief JV & Dunaiski PE

1 INTRODUCTION

One of the primary advances made with the development and implementation of the Eurocode Standards for structural design (CE 2002) is the extension of the general principles of structural reliability as formulated in ISO 2394 *General principles on the reliability of structures* (ISO 1998) into operational reliability-based partial factor limit states design procedures. The principles and procedures are formulated in general terms in EN 1990 *Basis of structural design* (CEN 2002, Gulvanessian *et al* 2002) and then applied in EN 1991 *Actions on structures* and EN 1992 – EN 1999 which specifies the materials related resistance requirements of structural design.

By capturing the principles of the reliability basis of structural design into the head Standard EN 1990, Eurocode not only strengthened its operational development into reliability-based limit states design procedures, but also provides the basis for unified treatment of actions on structures and materials based resistance. In addition to unification between actions and resistance, this approach includes foundations and earthquake resistance to provide consistency in the design of individual structures. It also results in consistency also between the parallel comparative performance requirements for different structures, and structures constructed from different materials, as diverse and ranging from steel to geotechnical materials.

The fundamental reliability-based methodology taken in Eurocode, at least at the conceptual level, was the only way in which the wide range of approaches and practice followed by Member States could be harmonised. Even where such harmonisation could not be achieved, such deviations were constrained to the National Determined Parameters, sometimes even to the range of values allowed for basic variables. Furthermore concerted efforts are made to improve the degree of consensus in the selection of NDP options and values presently and in the future.

This Chapter presents a review and analysis of the reliability basis of Eurocode, with an assessment made from the view of the revision of the present South African Loading Code SABS 0160-1989 (Amended) *The general procedures and loadings to be adopted in the design of buildings* into the proposed SANS 10160 *Basis of structural design and actions for buildings and industrial structures*. Since Eurocode has been selected as reference to the formulation of SANS 10160 such an assessment of the reliability basis of Eurocode serves as background to the revised South African Standard.

Following a presentation on the background to the development of standards for structural design, an overview of the reliability framework of Eurocode as formulated in EN 1990 is presented. The use of this framework in formulating application rules for action combinations, the specification of actions on structures, and general guidelines for materials-based resistance, is presented next. The theory of structural reliability is then applied in parametric analysis to determine appropriate partial factors. In conclusion the merit of the Eurocode reliability basis is assessed to derive guidelines for the formulation of SANS 10160.

2 BACKGROUND TO STANDARDS FOR STRUCTURAL DESIGN

The improved economy of structures designed using the principles of limit states procedures is underpinned by the theory of structural reliability in order to achieve the required levels of safety across the range of structures and conditions to which they are subjected.

Since structural reliability provides the theoretical and conceptual basis for the unification of the Eurocode structural design standards, an overview is provided in this chapter of the theory of structural reliability and its use to derive design procedures. First a brief overview is provided of the development of structural development standards into reliability-based partial factor limit states procedures.

2.1 General principles for reliability-based Limit States Design procedures

The principles for the application of the theory of structural reliability to derive design procedures are captured in the International Standard ISO 2394:1998 *General principles on reliability for structures*. This Standard is accepted internationally as the basis for the formulation of National Standards. These general principles are reformulated and developed into practical application rules in Eurocode EN 1990 *Basis for structural design*, which are then implemented as the design standards EN 1991 – EN 1999 for actions and structural resistance.

2.1.1 Design requirements

The reliability of the structure has to be guaranteed during the whole economically reasonable working life. In particular, the construction works must be designed and built in such a way that the loading likely to act during its construction and use does not cause:

- **collapse** of the whole or part of the work,
- **major deformations** to an inadmissible degree,
- **damage to other parts** of the works, equipment or installed devices,
- **disproportionate damage** in relation to the original cause.

The operational methods of structural design are further based on the concept of design situations and relevant limit states in conjunction with the partial factor method. A distinction is made between ultimate limit states and serviceability limit states. Limit states can be further differentiated into design situations to provide for all conditions that can reasonably be expected to occur during the life of the structure. In general, four types of the design situations are recognised: persistent situations for normal use; transient situations for temporary conditions; accidental situations for exceptional conditions; seismic situations which refer to earthquake events.

2.1.2 Limit states of structural performance requirements

The limit states denote particular circumstances beyond which the structural performance requirements are no longer satisfied:

The ultimate limit states are those associated with various forms of structural failure or states close to structural failure and may require consideration of:

- loss of equilibrium of the structure considered as a rigid body;
- excessive deformation or settlement, rupture, or the loss of stability.

The serviceability limit states are those associated with the criteria for the structure related to its use or function and may require consideration of:

- deformation or deflection;
- vibrations which limit the structural use;
- detrimental cracking.

2.1.3 Uncertainties

It is well recognised that construction works are complicated technical systems suffering from a number of significant uncertainties at all stages of execution and use. Some uncertainties can never be eliminated absolutely and must be taken into account when designing or verifying construction works. The following types of uncertainties can usually be identified, presented in the approximate order of decreasing knowledge and theoretical tools for analysis and implementation in design:

- **natural randomness** of actions, material properties and geometric data;
- **statistical uncertainties** due to a limited body of available data;
- **model uncertainties** caused by a simplification of actual conditions;
- **vagueness** due to inaccurate definitions of performance requirements;
- **gross errors** in design, execution and operation of the structure;
- **lack of knowledge** of the behaviour of new materials in real conditions.

The lack of available theoretical tools is obvious in the case of gross errors and lack of knowledge, which are nevertheless often the decisive causes of structural failures. To limit gross errors due to human activity a quality management system including the methods of statistical inspection and control may be effectively applied. Structural reliability forms the basis of contemporary systems of quality control and their operational techniques (ISO 1993; ISO 1997).

2.1.4 Structural reliability

The theory of structural reliability has been developed to describe and analyse uncertainties in a rational way and to take them into account in the design and verification of structural performance. In fact, the development of the whole theory was initiated by observed insufficiencies and structural failures caused by various uncertainties.

In Eurocode EN 1990:2002 it is required that *a structure shall be designed and executed in such a way that it will, during its intended life with appropriate degrees of reliability and in an economic way:*

- *remain fit for the use for which it is required; and*
- *sustain all actions and influences likely to occur during execution and use.*

The specification for reliability includes four important elements:

- **performance requirements** – the definition of the structural failure,
- **time period** – the specification of the required service-life T,
- **reliability level** – the specification of the probability of failure p_f,
- **conditions of use** – the restriction of input uncertainties.

2.2 Historical developments of design procedures

During their historical development the design methods that have had the objective of taking recognised uncertainties into account and of ensuring structural reliability have been closely linked to the available empirical, experimental as well as theoretical knowledge of mechanics and the theory of probability. Empirical methods for structural design gradually evolved into rationally based methods: permissible stresses; global safety factors; partial factor limit states methods; reliability-based design. The basis for these methods is briefly reviewed.

2.2.1 Permissible stress methods

The first worldwide design method for civil structures is the method of permissible stresses. It is based on the condition:

$$\sigma_{max} < \sigma_{per}, \text{ where } \sigma_{per} = \sigma_{crit} / k \quad (1)$$

where the coefficient k is assessed with regard to uncertainties in the determination of local load effect σ_{max} and of resistance σ_{per} and therefore may ensure with an appropriate level of security the reliability of the structure.

The main insufficiency of this method is perhaps the local verification of reliability (in the elastic range) and the impossibility of separately considering the uncertainties of basic quantities and uncertainties of computational models for the assessment of action effects and structural resistance. In this method, the probability of failure is controlled by one quantity only, the coefficient k.

2.2.2 *Global safety factor method*

The second widespread method of structural design is the method of global safety factor. It is based on the condition:

$$s = X_{resist} / X_{act} > s_0 \qquad (2)$$

according to which the calculated safety factor s must be greater than its specified value s_0 through the aggregate quantities of structural resistance X_{resist} and action effect X_{act}. Although the structural element or its section is considered, the probability of failure can again be controlled by one quantity only, the global safety factor s.

2.2.3 *Partial factor method*

At present, the most advanced operational method of structural design is the partial factor format based on the condition:

$$E_d\,(F_d, f_d, a_d, \theta_d) < R_d\,(F_d, f_d, a_d, \theta_d) \qquad (3)$$

where the action effect E_d and the structural resistance R_d are assessed according to the design values of basic quantities describing the action F_d, material properties f_d, dimensions a_d and model uncertainties θ_d. The design values of these quantities are determined (taking into account their uncertainties as well as the uncertainties of computational models) using their characteristic values (F_k, f_k, a_k, θ_k), partial safety factors γ, combination factors ψ and other measures of reliability. Thus a whole system of various partial factors and other reliability elements may be used to control the probability of structural failure.

Partial factor methods provide operational design procedures adopted in many national and international standards. General concepts of the partial factor method are described in International Standard ISO 2394-1998. The fundamental Eurocode EN 1990:2002 *Basis of Structural Design* describes details of accepted procedures based on a partial factor method in Section 6 *Verification by the partial factor method* and Annex A1 (Normative) *Application for buildings*. Additional background information and the theoretical basis of the method are described in Annex C (Informative) *Basis for partial factor design and reliability analysis*.

The partial factor method was introduced with limit states design in which differentiated requirements were set for safety (ultimate) and operational (serviceability) limits. From the view of reliability-based design the two limit states can be considered as basic reliability differentiation in terms of the consequences of the limits being exceeded.

2.2.4 *Probabilistic design methods*

Probabilistic design methods that are provided for in ISO 2394-1998 are based on the condition that during the service life of a structure (T) the probability of failure p_f does not exceed a specified target value p_T

$$p_f \leq p_T \qquad (4)$$

The probability of failure p_f is assessed using a computational structural model, defined through basic quantities $\boldsymbol{X}\,[X_1, X_2, \ldots, X_n]$ for actions, mechanical properties and geometrical data. The limit state of a structure is defined by the function for the safe state:

$$g(\boldsymbol{X}) \geq 0 \qquad (5)$$

Design directly based on probabilistic methods is allowed in EN 1990 Clause 3.5(5) as an alternative to the partial factor method. The JCSS Model Code provides consistent procedures for the design of structures from a probabilistic point of view (JCSS 2001).

2.2.5 *The Design Value Method*

The design value method, which is also called a *semi-probabilistic (level I) method* is a very important step from probabilistic design methods toward an operational partial factors method. The design value method is directly linked to the basic principle of EN 1990, according to which it should be verified that no limit state is exceeded when the design values of all basic variables are used in the models of structural resistance R and action effect E in terms of the design values E_d and R_d, *a structure is considered as reliable* when the following expression holds

$$E_d < R_d \tag{6}$$

where $$E_d = E\{F_{d1}, F_{d2}, \ldots a_{d1}, a_{d2}, \ldots \theta_{d1}, \theta_{d2}, \ldots\} \tag{7}$$

$$R_d = R\{X_{d1}, X_{d2}, \ldots a_{d1}, a_{d2}, \ldots \theta_{d1}, \theta_{d2}, \ldots\} \tag{8}$$

Here F is a general symbol for actions, X for material properties, a for geometrical properties, and θ for model uncertainties.

The design values E_d and R_d may be found by:

$$P(E > E_d) = \Phi(+\alpha_E\beta) \tag{9}$$

$$P(R \leq R_d) = \Phi(-\alpha_R\beta) \tag{10}$$

where β is the target reliability index, α_E and α_R, with $|\alpha| \leq 1$, are the values of the FORM weight (sensitivity) factors; taken in EN 1990 as $\alpha_E = -0{,}7$ and $\alpha_R = 0{,}8$.

2.3 Calibration

The assessment of various reliability measures (characteristic values, partial and combination factors) in the new structural design standards is partially based on probabilistic considerations, but to a great extent it is based on historical and empirical experience. Moreover, the choice of these reliability measures is affected in EN 1990 by the intention to simplify the calculation in practical design. Figure 1 shows possible calibration procedures indicated in EN 1990.

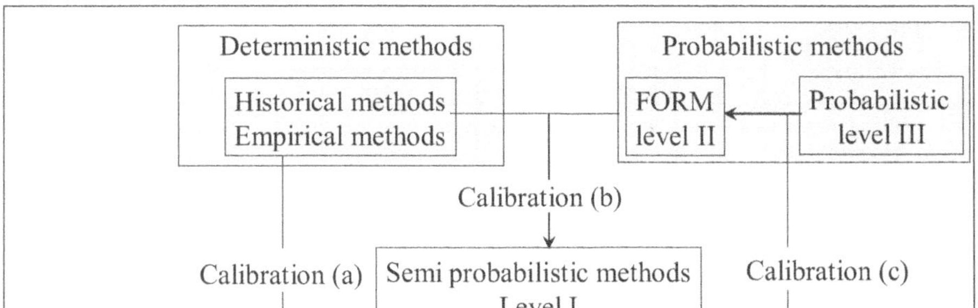

Figure 1 Alternative calibration procedures

The most frequently used calibration procedure (a) is based on past experience represented by historical and empirical design methods. However, reliability principles (calibrations (b) and (c)) represent very important theoretical bases enabling further improvement and generalization of existing design procedures. The theory of structural reliability makes it possible to extend the general methodology for new structural systems and construction materials.

3 RELIABILITY FRAMEWORK AND SPECIFICATION IN EUROCODE

The International Standard ISO 2394 *General principles on the reliability of structures* (ISO 1998) is globally accepted as providing the reliability basis for structural design on which structural design standards need to be based, as stated in its Introduction: *This International Standard constitutes a common basis for defining design rules relevant to the construction and use of the wide majority of buildings and civil engineering works, whatever the nature or combination of the materials used. … It has a conceptual character and it is of a fairly general nature.* Whilst Eurocode EN 1990 *Basis of structural design* is consistent with ISO 2394, its development of reliability-based partial factor Limit States design is taken further in order to provide operational design rules for application in EN 1991 – EN 1999.

3.1 Comparison of ISO 2394-1998 and EN 1990-2002

A comparison of the main features of ISO 2394 and EN 1990 is presented in Annex A. Although the two Standards are comparable in total length, the normative part of EN 1990 is more substantial. The main differences remain in the treatment of design verification procedures where ISO 2394 considers probability-based and partial factor procedures on an equal footing (including also the assessment of existing structures), EN 1990 develops the partial factor method into operational procedures. An important common feature is the inclusion of the reliability basis partial factor design in informative annexes, including design assisted by testing.

3.2 Design working life

The design working life of a structure is intended to relate to design and maintenance as summarised in Table 1. Included are also indicative values from EN 1990 which presents a slight elaboration of values from ISO 2394.

Table 1 Application and indicative values for design working life

<table>
<tr><th colspan="3">Application of Design Working Life (DWL)</th></tr>
<tr><td colspan="3">• Assumed period for which a structure or part of it is to be used for its intended purpose with anticipated maintenance but without major repair being necessary;
• Design: Actions (e.g. imposed, wind, earthquake) and deterioration of material (e.g. fatigue, creep, durability);
• Maintenance: Strategies and procedures for maintenance and renovation.</td></tr>
<tr><th>Category</th><th>Notional DWL (years)</th><th>. Examples</th></tr>
<tr><td>1</td><td>10</td><td>Temporary structures (e.g. scaffolding)</td></tr>
<tr><td>2</td><td>10-25</td><td>Replaceable structural parts, e.g. gantry girders, bearings</td></tr>
<tr><td>3</td><td>15-30</td><td>Agricultural and similar structures (e.g. buildings where people do not normally enter)</td></tr>
<tr><td>4</td><td>50</td><td>Building structures and other common structures (e.g. hospitals, schools etc.)</td></tr>
<tr><td>5</td><td>100</td><td>Monumental building structures, bridges, other civil engineering structures (e.g. churches)</td></tr>
</table>

3.3 Scope of operational procedures presented in EN 1990

The background and an interpretation of EN 1990-2002 are presented by Gulvanessian *et al* (2002) and Holický & Vrouwenvelder (2005). The EN 1990 operational procedures which derive from the principles, which also closely agree with ISO 2394 as indicated above, can be summarised in terms of the main elements as follows:

- **Reliability framework:** The various reliability elements are developed into a reliability framework consisting of
    - {reliability differentiation; limit states; design situations; basic variables}
- **Action combination schemes:** Appropriate action combination schemes are developed for the respective "cells" of the reliability framework
- **Design situations:** Provision is made and implemented for ranges of design situations
    - Safety: {transient; persistent; accidental; seismic (accidental); equilibrium; strength (structural; geotechnical)};
    - Serviceability: {irreversible; reversible; long-term}
- **Design verification:** Provision is made for the following variables and factors:
    - {basic variables (characteristic values); partial factors; combination factors (combination; frequent; quasi-permanent)}
- **Actions:** Standard range extended to include the following classes and types of actions:
    - {transient (execution); geotechnical; accidental (unidentified [integrity & robustness]; fire; impact; internal explosion); seismic}
- **Materials:** Standard range of structural materials
    - {concrete; steel; composite; timber; masonry; alloy aluminium} extended to include {geotechnical structures}
- **Scope of structures:** Various classes and types of structures include:
    - {generic; buildings; bridges; industrial (crane support & machinery; towers & masts; etc); containment (silos; liquid containing; pipelines); others}

3.4 EN 1990 reliability differentiation guidelines as compared to ISO 2394

EN 1990 establishes reliability classes RC1 – RC3 (also called consequence classes CC1 – CC3), with β values shown in Table 2 (indicating also the *reference class* reliability RC2):

- RC3 High consequence for the loss of human life; economic, social or environmental consequences very great;
- RC2 *Reference class* of medium consequences for most conventional structures;
- RC1 Low consequences for the loss of human life; small or negligible economic, social or environmental consequences.

Table 2 Target reliability levels (β) according to EN 1990

Reliability Class	**EN 1990 Minimum values for β**					
	Ultimate LS		Fatigue		Serviceability LS	
Reference period	1 year	50 years	1 year	50 years	1 year	50 years
RC3	5,2	4,3				
RC2	4,7	3,8		1,5 to 3,8	2,9	1,5
RC1	4,2	3,3				

The EN 1990 specification to achieve *reliability required for structures … by design … and by appropriate execution and quality management measures* (Clause 2.2(1)P) with *the choice of the levels of reliability for a particular structure should take account of the relevant factors* (Clause 2.2(3)) also provides that *different levels of reliability may be adopted inter alia for structural resistance; for serviceability* (Clause 2.2(2)). Guidelines for the application of reliability management in terms of reliability differentiation, provision for design supervision and inspection during execution, including adjustment of partial factors for resistance are presented in Annex B (Informative).

3.5 EN 1990 Nationally Determined Parameters

The fundamental alternative expressions for action combinations for transient and persistent design situations for the Ultimate Limit State are presented in symbolic terms in the normative Standard as either Expression (6.10) or Expressions (6.10a) and (6.10b). The Nationally Determined Parameter (NDP) choices are all presented in Annex A1 (Normative) *Application for buildings*. NDP choices include the expressions to be selected for the various design situations, for which no recommendations are made; and the partial action and combination factor values for which values are recommended.

3.6 Design assisted by testing

The informative EN 1990 Annex D elaborates design assisted by testing, providing procedures that may be followed when testing is used in the design process and in the verification of limit states. The procedures provide the basis for complying with the principle of Clause 5.2(2)P *Design assisted by testing shall achieve the level of reliability required for the relevant design situation.* Provision is made for the determination of either

- **representative or characteristic values** of actions, resistance or material properties; to be used together with appropriate partial factors;
- **design values** determined directly.

The procedures provide an important basis for activities related to design such as testing of materials or structural products, wind tunnel testing and soil tests.

4 DESIGN VERIFICATION FOR BUILDINGS

The design verification procedures are presented in EN 1990 Section 6 *Verification by the partial factor method* in symbolic terms for all the various limit states and design situations, and applied to buildings in Annex A1 (Normative) *Application for buildings*. The alternative combination schemes and the recommended partial factors for buildings are discussed here.

4.1 Action combination values

Representative values for actions consist of:

- **Characteristic** values (F_k)
- **Combination** value of a variable action ($\psi_0 Q_k$): Chosen so that the probability that the action effect values caused by the combination will be exceeded is approximately the same as when a single action is considered
- **Frequent** value of a variable action ($\psi_1 Q_k$): Determined so that the total time, within a chosen period of time, during which it is exceeded, is only a small given part of the chosen period of time; or the frequency of its excess is limited to a given small value
- **Quasi-permanent** value of a variable action ($\psi_2 Q_k$): Determined so that the total time, within a chosen period of time during which it is exceeded, is of the magnitude of half the chosen period

The recommended variable action combination values for buildings are listed in Table 3.

Table 3 Recommended combination factors for actions applied to buildings

Variable Action	ψ_0	ψ_1	ψ_2
Imposed loads on buildings			
- Category A/B: domestic residential, office areas	0,7	0,5	0,3
- Category C/D: congregation, shopping areas	0,7	0,7	0,6
- Category E: storage areas	1,0	0,9	0,8
- Category F/G: traffic area (vehicle < 30kN / >30kN)	0,7	0,7/0,5	0,6/0,3
- Category H: roofs	0	0	0
Wind loads on buildings	0,6	0,2	0
Temperature loads in buildings	0,6	0,5	0

4.2 Action combination schemes for various design situations

4.2.1 Design situations

The various combination schemes for actions are given in symbolic terms in Table 4 for the different design situations of the ultimate and serviceability limit states.

Table 4 Action combination schemes for design situations

Design Situation	Ultimate Limit State	
Persistent, Transient	$\sum_{j\geq 1}\gamma_{G,j}G_{k,j}\ \text{"+"}\ \gamma_P P\ \text{"+"}\ \gamma_{Q,1}Q_{k,1}\ \text{"+"}\ \sum_{i>1}\gamma_{Q,i}\psi_{0,i}Q_{k,i}$	[(6.10)]
	Alternative scheme:	
	$\sum_{j\geq 1}\gamma_{G,j}G_{k,j}\ \text{"+"}\ \gamma_P P\ \text{"+"}\ \gamma_{Q,1}\psi_{0,1}Q_{k,1}\ \text{"+"}\ \sum_{i>1}\gamma_{Q,i}\psi_{0,i}Q_{k,i}$	[(6.10a)]
	$\sum_{j\geq 1}\xi\gamma_{G,j}G_{k,j}\ \text{"+"}\ \gamma_P P\ \text{"+"}\ \gamma_{Q,1}Q_{k,1}\ \text{"+"}\ \sum_{i>1}\gamma_{Q,i}\psi_{0,i}Q_{k,i}$	[(6.10b)]
Accidental	$\sum_{j\geq 1}G_{k,j}\ \text{"+"}\ P\ \text{"+"}\ A_d\ \text{"+"}\ (\psi_{1,1}\ or\ \psi_{2,1})Q_{k,1}\ \text{"+"}\ \sum_{i>1}\psi_{2,i}Q_{k,i}$	
Seismic	$\sum_{j\geq 1}G_{k,j}\ \text{"+"}\ P\ \text{"+"}\ A_{E,d}\ \text{"+"}\ \sum_{i>1}\psi_{2,i}Q_{k,i}$ An importance factor γ_I is applied to the seismic action $A_{E,d}$.	
	Serviceability Limit State	
Characteristic, (Irreversible)	$\sum_{j\geq 1}G_{k,j}\ \text{"+"}\ P\ \text{"+"}\ Q_{k,1}\ \text{"+"}\ \sum_{i>1}\psi_{0,i}Q_{k,i}$	
Frequent (Reversible)	$\sum_{j\geq 1}G_{k,j}\ \text{"+"}\ P\ \text{"+"}\ \psi_{1,1}Q_{k,1}\ \text{"+"}\ \sum_{i>1}\psi_{2,i}Q_{k,i}$	
Quasi-permanent (Long-term)	$\sum_{j\geq 1}G_{k,j}\ \text{"+"}\ P\ \text{"+"}\ \sum_{i\geq 1}\psi_{2,i}Q_{k,i}$	

4.2.2 Recommended partial factors for ultimate limit states buildings

Recommended partial factors for the various limit states and design situations are summarised in Table 5; with $\xi = 0,85$. Application of the two alternative combination schemes {Expression (6.10) or Expression (6.10 a and b)} is presented (Brozzetti (2000).

A third alternative with $\gamma_{Q,a} = 0$ is revealed; as shown in the highlighted cell in Table 5. This option is quite similar to the present SANS 10160-1989 (Amended) scheme (Holický & Retief 2005).

Set A provides for static equilibrium, whilst Set B and Set C are defined in order to provide for geotechnical design, as discussed below. For the ultimate limit state in the accidental and seismic design situations the partial factors are specified as 1,0.

Table 5 Recommended partial factors for ultimate limit state combination sets

Set	Ultimate Limit State	Action: Permanent Unfavourable /Favourable	Action: Variable Unfavourable /Favourable
Set A	EQU: Static equilibrium for buildings		
	- Equilibrium only - with Resistance (instead of A & B)	1,10 / 0,90 1,35 / 1,15	1,50 / 0 1,50 / 0
Set B	STR/GEO: Alternative structural/geotechnical general equations		
Alt 1 Alt 2	- Equation (6.10) - Equation (6.10a) - Equation (6.10b)	1,35 / 1,00 1,35 / 1,00 ξ1,35 / ξ1,00	1,50 / 0 1,50 ψ_0 (= 1,05) / 0 1,50 / 0
Alt 3	- Equation (6.10a) modified - Equation (6.10b)	1,35 / 1,00 ξ1,35 / ξ1,00	0 1,50 / 0
Set C	STR/GEO: Geotechnical actions and ground resistance involved		
	- General equation	1,00	1,30 / 0

4.2.3 The serviceability limit state

For serviceability limit states the partial factors for actions should be taken as 1,0 except if differently specified in EN 1991 to EN 1999. Serviceability limit states in buildings should take into account criteria related, for example, to floor stiffness, differential floor levels, storey sway or/and building sway and roof stiffness. The serviceability criteria should be specified for each project and agreed with the client. The serviceability criteria may be defined in the National annex.

The serviceability criteria for deformations and vibrations shall be defined depending on the intended use in relation to the serviceability requirements and independently of the materials used.

4.2.4 Alternative Design Approaches for geotechnical design

Three alternative design approaches (Approach 1, 2 & 3) for the design of structural members (footings, piles, basement walls, etc.) (assigned as STR) involving geotechnical actions and the resistance of the ground (assigned as GEO) are introduced with various combinations of the use of Set

B and Set C together with alternative sets of material and resistance factors specified in EN 1997-1 *Geotechnical design – General rules*.

- Approach 1 alternatively applies Set B and Set C for structural and geotechnical actions; usually resulting in structural resistance governed by the Set B loading case and sizing of foundations governed by the Set C loading case.
- Approach 2 applies Set B to both structural and geotechnical actions.
- Approach 3 applies Set B to structural actions and Set C to geotechnical actions in a single load case.

Combination with the alternative sets of material factors (M1 & M2) and resistance factors (R1 – R4) according to EN 1997-1 are summarized in Table 6, where Set B & C are presented by A1 & A2 respectively.

Table 6 Sets of partial factors for alternative ULS design approaches

Design Approach (DA)		**Combination Scheme** General [Piles & Anchors]
Design Approach 1 Satisfy both:	Combination 1	*A1* "+" *M1* "+" *R1*
	Combination 2	*A2* "+" *M2* [*M1**] "+" *R1* [*R4*] (* *M2* for negative skin friction & lateral pile load)
Design Approach 2		*A1* "+" *M1* "+" *R2*
Design Approach 3		{*A1*(STR) or *A2*(GEO)} "+" *M2* "+" *R3*

An extract from the partial action (*A*), material (*M*) and resistance (*R*) factors recommended in EN 1997-1 Annex A (Normative) *Partial and correlation factors for ultimate limit states and recommended values* is provided in Table 7, which also provides some clarification of the intent, implications and differences for the alternative design approaches. Selection of the Design Approach is an NDP for EN 1997-1.

5 THEORETICAL RELIABILITY VERIFICATION IN EUROCODE

In accordance with the partial factor methods accepted in EN 1990 to EN 1999 the design values of the basic variables, X_d and F_d, are expressed in terms of their representative values X_{rep} and F_{rep}, which may be the characteristic X_k and F_k or nominal values X_{nom} and F_{nom}. The representative values X_{rep} and F_{rep} should be divided and/or multiplied, respectively, by the appropriate partial factors to obtain the design values X_d and F_d expressed as:

$$X_d = X_k/\gamma_M \quad (11)$$

$$F_d = \gamma_F F_k \quad (12)$$

where γ_M and γ_F denotes the materials and action partial factors, in most cases both > 1.

Table 7 Partial action, material and resistance factors recommended in EN 1997-1 Annex A

Action (STR), (GEO)		***A1***	***A2***		
Permanent	Unfavourable / Favourable	1,35/1,0	1,0/1,0		
Variable	Unfavourable / Favourable	1,5/0	1,3/0		
Material		***M1***	***M2***		
Angle of shear resistance (tan φ); effective cohesion		1,0	1,25		
Shear strength: Undrained, unconfined		1,0	1,4		
Density		1,0	1,0		
Resistance		***R1***	***R2***	***R3***	***R4***
Pile (driven/bored/CF Auger)	Base	1,0/1,25/1,1	1,1	1,0	1,3/1,3/1,45
	Shaft (compression)	1,0	1,1	1,0	1,3
	Total/combined (compression)	1,0/1,15/1,1	1,1	1,0	1,3/1,3/1,4
Spread foundation	Bearing	1,0	1,4	1,0	
Retaining	Sliding	1,0	1,1	1,0	
Equilibrium (EQU), Uplift (UPL) Hydraulic heave (HYD)		***A***			***M = M2***
		EQU	**UPL**	**HYD**	
Permanent	Unfavourable / Favourable	1,1/0,9	1,0/0,9	1,35/0,9	
Variable	Unfavourable / Favourable	1,5/0	1,5(/0)	1,5(/0)	
Angle of shear resistance (tan φ), effective cohesion					1,25
Undrained shear strength, tensile pile & anchorage resistance					1,4

5.1 General partial factors

Both partial factors should include model uncertainties, which may significantly affect the reliability of a structure. Design values for model uncertainties may be incorporated into the design expressions through the partial factors γ_{Ed} and γ_{Rd}, applied as follows:

$$E_d = \gamma_{Ed} E\{\gamma_{gj} G_{kj}; \gamma_P P; \gamma_{q1} Q_{k1}; \gamma_{qi} \psi_{0i} Q_{ki}; a_d \ldots\} \tag{13}$$

$$R_d = R\{\eta X_k / \gamma_m; a_d \ldots\} / \gamma_{Rd} \tag{14}$$

Here η denotes a conversion factor appropriate to the material property.

Reductions in the design values of variable actions are applied as ψ_0, ψ_1 or ψ_2 to simultaneously occurring accompanying variable actions. The following simplifications may be made to expressions (13) and (14), when required.

- On the loading side (for a single action or where linearity of action effects exists) :

$$E_d = E\{\gamma_{F,i} F_{rep,i}, a_d\} \tag{15}$$

- On the resistance, simplifications of expression (14) may be applied in the relevant material-oriented documents provided the level of reliability is not reduced.

For non-linear resistance and actions models, and multi-variable action or resistance models the above relations become more complex.

5.2 Partial factors for material properties

The partial factor for resistance γ_m is defined in equation (14) by fractiles X_k and X_d. Taking into account the general expression for fractiles of random variables the factor γ_m may be written as

$$\gamma_m = \frac{X_k}{X_d} = \frac{\mu_X + u_{0.05}\,\sigma_X}{\mu_X + u_p\sigma_X} = \frac{1 + u_{0.05}\,w_X}{1 + u_p w_X}, \quad p = \Phi(-0.8\beta) \tag{16}$$

where w_X denotes coefficients of variation of X, $u_{0,05}$ or u_p denotes 5%- p-fractile of the standardised random variable having the same probability distribution as the resistance X. In case of a lognormal distribution having the lower bound at zero, equation (16) may be written as

$$\gamma_m = \frac{X_k}{X_d} = \frac{\frac{1}{\sqrt{1+w_X^2}}\exp\left(u_{0.05}\sqrt{\ln(1+w_X^2)}\right)}{\frac{1}{\sqrt{1+w_X^2}}\exp\left(u_p\sqrt{\ln(1+w_X^2)}\right)} \cong \frac{\exp\left(u_{0.05}\times w_X\right)}{\exp\left(u_p \times w_X\right)}, \quad p = \Phi(-0.8\beta) \tag{17}$$

where u denotes the normal standardised variable.

The increase of γ_m with increasing the reliability index β is considerably greater for the normal distribution than for the lognormal distribution, particularly for coefficients of variation w_X greater than 0,10. Figure 2 shows the dependence of γ_m on w_X for normal (N), lognormal (LN, x_0 = 0) and lognormal distribution with the skewness (LN, α = 0,5) assuming β = 3,8.

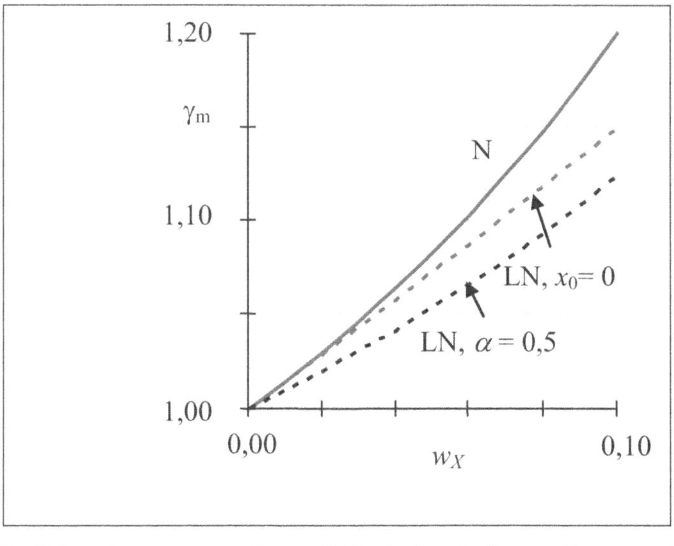

Figure 2 Partial factor γ_m versus the coefficient of variation w_X for normal and lognormal (x_0 = 0; skewness α = 0,5) distributions assuming β = 3,8

5.3 Partial factors for Permanent Action

Consider a permanent load G (self-weight) having a normal distribution. It is assumed that the characteristic value G_k is defined as the mean μ_G; with σ_G denoting the standard deviation and w_G the coefficient of variation; the sensitivity factor α_G = – 0,7. The design value G_d is then:

$$G_d = \mu_G - \alpha_G \times \beta \times \sigma_G = \mu_G + 0{,}7\times \beta \times \sigma_G = \mu_G(1 + 0{,}7\times \beta \times w_G) \tag{18}$$

The partial factor γ_G of G is given as:

$$\gamma_g = G_d / G_k \tag{19}$$

Taking into account of the fact that $G_k = \mu_G$ it follows from equations (5.7) and (5.8) that

$$\gamma_g = (1 + 0{,}7 \times \beta \times w_G) \tag{20}$$

Figure 3 shows the variation of the partial factor γ_g with the reliability index β for selected values of the coefficient of variation w_G = 0,05; 0,10; 0,15 and 0,20.

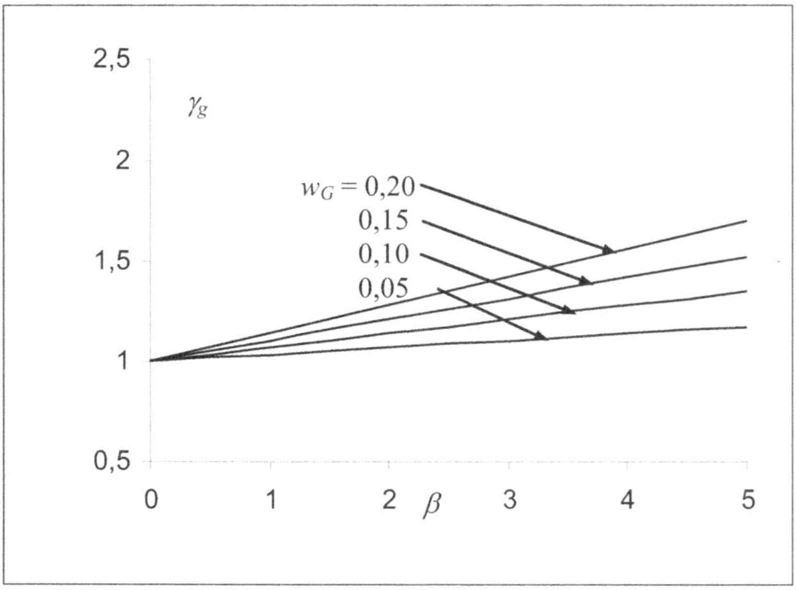

Figure 3 Variation of γ_g with β for a normal distribution of G

Consider a permanent load with a coefficient of variation w_G = 0,10; then γ_g = 1,28; if the model uncertainty γ_{Ed} = 1,05 is included, then:

$$\gamma_G = \gamma_g \, \gamma_{Ed} = 1{,}28 \times 1{,}05 = 1{,}33 \approx 1{,}35$$

Note that the value γ_G = 1,35 is recommended in EN 1990, corresponding to w_G = 0,10.

5.4 Partial factors for Variable Action

A similar procedure can be used for estimation of the partial factors γ_Q for variable loads Q. Assuming the Gumbel distribution with mean μ_G and coefficient of variation w_Q; the characteristic value is usually defined as 0,98 fractile of annual extremes (or extremes related to a certain basic reference period) and is given as

$$Q_k = \mu_Q \, (1 - w_Q \, (0{,}45 + 0{,}78 \ln(-\ln(0{,}98)))) \tag{21}$$

The design value Q_d related to the design working life or other reference period is given as

$$Q_d = \mu_Q (1 - w_Q (0{,}45 - \alpha_T 0{,}78 \ln(N) + 0{,}78 \ln(-\ln(\Phi^{-1}(-\alpha_E\beta)))) \quad (22)$$

where N denotes the ratio of the working design life, for example 50 years (or other reference period) to the basic reference period for the variable action (e.g. occupancy period for imposed loads or annual maximum for wind load), and α_T (usually taken = 1) is the time sensitivity factor, given by the ratio of w'_Q/w_Q, where w'_Q refers to the time variable part of w_Q.

Figure 4 shows the variation of $\gamma_q = Q_d / Q_k$ with the coefficients of variation w_Q for selected values of β assuming a Gumbel distribution of Q, and the period ratio $N = 10$ (the reference period 10 times greater than the basic reference period). For a variable load which has the coefficient of variation $w_Q = 0{,}3$ the partial factor $\gamma_q = 1{,}48$. If the model uncertainty $\gamma_{Ed} = 1{,}05$ is accepted, then it follows that:

$$\gamma_Q = \gamma_q \, \gamma_{Ed} = 1{,}48 \times 1{,}05 = 1{,}54 \approx 1{,}5$$

It appears that the value $\gamma_Q = 1{,}5$, recommended in EN 1990, is a reasonable approximation.

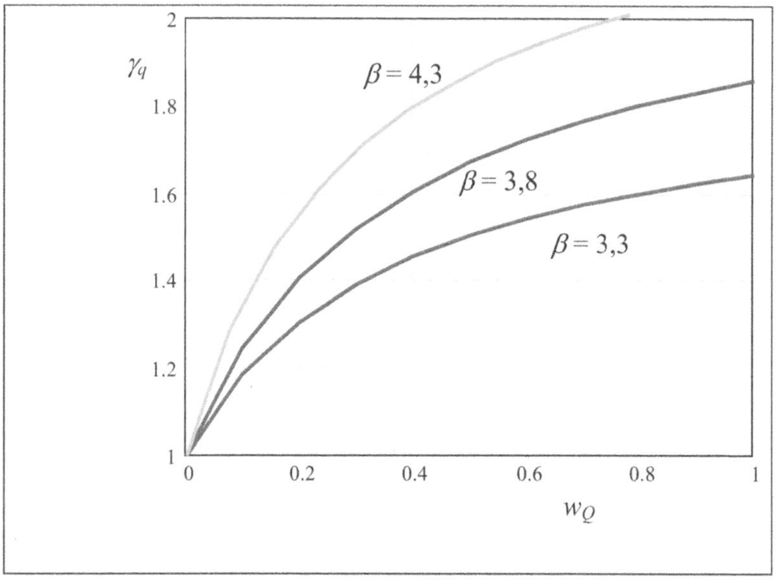

Figure 4 **Variation of γ_q with the coefficients of variation w_Q for selected values of β assuming a Gumbel distribution for Q, with N=10 and α_T = 1.**

5.5 Combination factor ψ_0

The combination rules provided in EN 1990 are based on the concept of one leading and various accompanying variable actions. Consider two variable actions, a leading action Q_1 having the basic reference period T_1 and an accompanying action Q_2 having the basic reference period T_2. The characteristic value Q_{2k} corresponds to the probability $\Phi(\alpha_E\beta)$ that the extremes corresponding to the basic reference periods (e.g. annual extremes) exceed Q_{2k}, thus

$$Q_{2k} = \mu_Q [1 - w_Q (0{,}45 + 0{,}78 \ln(-\ln(\Phi^{-1}(-\alpha_E\beta)))] \quad (23)$$

The representative (reduced characteristic) value $Q_{2,rep}$ corresponds with the increased probability $\Phi(0{,}4\alpha_E\beta)$ (additional factor 0,4 reduces β) that the extreme corresponding with the greater of the basic periods T_1 and T_2 exceeds $Q_{2,rep}$, thus

$$Q_{2,rep} = \mu_Q [1 - w_Q (0{,}45 + \alpha_T 0{,}78 \ln(N) + 0{,}78 \ln(-\ln(\Phi^{-1}(-0{,}4\alpha_E\beta)))] \quad (24)$$

In equation (24) N denotes the integer taken as a rounded value of the smaller of the ratios T/T_1 and T/T_2. The reduction (combination) factor ψ_0 is then given as

$$\psi_0 = Q_{2,\text{rep}}/\ Q_{2\text{k}} \tag{25}$$

where $Q_{2\text{k}}$ and $Q_{2,\text{rep}}$ are given by (23) and (24). Figure 5 shows the variation of ψ_0 with the coefficient of variation related to the reference period T, for β=3,8, for N = 7 and selected time sensitivity factors α_T.

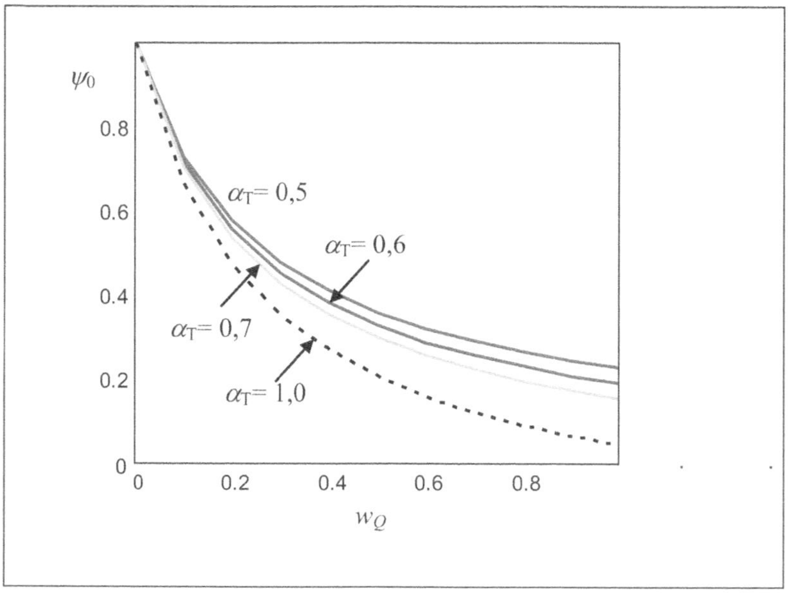

Figure 5 **Variation of ψ_0 with w_Q related to the reference period T; for β=3,8; N = 7 and selected time sensitivity factors αT**

Consider a combination of an imposed load Q_1 having the basic reference period T_1 = 7 years and an accompanying action (wind) Q_2 having the basic reference period T_2 = 1 year. Consider a time sensitivity factor α_T = 0,5. Considering the reference period T = 50 years then N = 50/7≈7. It follows from Figure 5 that for β=3,8, w_Q=0,35 the factor $\psi_0 \approx 0{,}46$ (for α_T = 1, ψ_0 is less than 0,4).

5.6 Combination factor ψ_1 and ψ_2

Factors ψ_1 and ψ_2 are defined in such a way that the corresponding representative $Q_{\text{rep}} = \psi_1 Q_\text{k}$ or $\psi_2 Q_\text{k}$ are exceeded with the probability 0,01 or 0,50 respectively. Assuming Gumbel distribution the reduction factors may be determined as

$$\psi_{1(2)} = \frac{1 - w\{0{,}45 + 0{,}78\ln[-\ln(\Phi(1 - \frac{\eta_{1(2)}}{q}))]\}}{1 - w\{0{,}45 + 0{,}78\ln[-\ln(\Phi(0{,}98))]\}} \tag{26}$$

where $\eta_{1(2)}$ is the required probability (0,01 or 0,50 respectively) and q is the probability of variable action Q being non-zero

Consider the following informative examples of frequent and quasi-permanent values of different variable actions Q. Recall that the probability ρ □= $1-\eta/q$, where η is a fraction of the reference period (0,01 or 0,5) during which Q_1 or Q_2 are exceeded, q denotes the probability of Q being non zero.

- Imposed load: It is assumed that the coefficient of variation $w_Q = 1{,}1$ (approximates a point in time distribution), short term imposed load is supposed to be on about 18 days a year, then $\rho_1 = 1-\eta_1/q_1 = 1-0{,}01/0{,}05 \sim 0{,}8$ and it follows from Figure 5 that $\psi_1 \sim 0{,}5$, long term load is almost always on, then $\rho_2 = 1-\eta_2/q_2 = 1-0{,}5/1 \sim 0{,}5$ and it follows from Figure 5 that $\psi_2 \sim 0{,}2$,
- Wind load: The assumed coefficient of variation is $w_Q = 0{,}5$, the load is on about 10×8 hours a year $\rho_1 = 1-\eta_1/q_1 = 1-0{,}01/0{,}009 < 0$ (not applicable), $\psi_1 \sim 0{,}0$, for quasi-permanent value $\rho_2 = 1-\eta_2/q_2 = 1-0{,}5/0{,}009 < 0$, (not applicable) and $\psi_2 \sim 0{,}0$,

Based on these results, factors ψ_0, ψ_1 and ψ_2 recommended in EN 1990 (see §4.1) seem to be rather conservative and should be revised taking into account local conditions.

6 CONCLUSIONS

The following conclusions are made on the basis of structural design applied in Eurocode through the head Standard EN 1990-2002 in terms of its relevance to the revision of SANS 10160 in particular and South African structural design standards in general:

- **International harmonisation:** The consistency between EN 1990 and ISO 2394 provides a sound basis for harmonisation by referencing to the Eurocode Standard, not only to European Member States, but also to international practice based on the International Standard.

- **Unified design:** The extensive development of reliability-based partial factor Limit States Design in EN 1990 as the basis for unification with the treatment of actions on structures and structural resistance through EN 1991 – EN 1999 (Sedlacek *et al* 1996) provides the platform for the consistent treatment of actions in the present revision of SANS 10160 and the application of unified procedures with the future revision of materials-based design standards.

- **Reliability framework:** The extended reliability framework set up in EN 1990 provides the basis for an extended scope of design situations and actions that can be provided for, whilst maintaining consistent levels of reliability as based on rational analysis and calibration; specifically providing operational procedures for transient (execution of structures); persistent (equilibrium, strength, geotechnical) and accidental (robustness, seismic) situations.

- **Parameter selection:** Although the selection of Nationally Determined Parameters is constrained and managed, the options are sufficiently wide to accommodate present South African practice whilst maintaining effective consistency with Eurocode in a revised SANS 10160. It also provides the basis for improvement of the consistency of reliability (SAKO 1999; BRE 2003; Holický & Retief 2005).

It is therefore concluded that Eurocode EN 1990-2002 *Basis of structural design* represents a significant advance in presenting rational structural design procedures, that could serve as useful reference in the revision of the present South African Standard SABS 0160-1989 (Amended).

7 REFERENCES

BRE (2003). UK Building Research Establishment *Review of SAKO Report.* Gulvanessian H – personal communication.

Brozzetti J, Sedlacek G (2000). *Background to the combination rules (6.10) or (6.10a) and (6.10b) in prEN 1990 Basis of design.* Document CEN/TC 250 N419. CEN/TC 250 Secretariat.

CE (2002). European Commission Guidance Paper L *Application and use of Eurocodes.* Document CONSTRUCT 01/483 Rev.1. Brussels.

CEN (2002). European Committee for Standardisation EN 1990:2002 *Eurocode – Basis of structural design.* Brussels.

Gulvanessian H, Calgaro J-A & Holický M (2002). *Designer's guide to EN 1990 Eurocode: Basis of structural design.* Thomas Telford. 192 pp.

Holický M & Retief JV (2005). *Reliability Assessment of Alternative Load Combination Schemes for Structural Design.* Journal SAICE, Vol. 25, No 1, pp 15 – 20, 2005.

Holický M & Vrouwenvelder T (2005). *Basic concepts of structural reliability.* Chapter I in Implementation of Eurocodes Handbook 2 *Basis of structural reliability and risk engineering.* Leonardo da Vinci Pilot Project.

ISO (1993). International Standards Organisation. ISO 3534 *Statistics- Vocabulary and Symbols*: Part 1: Probability and general statistical terms, Part 2: Statistical quality control.

ISO (1997). International Standards Organisation. ISO 12491 *Statistical methods for quality control of building materials and components.* 1997.

ISO (1998). International Standards Organisation. ISO 2394:1998 *General Principles on Reliability for Structures.* International Standard. [Also issued as SANS 2394:2004]

JCSS (2001). Joint Committee on Structural Safety. *Probabilistic model code.* http://www.jcss.ethz.ch/.

SABS (1994). SA Bureau of Standards, Pretoria. *Code of Practice: The general procedures and loadings to be adopted in the design of buildings.* SABS 0160:1989 (Amended 1990, 1991, 1994).

SAKO (1999). Joint Committee of NKB and INSTA-B. *Basis of Design of Structures. Proposal for Modification of Partial Safety Factors in Eurocodes.* NKB Committee and Work Report 1999:01 E, pp. 55.

Sedlacek G, Brozzetti J, Hanswille G, Litzner HU, Canisius TDG (1996). *Relationship between Eurocode 1 and the "material" oriented Eurocodes.* IABSE Colloquium Background and application of Eurocode 1. Delft.

ANNEX A COMPARISON OF THE SCOPE OF ISO 2394 AND EN 1990

ISO 2394	# of pages		EN 1990
INTRODUCTORY AND GENERAL SECTIONS			
1-3 Scope, Definitions, Symbols **4 Requirements and concepts** **5 Principles of limit states design** **6 Basic variables** **7 Models**	21	37	**1-2 General; Requirements** **3 Principles of limit states design** **4 Basic variables** **5 Structural analysis and design assisted by testing**
DESIGN VERIFICATION PROCEDURES			
8 Principles of probability-based design – General; – Systems reliability vs. element reliability; – Specified degrees of required reliability; – Calculation of failure probabilities; – Implementation	2,5	-	Provided for in Clause 3.5(5)
9 Partial factors format – Design conditions and values; – Action characteristic, combination, frequent, quasi-permanent values; – Characteristic values · Materials · Including soils; · Geometry – Load cases and load combinations; – Action effects and resistances; – Fatigue; – Calibration	4	19	**6 Verification by the partial factors method** **Annex A1 (Normative) Applications for buildings** – Design values · Actions, effects; · Material; · Geometry, · Resistance – Ultimate LS · Static equilibrium & resistance; · Combination of actions; · Partial factors for actions & combinations; · Materials & products; – Serviceability LS · Verification; · Criteria; · Combinations; · Partial factors for materials
10 Assessment of existing structures – Relevant cases; – Principles of assessment; – Basic variables; – Investigation; – Assessment of damage	2,5		
INFORMATIVE ANNEX			
A Quality management and quality assurance **B Examples of permanent, variable and accidental actions** **C Models for fatigue** **D Design based on experimental models** **E Principles of reliability-based design** **F Combination of actions, estimation of a-values** **G Example of a method of combination of actions**	43	30	**B Management of structural reliability for construction works** **D Design assisted by testing** **C Basis for partial factor design and reliability analysis**
TOTAL	**73**	**86**	

Parts 2, 7 & 8 – General Actions:

Self-weight & Imposed Loads; Thermal Actions; Actions during Execution

2-1 Review of Provisions for General Actions in SANS 10160

Retief JV & Dunaiski PE

1. INTRODUCTION

The general actions on buildings which are reviewed in this chapter include provisions for self-weight of the structure and non-structural components of the building; various imposed loads on floors, roofs and horizontal loads on balustrades and partition walls; thermal actions caused by temperature gradients and changes; actions during execution of the structure, including construction, fabrication and erection. These provisions are presented respectively in SANS 10160-2 *Self-weight and imposed loads*; SANS 10160-7 *Thermal actions* and SANS 10160-8 *Actions during execution*.

Although the provisions for self-weight and imposed loads represent a revision of the SABS 0160:1989 stipulations for these basic sets of loads, SANS 10160-2 is primarily based on reference to the Eurocode EN 1991-1-1:2002. The revised provisions include the introduction of a number of new loads and classes from EN 1991-1-1 *Actions on structures: General actions – Self-weight, densities and imposed loads*. Provisions for thermal actions represent the introduction of a new class of action from EN 1991-1-5:2004 *Actions on structures: General actions – Thermal actions*, considering only the sections relevant to building structures and omitting the substantial section on bridges. The introduction of the provision for actions during the transient situation during the execution of the structure from EN 1991-1-6:2005 *Actions on structures: General actions – Actions during execution* represents the consideration of an important stage in the design life of a structure.

Eurocode treats wind actions on structures also as a *general action*. Provisions for wind actions are however given in SANS 10160-3 and discussed in Chapter 3-1 and Chapter 3-2.

Another Eurocode *general action* is the provision for accidental actions. EN 1991-1-7:2006 *Actions on structures: General actions – Accidental actions* provides the general procedures for the treatment of accidental actions, together with specific procedures for actions due to impact and internal explosions. As indicated in Chapter 1-1, these two types of accidental actions are considered to be too special in character to warrant inclusion in SANS 10160. Only the basis of design for the accidental design situation was taken over from Eurocode and incorporated into SANS 10160-1, as discussed in Chapter 1-2.

The special case of an accidental design situation which is also treated as a Eurocode *general action* provides for actions on structures exposed to fire. Such actions are however not included in SANS 10160. This important category of action requires the availability of unified provisions for materials-based design procedures, as in Eurocode. A concerted effort would be required to introduce provisions for actions exposed to fire together with at least one materials-based standard, preferably together with at least the standards for structural concrete and steel.

2. ACTIONS DUE TO SELF-WEIGHT

The actions on a structure due to its self-weight, and those of the non-structural parts of the building are the least problematical actions to be considered. They are however not without complications, and therefore require careful consideration in the design process. Although the variability of self-weight is relatively small, differences do occur between the various classes of self-weight. There is also evidence that it is consistently underestimated. Furthermore the uncertainty of modelling the load effect needs to be taken into account. Although measures are taken to improve the strength-to-weight ratio of structural systems on the one hand, the tendency towards larger structures with longer spans increases the contribution of self-weight to the total load on the other hand.

Self-weight represents the dominant component of the more general class of *permanent actions* on the structure. Although this class of actions clearly refers to the temporal nature of the action, it also implies limited variability. This section only considers proper self-weight permanent actions; other types of permanent action such as geotechnical and hydrostatic permanent actions are excluded here – see Chapter 5.1.

Improved rationality of the treatment of loads as derived from principles of structural reliability is required not only for variable or accidental actions, but also for permanent actions, including self-weight. Such an approach is in contrast to traditional design procedures where relatively large partial factors for self-weight were applied simply as part of the overall safety provisions.

The implementation of the dual action combination function (STR & STR-P) as discussed in Chapter 1-2 is one step taken towards improved treatment of self-weight loads. Eurocode introduces an elaborate scheme for the treatment of permanent actions to be considered across the wide scope of structures and situations. The need for such an elaborate scheme for the limited scope of SANS 10160 was one of the issues that needed attention.

In this section the conceptual basis of actions due to self-weight and its specification in Eurocode EN 1991-1-1:2002, which complies with EN 1990:2002, is discussed as background to its provision in SANS 10160-2, which is in turn complies with SANS 10160-1.

2.1 Conceptual Basis for Self-weight

Models for the probabilistic treatment of self-weight loads are presented in the CIB Report 115 (CIB 1989a) as part of a series of reports on actions on structures. These self-weight models formed the basis for the specifications for self-weight of the JCSS Model Code (JCSS 2002). Although the models are set up in probabilistic terms, their qualitative description provides a useful conceptual basis for actions due to self-weight.

An appreciation of the processes and mechanisms that contribute to the conversion of self-weight into load effects, and the associated uncertainties, assists in the formulation and use of the simplified design procedures for self-weight. Such insight will be particularly useful for identifying situations where the procedures are not sufficiently reliable, or conversely where it is justified to reduce the conservatism which should result from simplification.

The discussion below is extracted mainly from the CIB Report 115 (CIB 1989a).

2.1.1 *General model for self-weight loads and effects*

The elements of the self-weight load effect (E) are represented in Equation (1) below in terms of the weight density of the material (γ) over the volume element (dV) and the influence function (I) for the specific load effect:

$$E = \int_V I \gamma \, dV \qquad (1)$$

Uncertainties in self-weight load effects therefore derive from the uncertainties or variability during the design life of the structure of the densities of the materials and their distribution across the structure; the geometry of the structure which is defined in terms of nominal dimensions and incomplete specification during the design stage; modelling of the structural behaviour. Contributions to the variability of self-weight load effects arise from fluctuations in weight densities of the material, the dimensions of the structure, environmental factors such as humidity; the addition of unforeseen protective material. Differences between the information used during design, and the actual construction, use and maintenance of the structure also contribute to the variability.

2.1.2 *Classes of actions due to self-weight*

The main classification of actions due to self-weight consists of the differentiation between structural and non-structural building elements and materials. Since the structural material forms an inherent part of the load carrying system, it does not change over the lifetime of the structure and has a probability of 1 to occur at any point in time. Depending on the type of structural material, there is good control over the dimensions of the structural element and other properties defining the density of the construction material. The load produced by the weight of non-structural components may however change over the lifetime of the structure due to maintenance, repair and remodelling with a consequent probability of less than one at any point in time. The important issue of considering the weight of building components as variable loads during execution is treated separately in Eurocode EN 1991-1-6 and in SANS 10160-8 (see Section 6 of this chapter).

The characteristics of actions due to self-weight of structural elements are clearly dependent on the type of structural material. In the CIB models the following characterisation is given:

Concrete structures: As a rule, the self-weight of a concrete component is a substantial part of the total load. Influencing factors include the cross section dimensions; the weight density of the concrete, including the bulk density of the aggregates; the weight of reinforcement, including deviations from the weight assumed in the design.

Masonry structures: The self-weight of a masonry structure is almost always a substantial part of the load on the structure. Influencing factors include the dimensions and bulk density of bricks; the properties of mortar and joints; moisture contents at production and in the ultimate dried state.

Metal structures: The self-weight of the structural component of a metal structure is generally small compared to the supported variable loads. The variability of the structural element proper is generally small because of small variances in cross section and virtually constant density. The weight of connections also contributes to the self-weight. Deviations from the design values derive primarily from simplified estimates.

Timber structures: Timber structures are generally used to carry low loads; in which case the self-weight component is important. The determination of self-weight depends on the weight density of the timber, including moisture contents; dimensions of parts; connecting elements and metal parts.

2.1.3 *Sources of uncertainty*

In addition to the values and variability of self-weight that derive from the physical properties of the structure discussed above substantial deviations may occur during the design, construction or use of the building, as characterised by the CIB Report:

Design: Design errors may occur due to inadequate simplification of the design by underestimating the significance of self-weight; insufficient accuracy in the estimates of weights, dimensions and densities; neglecting some loads; lack of data at the time of design.

Execution: Deviations may occur during the manufacture, transport, erection or construction of the building related to changes in materials used; the dimensions as erected or constructed; all these deriving mostly from insufficient knowledge, experience and control during construction, or shortage of the specified materials.

Use: During the use of the structure, the self-weight loads may change due to insufficient consideration of the structural implications of modifications, reconstruction and maintenance due to changes in use or additions to the building; functional failure of water-proofing increasing the moisture content.

2.2 Specification of Actions due to Self-weight in EN 1991-1-1

The general specification of the characteristic value of permanent actions (G_k) given in EN 1990 Clause 4.1.2 is to provide a single value, unless the variability of G cannot be considered to be small (coefficient of variation $V_G < 0{,}05$ to $0{,}10$, depending on the type of structure) or where the structure is sensitive to variations in G; in which case upper and lower values need to be determined.

For the self-weight of a structure, a single characteristic value is calculated on the basis of *nominal dimensions* and *mean unit masses*.

In the IStructE *Manual for the design of building structures to Eurocode 1* (IStructE 2009) this is interpreted as requiring the use of the dual scheme where the limit $V_G < 0,10$ generally applies, and when self-weight contributes to equilibrium, the limit $V_G < 0,05$; upper and lower values for G then need to be determined when these limits are exceeded. According to the *Designers Guide to EN 1990* (Gulvanessian *et al* 2002) values can be determined by using Equation (2). For the two limiting values of V_G this implies a decrease / increase of 16% and 8% respectively for the lower and upper values for G_k.

$$G_{k,\inf/\sup} = \mu_G \pm 1,64\sigma_G = \mu_G(1 \pm 1,64V_G) \qquad (2)$$

These requirements are then taken over in EN 1991-1-1 in specifying the self-weight of structural and non-structural elements as a single permanent fixed action as calculated from nominal dimensions and mean unit masses. The requirement to determine upper and lower values for self-weight loads under certain circumstances is maintained. Non-structural elements, fixed services and machinery, earth pressure and ballast, and manufactured building systems are all included.

Nominal densities of construction and stored materials are provided in an informative annex, together with the provisions for determining the densities of materials not listed, in accordance with EN 1990 requirements.

2.3 Specification of Actions due to Self-weight in SANS 10160-2

The specification of actions due to self-weight for buildings from EN 1991-1-1 has generally been maintained in SANS 10160-2, with the main difference being the simplification of the requirements due to the scope of SANS 10160 being limited to buildings. The tables of mass densities of structural and stored materials presented in an informative annex are extended substantially from that given in SABS 0160:1989, as based on the similar annex in EN 1991-1-1.

An important simplification of SANS 10160-2 as compared to the referenced EN 1991-1-1 is that the need to differentiate into upper and lower values for self-weight for certain situations has been omitted. Although it is judged that such differentiation is not warranted for the restricted scope of building structures, designers should be sensitive to exceptional situations where such a simplification may not be reasonable. Although EN 1990 and EN 1991-1-1 does not give any further guidance other than indicating the need for such differentiation, the Eurocode procedures and practice could be followed when such conditions are applicable in a design situation. As indicated above the IStructE Manual provides useful guidance.

2.3.1 *Classification and design situations*

In accordance with the requirements of Part 1 self-weight is formally classified as a permanent action in the general sections of Part 2. Various design situations are identified for which the critical situation should be implemented in the verification, such as the addition or removal of certain non-structural elements, new coatings, water levels or moisture contents.

An important stipulation is that the total self-weight of structural and non-structural members shall be taken into account as a *single action* in a combination of actions.

2.3.2 *Densities of materials*

A practical aspect of the revised specification of self-weight is the presentation of Annex A (Informative) which provides an extended list of weight densities of materials ranging from various structural and building materials to classes of stored materials and their angle of repose where relevant. The annex represents a substantial extension of the tables from SABS 0160 with the inclusion of additional information from EN 1991-1-1. It is noted that the nominal values of weight densities may be used to determine characteristic values of self-weight.

2.3.3 *Non-structural elements*

A convenient indication is provided of the types of non-structural elements that need to be included, such as roofing, surfacing and coverings, partitions and linings, hand rails and barriers etc; a similar listing of fixed services is provided, such as equipment for lifts, heating, ventilation and air conditioning, electrical distribution, pipes and their contents.

3. CONCEPTUAL BASIS FOR IMPOSED FLOOR LOADS

The revision of SABS 0160:1989 afforded the opportunity to assess the provisions for imposed floor loads on building structures in terms of the scope of loads which is provided, together with the classification of occupancies; and finally the specified minimum values to be applied.

Although numerous load surveys have been performed over many years, with extensive efforts to convert the observations into floor load models, there is not sufficient information available to present comprehensive specifications for imposed floor loads for buildings across the practical range of design situations on a rational basis. Specified imposed floor loads are consequently generally based on judgement by standards committees and presented in convenient format at conservative values.

It is nevertheless instructive to consider the mechanisms which contribute to floor loads, and the resulting models that are used to represent such loads, together with the relevant parameters incorporated in the models for the cases that are available. These load models provide some insight on the mechanisms and factors underlying the somewhat simplified load specifications incorporated in the codes. Such awareness assists code committees in reducing the degree of subjectivity in specifying the imposed loads. It will also assist the designer in applying judgement to identify situations where the code specifications may either not be sufficient or where relaxation may be justified due to special conditions applying to any specific situation.

This section will summarise some information which is available in the form of probabilistic models which could be used in structural reliability modelling. The procedures followed in the formulation of imposed loads through experience-based judgement are also reviewed. Until the stipulated loads can be based extensively on rational probabilistic structural mechanics models, they should complement the judgement-based procedures.

3.1 Theoretical Models for Imposed Floor Loads

Two useful model codes for imposed floor loads that could be applied for the purpose of understanding imposed load mechanisms are the CIB Report 116 *Actions on structures – Live loads in buildings* (1989b) which incorporates an extensive body of information from load surveys and models and the JCSS Model Code (2002) which applies this information to provide probabilistic models and parameter data for use in structural reliability modelling. Although the information is still not sufficient to form the basis for standardised specification, it provides not only insight into the qualitative processes of imposed loads, but also some quantitative indication of magnitudes and variability.

The various processes that have an influence on the structural characteristics of imposed loads on building floors are summarised in the CIB Report 116 to consist of the following mechanisms:

- **Sustained loads:** Loads caused by furniture and equipment, with slow variations over time, but rapid changes associated with special occurrences such as changes in occupancy or tenants, or changes in the use of one or more rooms. Live loads due to storage tend to increase over time of occupancy. There may be a concentration of furniture or heavy equipment along walls, with movable partition walls themselves treated as live loads.
- **Intermittent loads:** Loads associated with persons and light equipment are generally random in space, with fairly rapid variations in time and duration, occurring for a relatively small part of the time. The characteristics of the intermittent loads and the relative ratio of sustained and intermittent loads may vary considerably, depending on the type of building and occupancy.
- **Special situations:** The load may be increased considerably during special situations, such as the gathering of people during planned events such as parties or meetings; crowding of people under emergency situations; the piling up of furniture in one area, where it is concentrated over a smaller area.

The spatial distribution and variability of the various loads are represented in terms of the equivalent uniformly distributed load (EUDL) for the various load effects. Appropriate data are derived from floor load surveys. Area effects are taken into account by converting EUDL values to specific reference areas. CIB Report 116 provides an assessment of such surveys and their interpretation in term of the various load mechanisms for typical occupancies, trends over time and comparisons between countries. The life-time maximum imposed load for a number of occupancies are given in conclusion in CIB R116, as shown in Table 1.

Table 1 Imposed load values derived from probability models of load processes

Imposed Load	CIB R116		Eurocode Background			
	Mean kN/m²	Standard Deviation kN/m²	Area m²	p_k kN/m²	Combination Factor	
					Characteristic ψ_0	Long-term ψ_2
Residential building	1,73	0,34	10	1,75	0,51	0,23
			50	0,87	0,69	0,32
Hotel	2,20	0,30	10	2,30	0,54	0,09
			50	0,90	0,72	0,26
Hospital			10	0,80	0,58	0,43
			50	0,55	0,31	0,56
Office building	2,64	0,49	10	1,90	0,44	0,27
			50	0,95	0,68	0,50
Commercial building	2,73	0,30	10	2,10	0,45	0,14
			50	1,00	0,66	0,31
School	1,63	0,20	10	2,20	0,50	0,23
			50	1,30	0,67	0,37

An updated set of probabilistic models is presented in the JCSS Model Code (2002), together with parameters for sustained loads and intermittent loads for various types of occupancies, and loads due to crowding of people. The parameters for the models are summarised in Table 2.

The magnitudes of mean and standard deviation for the respective load mechanisms give an indication of the relative importance of each mechanism and its variability. The variability of the sustained load is dominated by the contribution of the area-dependent component (σ_u), particularly for cases where the effective area is at about the reference area (A_0). The short duration of the intermittent load also indicates that the arbitrary point-in-time value is dominated by the sustained load, which has typically only five to ten occupancy changes during a 50 year design life. The large mean and standard deviation for the intermittent load due to the concentration of people confirms the importance of crowding for occupancies where this load mechanism is relevant.

Table 2 Parameters for imposed floor load models (JCSS 2001)

Occupancy	A_0	Sustained Load				Intermittent Load			
		m_q	σ_v	σ_u	$1/\lambda$	m_p	σ_U	$1/v$	d_p
	[m^2]	[kN/m^2]			[y]	[kN/m^2]		[y]	[d]
Office	20	0,5	0,3	0,6	5	0.2	0,4	0,3	1-3
Lobby	20	0,2	0,15	0,3	10	0,4	0,6	1,0	1-3
Residence	20	0,3	0,15	0,3	7	0,3	0,4	1,0	1-3
Hotel room	20	0,3	0,05	0,1	10	0,3	0,4	0,1	1-3
Patient room	20	0,4	0,3	0,6	5-10	0,2	0,4	1,0	1-3
Laboratory	20	0,7	0,4	0,8	5-10				
Library	20	1,7	0,5	1,0	>10				
Classroom	100	0,6	0,15	0,4	>10	0,5	1,4	0,3	1-5
Retail – first floor	100	0,9	0,6	1,6	1-5	0,4	1,1	1,0	1-14
Retail – upper floor	100	0,9	0,6	1,6	1-5	0,4	1,1	1,0	1-14
Storage	100	3,5	2,5	6,9	0,1-1,0				
Industrial light	100	1,0	1,0	2,8	5-10				
Industrial heavy	100	3,0	1,5	4,1	5-10				
People - concentration	20					1,25	2,5	0,02	0,5

An assessment of probabilistic models as basis for the determination of imposed loads for inclusion in Eurocode is reported by Sedlacek & Gulvanessian (1996). Values for characteristic values of imposed loads are presented in Table 1. Probability based values for the combination value (ψ_0) and frequent combination value (ψ_2) are also shown in Table 1. A marked reduction in imposed load occurs when the tributary area increases from 20m^2 to 50m^2. It is also notable that the combination values are significantly lower than those specified in EN 1990 (see Chapter 1-4), specifically for the smaller reference area. It was however concluded by Sedlacek & Gulvanessian that the lack of sufficient information precluded the application of such rational procedures to determine specified values.

3.2 General Specification Practice

The general practice followed in the specification of imposed loads is to devise a convenient scheme of classification to define the various processes related to the characteristics of the load. Elements of such a classification system used for imposed loads include the occupancy of the building area, from general classes such as domestic, residential, institutional, business, commercial, industrial, etc to specific zoning such as balconies and corridors; structural elements such as floor, roof or partition wall; distinctive activities such as the possibility of crowding, accumulation of material, forklift activities, etc.

The two issues to be considered are therefore (i) the definition of the situation and (ii) the related load values. Although the two issues are interrelated, and therefore need to be considered together in drawing up the specifications, it is convenient first to consider them separately.

3.2.1 *Imposed load source classification scheme*

The two elements of an effective classification scheme for imposed loads are:

- that conditions for which distinctive values for the load can be identified and assigned;
- that the respective conditions can then be arranged into a logical and convenient scheme which allows for a simple and unambiguous selection of a load which corresponds to the respective areas or components of the structure to be designed. A more refined classification scheme allows for better provision for the mechanisms

that contribute to the imposed load and consequently the assignment of proper and appropriate characteristic values for the stipulated loads.

The logical arrangement of the classification scheme then also conveys additional information to the user on the pertinent characteristics of the load mechanism. It should be noted that a more elaborate scheme for the stipulation of imposed loads does not result in complexity in using it; on the contrary, it increases the clarity of the scheme and the convenience in using it.

Different approaches taken by SABS 0160:1989 and AS 1170.1-1989 arguably represent two extremes of the way in which imposed loads are presented: In SABS 0160 all the occupancies for which the same imposed floor load value apply are listed in a very compact manner. In AS 1170.1 separate tables are given for eight building types, with a further lower level of classification of the various parts of such a building, arranged alphabetically. The concise arrangement of SABS 0160 given in Table 3 can be compared with the elaborate scheme used in AS 1170.1 as shown in Table 4, showing extracts of the respective stipulations in the two tables.

Table 3 Extract from scheme for imposed floor loads used in SABS 0160

Occupancy class of building or floor zone (description of room or floor use)	Uniformly distributed imposed floor load, kN/m^2	Concentrated load kN
All rooms in a dwelling unit and a dwelling house including corridors, stairs and lobbies to a dwelling house Bedrooms, wards, dormitories, private bathrooms and toilets in educational buildings, hospitals, hotels and other institutional occupancies Access catwalks in buildings	1,5	1,5
Classrooms, lecture theatres X-ray rooms, operating theatres	2,0	5,0
Garages and parking areas for vehicles of gross weight less than 25 kN excluding garages where mechanical parking or stacking devices are employed	2,0	10,0
Offices for general use	2,5	9,0
Offices with data-processing and similar equipment	3,0	
Cafés, restaurants Dining rooms, dining halls, lounges, kitchens, communal bathrooms and toilets in educational buildings, hotels and offices Entertainment, light industrial and institutional occupancies	3,0	5,0
Assembly halls, theatres, cinemas, sports complexes, grandstands, all with fixed individual seating	4,0	3,0
Light laboratories Sales and display areas in retail shops and departmental stores Banking halls	4,0	5,0
Etcetera		

3.2.2 *Selection of specified imposed load values*

Although the conceptual models for imposed loads as discussed above provide some background and insight into the way in which mechanical and probabilistic models can be devised, specified values of characteristic loads are essentially based on experience and judgement. Comparison with other internationally recognised standards broadens the base for selecting appropriate values. At the same time this is a self-referencing process that could lead to creeping conservatism if the process of selecting the values is not managed and moderated properly.

An extensive comparison between the SABS 0160:1989 values for imposed loads and those of AS 1170.1-1989, BS 6399: Part 1: 1996, ASCE-7:95 and ENV 1991-1-2 was made at the early stage of the development of SANS 10160 (Retief *et al* 2001). Similarly a comprehensive comparison of imposed load values amongst Eurocode Member States is reported by Hemmert-Halswick *et al* (1988).

In EN 1991-1-1 ranges of values are given from which Member States are allowed to select values as NDP options. In most cases recommended values are indicated. In cases where Member

States decide to elaborate the classification scheme for imposed loads, as illustrated above for the UK, the allowable ranges and recommended values lose much of their relevance.

Table 4 Extract from scheme for imposed floor loads used in AS 1170.1

Occupancy		Uniformly distributed imposed floor load, kN/m²
1 HOUSES		
1.1	General	1.5
1.2	Balconies – 1 m or more above ground – others	See clause x 1.5
1.3	Stairs and landings	3.0
1.4	Parking, including driveways and ramps	3.0
Etcetera		
2 RESIDENTIAL AND APARTMENT BUILDINGS (flats, hotels, motels, boarding houses, residential clubs)		
2.1	Communal assembly areas with fixed seating	3.0
2.2	Communal assembly areas without fixed seating, such as dance areas, bars, ...	5.0
Etcetera		
3 GARAGES AND PARKING		
3.1	Balconies	Same as areas ...
3.10	Toilet and bath rooms	2.0
Etcetera		
4 INDUSTRIAL (workshops, factories, warehouses)		
4.1	Balconies	
Etcetera		
8 RETAIL PREMISES (shops, department stores, supermarkets, etc)		
8.1	Balconies	Same as ...
8.5	Dining rooms and cafeterias	2.0
8.16	Vaults and strong rooms	5.0

An alphabetical arrangement of occupancy types is followed in ASCE-7 and also in the Canadian loading code, as illustrated in Table 5.

In the case of Eurocode EN 1991-1-1 a scheme for floor loads is used which is based mainly on public exposure for buildings of residential, social, commercial or administrative functions for imposed floor loads. The degree to which people may congregate is then used for further subdivision of the occupancies. The definition of the occupancy classification is given in Table 6. Allowable ranges and recommended values for imposed loads are tabulated separately, to be determined as NDP in the National Annex (NA) of each Member State. The same scheme is applied to accessible roofs and horizontal loads on parapets and partition walls. Individual clauses are assigned to the various imposed loads related to industrial activities and parking garages.

In the British National Annex to EN 1991-1-1 (BS EN 1991-1-1 NA) an additional level of sub-categories is added as illustrated in Table 7. The refined classification allows for improved differentiation in the selection of occupancies, which is particularly effective for Category C areas of public areas where people may congregate. It is shown below that the net effect is that areas are identified which mostly lead to lower load values, rather than the identification of cases where higher loads are required.

4. SPECIFICATION OF IMPOSED LOADS IN SANS 10160-2

Since imposed loads form a key part of SABS 0160:1989 these loads have received substantial attention during the early phases of the review process as outlined in Chapter 1-1. The results from a comparative review of imposed floor load values specified in SABS 0160, the values given in a number of representative International Standards and the selected revised values formed the most significant body of information that was carried over to the stage when Eurocode was selected as reference to the revised standard, with specific reference to EN 1991-1-1:2002.

The most important ways in which SANS 10160-2 refers to EN 1991-1-1 for imposed loads on building structures are that it follows its layout format and introduces a number of imposed load mechanisms, particularly relevant to industrial buildings and processes, and helicopter landing pads on top of buildings. The specified values from the previous version of Eurocode, ENV 1991-2-1 were but part of a general review of international practice referred to above.

4.1 Scope of Imposed Loads and Procedures

The scope of imposed loads and various related procedures employed in SANS 10160-2 and the relevance of EN 1991-1-1 are summarised in Table 8. Although the bulk of imposed load classes, namely floor and roof loads and horizontal loads on partitions and balustrades can be considered as being carried over from SABS 0160, the more logical differentiation between general and industrial buildings derives from EN 1991-1-1.

Table 5 Alphabetical arrangement of occupancies for floor loads

Extract from ASCE-7 Scheme		
Occupancy or use	**Uniform psf (kN/m^2)**	**Concentration lb (kN)**
Apartment (see Residential)		
Access floor systems		
Office use	50 (2.4)	8 000 (8,9)
Computer use	100 (4.79)	8 000 (8,9)
Armories and drill rooms	150 (7.18)	
Assembly areas and theatres		
Fixed seats (fastened to floor)	60 (2.87)	
Lobbies	100 (4.79)	
Movable seats	100 (4.79)	
Etcetera		
Extract from Canadian Scheme		
Use of Area of Floor or Roof		**Specified load (kPA)**
Assembly areas a) Except for those areas listed under (b) and (c), assembly areas with or without fixed seats including Arenas, Auditoria, Churches, ... Theatres and other areas with similar use		4.8
b) Assembly areas with fixed seats that have backs over at least 80% of assembly area for: Churches, Courtrooms, Lecture Halls, Theatres		2.4
c) Class rooms with or without fixed seats		2.4
Attics: Accessible by a stairway in *residential occupancies* only		1.4
Attics: Having limited accessibility so that there is no storage of equipment or material		0.5
Balconies: Exterior		4.8
Balconies: Interior and *mezzanines* that could be used by assembly of people as viewing areas		4.8
Etcetera		
Specified Concentrated Live Loads on an Area of Floor or Roof		**Load, kN**
Roof surfaces		1.3
Floors of classrooms		4.5
Floors of offices, manufacturing *buildings*, hospital wards and *stages*		9.0
Etcetera		

Table 6 Eurocode scheme for the specification of imposed floor loads

Category	Specific Use	Example
A	Areas for domestic and residential activities	Rooms in residential buildings and houses; bedrooms and wards in hospitals; bedrooms in hotels and hostels; kitchens and toilets.
B	Office areas	
C	Areas where people may congregate (with the exception of areas defined under category A,B and D	**C1**: Areas with tables etc. *e.g.* areas in schools, cafés, restaurants, dining halls, reading rooms, receptions
		C2: Areas with fixed seats, *e.g.* areas in churches, theatres or cinemas, conference rooms, lecture halls, assembly halls, waiting rooms, railway waiting rooms
		C3: Areas without obstacles for moving people, *e.g.* areas in museums, exhibition rooms, etc. and access areas in public and administration buildings, hotels, hospitals, railway station forecourts
		C4: Areas with possible physical activities, *e.g.* dance halls, gymnastic rooms, stages
		C5: Areas susceptible to large crowds, *e.g.* in buildings for public events like concert halls, sports halls including stands, terraces and access areas and railway platforms
D	Shopping areas	**D1:** Areas in general retail shops
		D2: Areas in department stores

Table 7 Extract from British scheme for the specification of imposed floor loads

BS EN 1991-1-1 NA:2005 Scheme				
Category	**Specific Use**	**Sub-Category**	**Example**	
A	Areas for domestic and residential activities	A1	All usages within self-contained dwelling units (unit occupied by single family, modular student accommodation unit with a secure door and comprising not more than six single bedrooms and internal corridor) Communal areas (including kitchens) in blocks of flats with limited use.	
		A2	Bedrooms and dormitories except those in self-contained single family dwelling units, in hotels, motels	
		A3	Bedrooms in hotels, motels; hospital wards; toilet areas	
		A4	Billiard/snooker rooms	
B	Office areas	B1	General use other than in B2	
		B2	At or below ground floor level	
C	Areas where people may congregate (except areas defined under category A, B and D	C1 Areas with tables	C11	Public, institutional and communal dining rooms and lounges, cafes and restaurants
			C12	Reading rooms with no book storage
			C13	Classrooms
		C2 Areas with fixed seats	C21	Assembly areas with fixed seating
			C22	Places of worship

Table 8 Reference of SANS 10160-2 scope of imposed loads to EN 1991-1-1

Elements of SANS 10160-2 Scope	Reference to EN 1991-1-1
Classification of imposed loads	New from EN 1991-1-1
Floor loads for residential, social, commercial and administration areas (Classes A, B, C & D)	Classification based on EN 1991-1-1 Values from wide review
Movable partitions as uniform load	From EN 1991-1-1
Area reduction Reduction for multi-storeys	Maintain SABS 0160 procedure No provision for multi-storey reduction
Floor loads for industrial use and storage (E1 – E4)	New from EN 1991-1-1
Actions induced by forklifts (FL1 – FL6), including dimensions	New from EN 1991-1-1
Actions induced by transport vehicles & devices for maintenance	New from EN 1991-1-1 (nominal)
Garages and vehicle traffic areas (F; G), excluding bridges	Additional class from EN 1991-1-1
Vehicle barriers and parapets for car parks	Informative annex not included
Roof loads (Classes H; J; K & L) Including area reduction	Revision of SABS 0160
Helicopter roof landing pads (HC1; HC2)	New from EN 1991-1-1
Horizontal loads on parapets, partitions walls and guardrail barriers	Based on SABS 0160 & EN 1991-1-1
Impact test	Noted procedure updated

4.2 Format and Classification Scheme

A summary of the classification scheme employed in SANS 10160-2 is given in Table 9, from where it can be compared particularly to the levels of differentiation used in EN 1991-1-1:2002 and BS EN 1991-1-1 NA:2005 as shown in Tables 6 and 7 above. Table 9 also provides an extract of all the various imposed load classes given in SANS 10160-2, albeit in an abbreviated format.

The classification used in SANS 10160-2 clearly follows the EN 1991-1-1 scheme, but with somewhat more elaborate differentiation for the residential to administration categories. The differentiation is however not as elaborate as that of the BS EN 1991-1-1 NA.

In SANS 10160-2 the arrangement of stipulations for industrial occupancies and other imposed loads as given in EN 1991-1-1 is generally followed.

4.3 Characteristic Imposed Load Values

The two diverse sources of information on which the determination of the characteristic value of the imposed load for a given properly demarcated situation could be based are the probability models which are derived from load surveys and structural mechanical models for load effects on the one hand, and international standards practice on the other hand. As indicated above the theoretical modelling is not sufficiently mature to be used in standards development. The implication is that the values reflected in leading international standards are bound to be not entirely consistent, with elements of subjectivity and differences in levels of reliability, or degrees of conservatism, imbedded in the selected values. Nevertheless guidance from such standards is the only feasible way to obtain reasonable values for characteristic values.

4.3.1 *Survey of international practice for specified imposed load values*

The values for imposed floor loads given in SABS 0160:1989 were compared to a number of structural standards relevant to South African practice (Retief *et al* 2001) as summarised in Table 10. From the comparison it was concluded that the SABS 0160 values were systematically lower than practice followed elsewhere for both general types of occupancy and for areas for special use.

It was concluded that the specified values should be aligned with at least one recognised standard, with proper justification where deviations are still allowed. Differences in specified values amongst the various standards give an indication of some lack of consistency internationally. It would for instance not be justified to select the highest load from the various standards, since this would lead to systematic excessive conservatism.

Table 9 SANS 10160-2 imposed load categories

Residential, social, commercial and administration areas			
Category	**Specific use**	**Sub-C**	**Example**
A	Areas for domestic and residential activities	A1	All rooms in dwelling unit and dwelling house, including corridors and lobbies.
		A2	Bedrooms, wards, dormitories, private bathrooms and toilets in hospitals, hotels, hostels and other institutional residential occupancies.
		A3	Stairs and escape routes in residential occupancies for example, serving hospitals, hotels, hostels and other institutional residential occupancies.
		A4	Balconies accessible to domestic and residential occupancy areas.
B	Public areas (not susceptible to crowding)	B1	Office areas for general use.
		B2	Public libraries, excluding stack areas.
		B3	Kitchens, communal bathrooms and toilets in educational buildings, hotels, office buildings and other institutional occupancies.
		B4	Light laboratories, operating theatres, X-ray rooms.
		B5	Filing and office storage areas, stack areas in libraries and archives.
C	Public areas where people may congregate (with the exception of areas defined under category A, B and D)	C1	Areas with movable furniture, tables etc. e.g. classrooms, areas in schools, cafes, restaurants, dining halls, reading rooms.
		C2	Areas with fixed seats, e.g. areas in churches, theatres or cinemas, conference rooms, lecture halls, assembly halls, waiting rooms, railway waiting rooms; grandstands with fixed individual seating.
		C3	Areas without obstacles for moving people, all without fixed individual seating e.g. assembly halls & areas, sport complexes, grandstands, exhibition rooms, etc. and access areas in public buildings, hotels, hospitals, airports terminals; stairs, corridors, landings; cantilever balconies accessible to the public.
		C4	Areas with physical activities, e.g. dance halls, gymnastic rooms, stages.
		C5	Areas susceptible to large crowds, e.g. in buildings for public events like concert halls, exhibition halls, sports halls including stands, terraces, escape routes.
D	Shopping	D	Areas in general retail shops and department stores.

Imposed loads on floors due to industrial use and storage		
Category	**Specific use**	**Example**
E1	Light industrial use	Production rooms such as workshops with lightweight equipment
E2	Industrial use	Production rooms such as workshops in works and factories.
E3	Areas susceptible to accumulation of goods	Areas for storage use including storage of books & documents.
E4	Access ladders and walkways	Maintenance walkways in buildings.

Axle loads of forklifts according to classes FL						
Class FL1-FL6	Net weight	Hoisting load [kN]	Axle width	Overall width	Overall length	Axle load [kN]

Imposed loads on traffic and parking areas in buildings		
Categories	**Specific use**	**Examples**
F	Traffic and parking areas: light vehicles ($\leq$ 25 kN)	Garages, parking areas, parking halls.
G	Traffic and parking areas: medium vehicles (25 - 160 kN)	Access routes; zones accessible to fire engines

Imposed loads on roofs (minimum values)			
Category	**Specific use**	**Examples**	
H	Inaccessible roofs	H1	Inaccessible roofs during construction
		H2	For normal maintenance and repair
J	Accessible flat roofs	Where access is provided in addition to access necessary for maintenance, excluding occupancy according to categories A to D	
K	Accessible flat roofs	Where access is provided according to occupancy categories A to D.	
L	Special services	Helicopter landing areas	

Imposed loads on roofs of category L for helicopters		
Helicopter class HC1 & HC2	Take-off load of helicopter Q [kN]	Dimension: loaded area [m^2]

4.3.2 *Approach followed in selection of values*

The general approach that was followed in selecting appropriate values for imposed loads for SANS 101060-2 was to compare the SABS 0160 values primarily to the values given in ASCE 7-95 and ENV 1991-2-1 as leading international standards. Some consideration was also given to BS 6399-1:1966 as a more recent version of the standard from which SABS 0160 was derived. The Australian Standard AS 1170.1-1989 was similarly considered. Values were selected to agree at least with one of the two primary references, with some consideration of the other two complementary standards.

The selected characteristic imposed load values were then cast into the classification system based on EN 1991-1-1, with some modification and refinement as indicated above. Some superficial comparison was made to the EN 1991-1-1 recommended values and optional ranges when they became available. The values for new classes of imposed loads introduced from EN 1991-1-1 were mainly maintained in SANS 10160-2.

Reference is also made above to the more refined scheme used in BS EN 1991-1-1 NA: 2005 which provides for a more elaborate subdivision of occupancy classes. It is interesting to note that EN 1991-1-1 seems to be a simplified version of BS 6399-1 for imposed floor loads, which is then subsequently re-extended in the BS National Annex.

Table 10 Comparison of SABS 0160 imposed loads to international standards

<table>
<tr><th>OCCUPANCY</th><th>SABS</th><th>ASCE</th><th>ENV</th><th>BS</th><th>AS</th></tr>
<tr><td>Dwelling house/unit</td><td rowspan="4">1.5</td><td>1.4</td><td rowspan="3">2</td><td>1.5</td><td>1.5</td></tr>
<tr><td>Bedrooms, wards, dormitories, etc in hospitals, hotels</td><td rowspan="3">1.9</td><td>2</td><td>2</td></tr>
<tr><td>Corridors, lobbies, landings to dwelling house</td><td rowspan="2">1.5</td><td rowspan="2">3</td></tr>
<tr><td>Stairs to dwelling house</td><td>3</td></tr>
<tr><td>Classrooms, lecture theatres</td><td rowspan="3">2</td><td>1.9</td><td>3</td><td>3</td><td>3</td></tr>
<tr><td>Operating theatres, x-ray rooms</td><td>2.9</td><td>3</td><td>2</td><td>3</td></tr>
<tr><td>Reading rooms in libraries</td><td>2.9</td><td>3</td><td>2.5</td><td>2.5</td></tr>
<tr><td>Garages, parking areas: < 25 kN gross weight</td><td>2</td><td>2.4</td><td>2</td><td>2.5</td><td>3</td></tr>
<tr><td>Offices for general use</td><td>2.5</td><td>2.4</td><td>3</td><td>2.5</td><td>3</td></tr>
<tr><td>Offices with data processing equipment</td><td>3</td><td></td><td></td><td>3.5</td><td></td></tr>
<tr><td>Cafes, restaurants, dining rooms lounges</td><td rowspan="5">3</td><td>4.8</td><td>3</td><td></td><td>2</td></tr>
<tr><td>Kitchens, laundries in hotels, offices, educational etc</td><td></td><td>2</td><td>3</td><td>4</td></tr>
<tr><td>Communal bathrooms, toilets in hotels, offices, etc</td><td>1.9</td><td>2</td><td>2</td><td>2</td></tr>
<tr><td>Entertainment areas</td><td>3.6</td><td></td><td>3</td><td></td></tr>
<tr><td>Light industrial</td><td>6</td><td></td><td>2.5</td><td>4</td></tr>
<tr><td>Assembly areas; fixed seating in residential buildings</td><td rowspan="3">4</td><td>2.9</td><td>4</td><td>4</td><td>3</td></tr>
<tr><td>Assembly halls, theatres, sport complex; fixed seats</td><td>2.9</td><td>4</td><td>4</td><td>4</td></tr>
<tr><td>Grandstands with fixed seating</td><td>4.8</td><td></td><td>5</td><td>5</td></tr>
<tr><td>Retail shops, department stores: sales and display</td><td rowspan="3">4</td><td>4.8</td><td rowspan="2">5</td><td rowspan="2">4</td><td rowspan="2">5</td></tr>
<tr><td>- upper floors</td><td>3.8</td></tr>
<tr><td>Light laboratories, banking halls</td><td></td><td></td><td>3</td><td>3</td></tr>
<tr><td>Assembly halls, sport complex; without fixed seats; stairs, corridors, landings of grandstands; public assembly areas, cantilever balconies.</td><td rowspan="2">5</td><td>4.8</td><td rowspan="2">5</td><td>5</td><td>5</td></tr>
<tr><td>Stages to assembly halls, theatres</td><td>7.2</td><td>7.5</td><td>7.5</td></tr>
<tr><td>Filing and storage: offices, hotels, institutions</td><td rowspan="4">5</td><td></td><td>6</td><td>5</td><td>5</td></tr>
<tr><td>Stack rooms: books, stationary</td><td></td><td></td><td>2.4/m</td><td>4/m</td></tr>
<tr><td>Shelved areas in libraries</td><td>7.2</td><td>6</td><td>4</td><td>3.3/m</td></tr>
<tr><td>Exhibition halls</td><td></td><td>5</td><td>4</td><td></td></tr>
<tr><td colspan="2">Comparison: International Standards to SABS 0160</td><td colspan="2">HIGHER</td><td colspan="2">LOWER</td></tr>
</table>

4.4 Assessment of SANS 10160-2 Characteristic Imposed Load Values

Since other standards specifying imposed loads are the primary sources of information on which SANS 10160-2 values are based, comparison with the values of relevant standards should also serve as basis for the assessment of the selected values.

4.4.1 *Comparison of floor and roof to SABS 0160*

A comparison with the SABS 0160 values gives an indication of the changes that would result from the use of the new stipulations. Due to differences in the respective category schemes, direct comparisons cannot be made in all cases. The SANS 10160-2 scheme was used as basis for the comparisons presented here, requiring some interpretation to select the equivalent category and value from SABS 0160. The comparison is shown graphically in Figure 1 for the uniformly distributed load (q_k) and Figure 2 for the concentrated load (Q_k).

There is an overall agreement between the revised minimum imposed load values and that stipulated in SABS 0160 for both q_k and Q_k. As indicated above the differences derive mainly from considering practice used in EN 1991-1-1 and ASCE-7. The following differences between SABS 0160 and SANS 10160-2 should be noted:

Uniformly distributed load q_k:

- Institutional occupancies are treated separately from domestic dwelling units, with an increased load for sleeping areas.
- Light laboratories, operating theatres and X-ray rooms are classified together (at q_k = 3,0), where they are classified separately in SABS 0160 (at q_k = 4,0 and 2,0 respectively).
- Areas for filing, office and industrial storage and areas where material may accumulate is stipulated at 2,5 kN/m stack height, in addition to the minimum value of 5 kN given in SABS 0160.
- Classrooms are classified together with other areas with movable furniture (cafes, restaurants, etc), with an increase for q_k = 3,0.
- For shopping areas q_k is increased from 4,0 to 5,0 kN/m^2, without differentiating between retail shops and department stores.
- A new sub-category for traffic loads (excluding bridges) for vehicles between 25 kN and 150 kN is introduced from EN 1991-1-1.
- Imposed loads on inaccessible roofs are specified for < 3 m^2 and > 15 m^2, with linear interpolation in between. The following modifications from SABS 0160 were applied (i) Execution treated separately (ii) Values derived from Retief & De Villiers (2005):
  - <u>For areas > 15 m^2:</u> Decreased from 0,30 kN/m^2 to 0,25 kN/m^2 for both construction and maintenance
  - <u>For areas < 3 m^2:</u> When considering
    - Construction phase: Increase from 0,50 kN/m^2 to 0,75 kN/m^2,
    - Maintenance only: Maintained at 0,50 kN/m^2

Concentrated load Q_k:

- A3 Stairs and escape routes for institutional occupancies: the Q_k value is above the EN 1991-1-1 recommendation.
- B1 Office areas: Q_k taken at the EN 1991-1-1 recommended value.
- C4 Areas with possible physical activities: a value for Q_k within the EN 1991-1-1 range is taken, although it is below the recommended value (no value is specified in ASCE-7). Such occupancy is not explicitly indicated in SABS 0160.
- C5 Areas susceptible to large crowds: a value for Q_k just above the EN 1991-1-1 recommendation of 4,5 kN is taken, which is substantially above the SABS 0160 value.

- E Industrial activities: Although the Q_k values are below the EN 1991-1-1 recommendations, it should be noted that these are *minimum* values, to be read together with the clauses requiring consideration of all the activities to take place, together with the associated equipment and materials, as given in SABS 0160.
- F Traffic load for vehicles below 25 kN gross weight: Q_k is increased from the SABS 0160 value to mid range of EN 1991-1-1 but below its recommended value of 20 kN.
- G Traffic loads for vehicles from 25 kN to 150 kN gross weight: Q_k is stipulated in accordance with EN 1991-1-1.

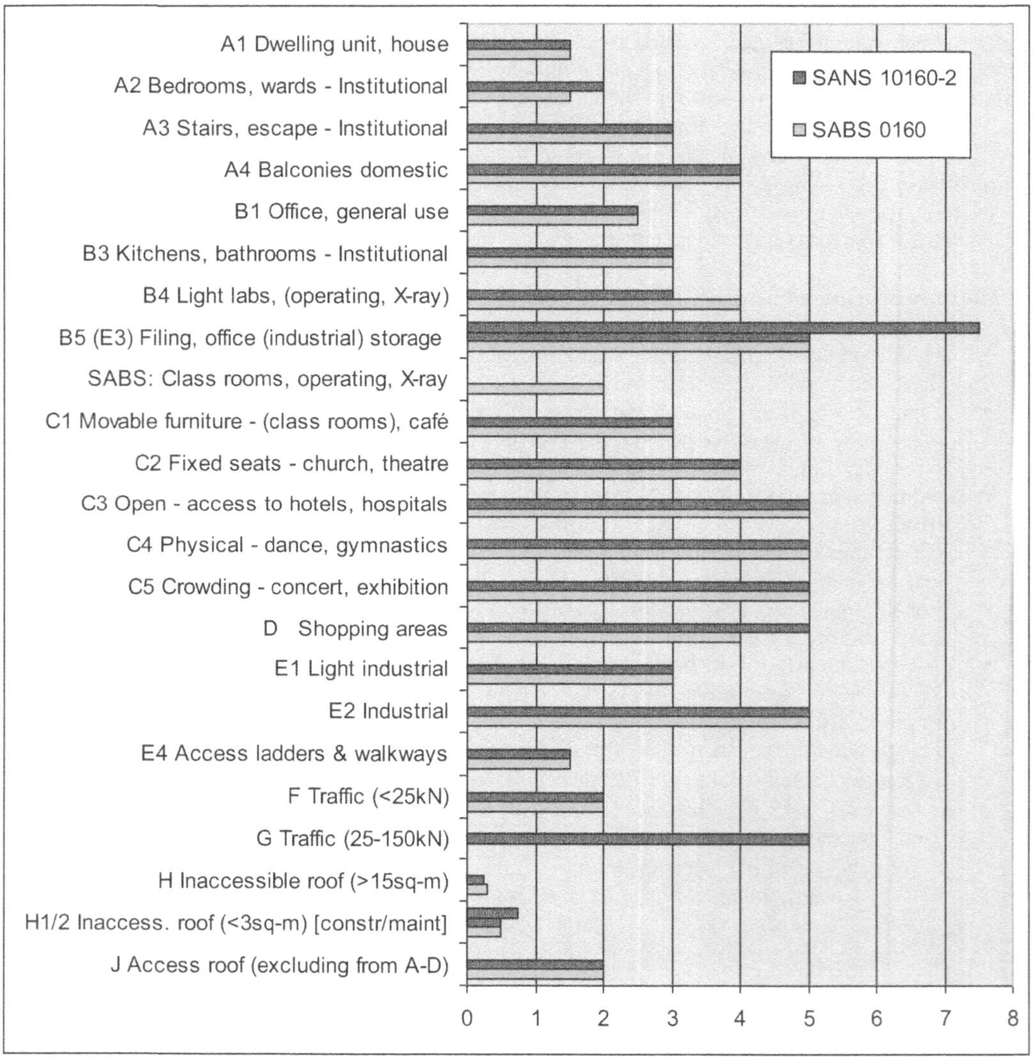

Figure 1 Comparison: SANS 10160-2 & SABS 0160 distributed floor load (q_k) [kN/m^2]

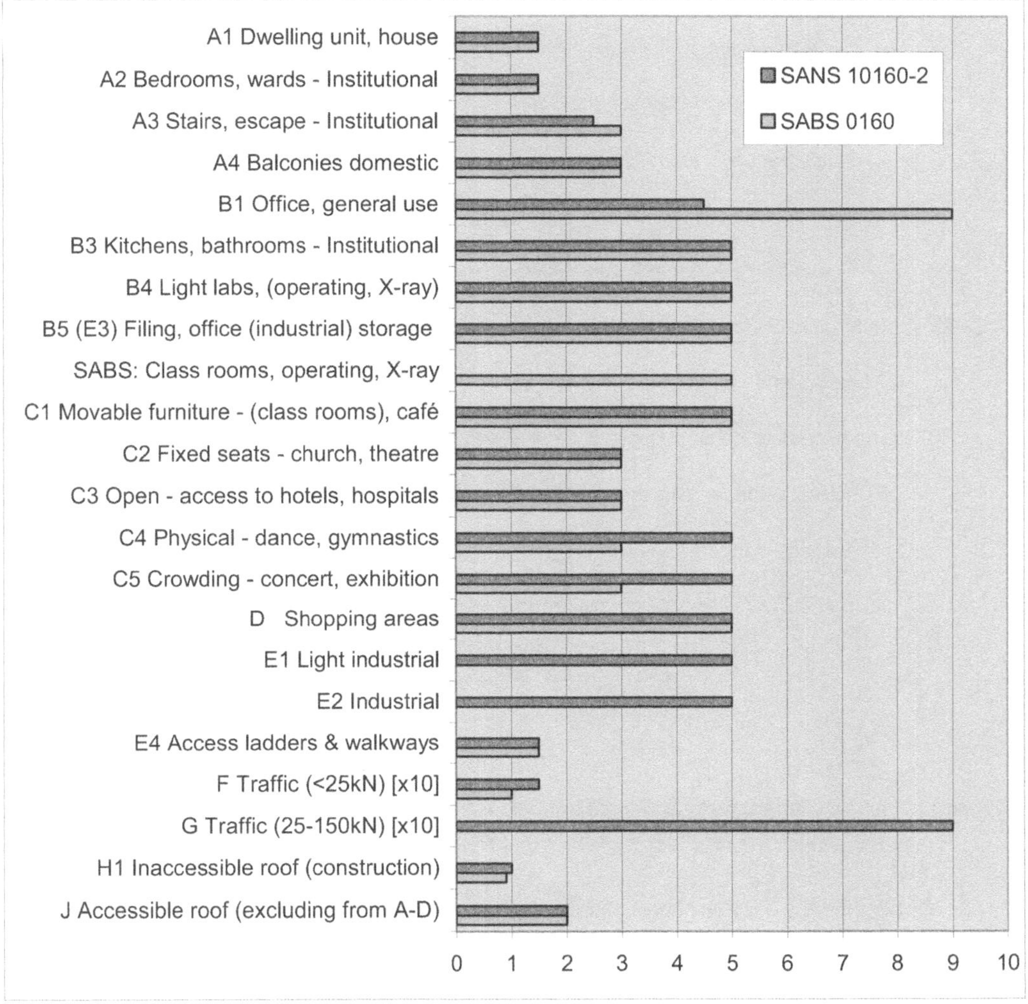

Figure 2 Comparison: SANS 10160-2 & SABS 0160 concentrated floor load (Q_k) [kN]

4.4.2 ***Comparison of SANS 10160-2 representative floor loads to allowable range of EN 1991-1-1 values***

A comparison between the values for minimum uniformly distributed imposed floor loads stipulated in SANS 10160-2 and the allowable range and recommended values given in EN 1991-1-1 is presented graphically in Figure 3. The following observations can be made about the SANS 10160-2 values:

- In all cases the SANS 10160-2 values fall within the EN 1991-1-1 range, in seven of the twelve cases shown they are at the recommended value, of which six are at the upper limit of the range.
- For dwelling units the lower limit is selected, rather than the recommended upper limit.
- Conversely for residential balconies the upper limit is selected, rather than the recommended lower limit.
- No differentiation is made between general shopping areas and department stores in SANS 10160-2; EN 1991-1-1 differentiates in terms of the recommended value.

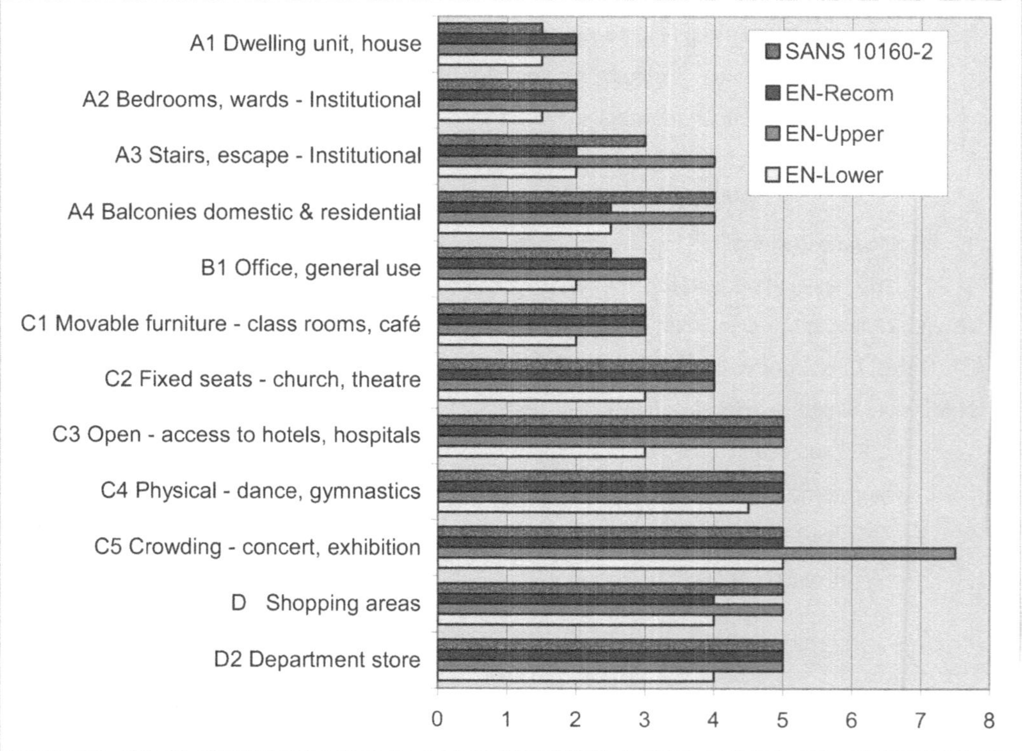

Figure 3 **Comparison: SANS 10160-2 & EN 1-1-1 allowable and recommended uniform imposed floor load values (q_k) [kN/m^2]**

4.4.3 ***BS EN 191-1-1 NA floor loads***

A comparison between the SANS 10160-2 stipulated values and the values given in the BS National Annex to EN 1991-1-1 is shown graphically in Figure 4. In most cases the SANS 10160-2 values agree with the highest value of that specified for the BS EN 1991-1-1 NA differentiated sub-occupancy zone. Such refined classification clearly allows for the specification of lower imposed load values. The cases where higher values are stipulated are for heavy duty walkways and stages in public assembly areas, which can clearly be justified. These specifications could be considered in a future revision of SABS 10160-2. In the mean time there is sufficient consistency between the two standards for consideration of BS EN 1991-1-1 NA values for local design for special cases.

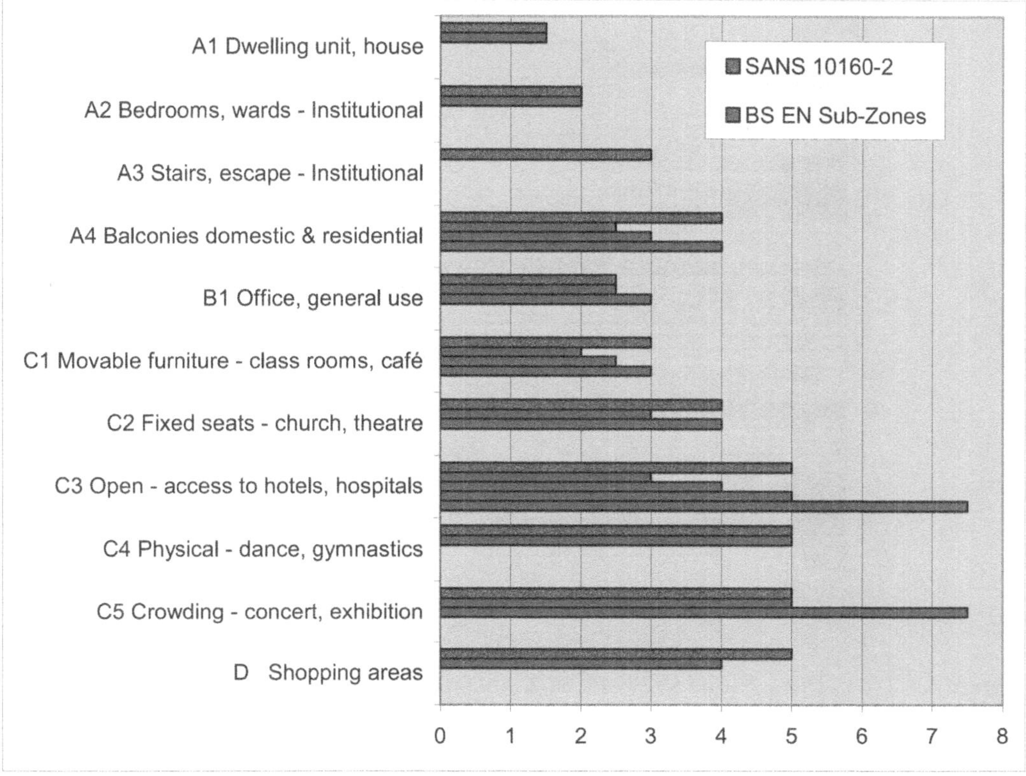

Figure 4 **Comparison: SANS 10160-2 & BS EN 1-1-1 NA uniform floor loads (q_k) [kN/m^2]**

4.4.4 *Horizontal loads on partition wall and parapets*

The values of horizontal loads on partition walls and parapets are given in terms of the occupancy classification in SANS 10160-2, similar to the scheme used in EN 1991-1-1. The values stipulated in SANS 10160-2 are compared in Figure 5 to the range and recommended values given in EN 1991-1-1, expressed as linearly distributed loads (kN/m). The ranges of values specified for the various sub-categories defined in the BS EN National Annex are also indicated in Figure 5.

In all cases the SANS 10160-2 values fall within (five times) or exceed (four times) the EN 1991-1-1 range, agreeing with the recommended value of four times out of the nine occupancy classes. The BS EN NA values show the same tendency as before, where the sub-categorization leads to the relaxation for certain occupancies; the exception is the specification of a higher value for restaurants and cafés (C1).

A concentrated horizontal load of Q_k = 1,0 kN is specified for all cases in SANS 10160-2; no guidance is given in EN 1991-1-1, allowing it to be specified as NDP.

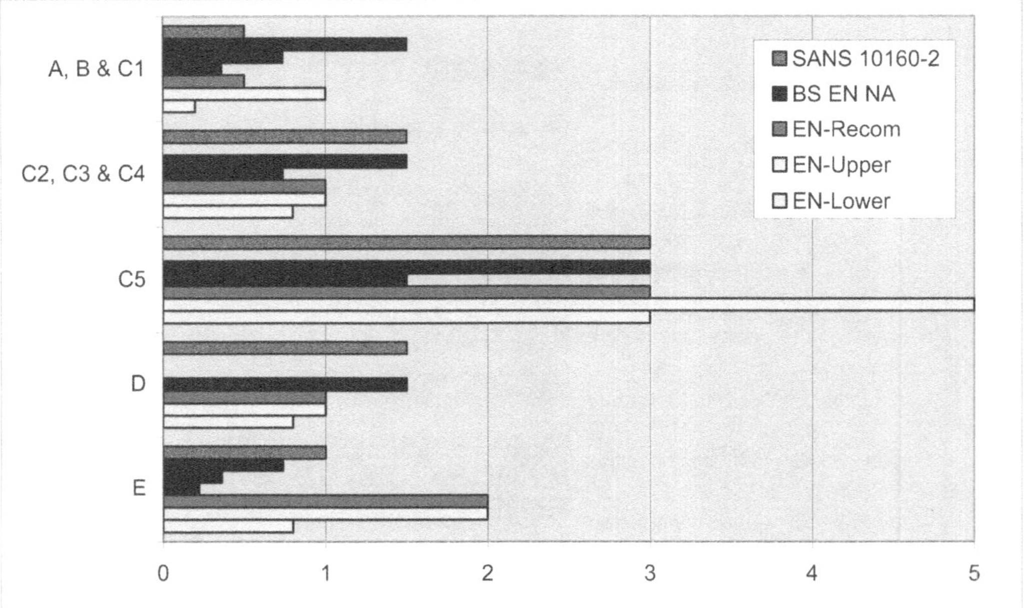

Figure 5 **Comparison: SANS 10160-2, EN 1-1-1 & BS EN 1-1-1 NA for distributed horizontal loads [kN/m]**

For guardrails around elevated parking areas a load of 30 kN over a length of 1,5 m is specified for vehicles < 25 kN in SANS 10160-2; EN 1991-1-1 refers to the informative Annex B which presents a procedure to calculate impact forces for all classes of vehicle, giving a value of 150 kN for vehicles up to gross mass of 2 500 kg on impact against a rigid barrier.

4.4.5 ***Various imposed loads and procedures***

A summary of the range of imposed loads and associated procedures and the relation between SANS 10160-2 and en 1991-1-1 is given in Table 5. New loads introduced from EN 1991-1-1 include loads caused by forklifts; general requirements for transport vehicles and maintenance devices; vehicle loads for vehicles with a gross weight of 25 – 160 kN; helicopter loads on roof-top landing pads. The following imposed loads and procedures should be noted:

Removable partitions: The loads from removable partitions from EN 1991-1-1 are used, changing the range of values for a uniform load from 1 – 1,5 kN/m^2 in SABS 0160 to 0,5 – 1,2 kN/m^2 in SANS 10160-2 for partitions with a linear weight up to 3 kN/m.

Area reduction for distributed loads: Since the reduction schemes for distributed loads for both contributing areas and number of storeys are based on the characteristic combination value ψ_0 in EN 1991-1-1, the SABS 0160 scheme for area is maintained in SANS 10160-2 whilst no separate provision is made for multiple storey effects.

Forklifts: The loads due to the use of forklifts in industrial areas and geometry of application are given for six classes of forklifts with a hoist load capacity of up to 80 kN. Provision is made for dynamic effects and horizontal loads.

Loads on inaccessible roofs: Separate treatment of execution (construction) and use/maintenance from Part 1 design situations (see also SANS 10160-8); SABS 0160 values adjusted based on an investigation reported by Retief & De Villiers (2005). See also Section 4.4.1 of this chapter.

Roof-top helicopter landing pads: Loads on roof-top helicopter landing pads are given as a load due to special services, for two classes of helicopter, with take-of loads of < 20 kN and 20 – 60 kN respectively. The geometry of the loaded area and a factor for dynamic effects are stipulated.

Testing for impact resistance: Requirements for impact resistance of large glazed areas and acceptable testing procedures are carried over from SABS 0160, with some modification to the formulation of the clauses and acceptable testing procedures.

5. SANS 10160-7 THERMAL ACTIONS

Provisions for the determination of characteristic thermal actions on building structures in SANS 10160 Part 7 *Thermal actions* represent an extension of the scope of SABS 0160:1989. Principles and rules are provided for calculating thermal actions, classified as variable and indirect actions, on buildings and their structural elements.

The new standard SANS 10160-7 is based on the sections of Eurocode EN 1991 *Actions on structures* Part 1-5 *General actions – Thermal actions* or EN 1991-1-5:2004. In accordance with the scope of SANS 10160, only the sections of EN 1991-1-5 relevant to building structures are included in Part 7, excluding the substantial part referring to bridges. Provisions for local characteristic (50 year return period) maximum and minimum shade air temperatures are based on the information from TMH-7.

The procedures include the following main elements:

Characteristic environmental temperatures: Environmental temperatures are determined from isotherms of maximum and minimum shade air temperatures for a 50 year return period at sea level; adjustment rates for height above sea level; adjustment for different return periods.

Representation of actions: Temperature distributions and changes and the associated thermal strains are split into uniform distributions; linear variations about the two principal cross section axes; a non-linear distribution causing a self-equilibrating stress distribution.

Radiation effects: The effects of radiation on exposed surfaces are taken into account by considering the season,, surface colour, shade air temperature and the orientation of the surface.

6. SANS 10160-8 ACTIONS DURING EXECUTION

Execution of the structure is a general term which refers to conditions and activities which apply to the incomplete structure such as construction, erection, fabrication, temporary structures, structural alterations and refurbishment, partial or complete demolition. In accordance with SANS 10160-1 it is classified as a *transient design situation*. Inclusion of Part 8 into SANS 10160 is derived from Eurocode EN 1991 *Actions on structures* Part 1-6 *General actions – Actions during execution* or EN 1991-1-6:2005.

The execution of the structure represents an important phase in the life cycle of a structure during which a significant number of failures do occur. Although structural failure during execution is closely related to requirements for occupational health and safety at the construction site, the requirements are not directed towards compliance with such regulations as such, but primarily towards the performance and behaviour of the structure due to the related transient conditions and loads.

6.1 Main Elements of Design Situations and Actions during Execution

The main elements of measures to ensure the acceptable and safe performance of the structure during the execution phase include the behaviour of the partially completed structure, as exposed to the transient and construction loads, with temporary support taken into account. Structural performance should be assessed for the ultimate limit state, considering transient and accidental design situations and robustness of the partial structure; applying suitable criteria for the serviceability limit state for irreversible, reversible and the early manifestation of long-term effects.

This process needs to be taken through the various phases of construction by validating compliance for all critical phases of the partial structure and its resistance to the transient and construction loads, taking account of or providing the necessary temporary support.

6.2 Requirements from SANS 10160 Part 8 & Part 1

In SANS 10160-8 these main elements for structural performance during execution are set up systematically by defining their main components and setting requirements for how these elements are to be treated. However, since the range of situations exceeds that of the completed structure by undefined orders of magnitude, the requirements can only be expressed in general terms. The structural engineer responsible for structural performance during execution is therefore obliged to apply the principles of Part 1 in selecting suitable design situations, suitable action and material or resistance values and even adjusted partial factors for use in design verification. The importance of suitable experience and quality management procedures also derive from using Part 1 as general guideline.

6.3 Management of the Execution Process

The most important measure available to manage structural performance during execution is the use of the control of the process, in contrast to the lack of direct control by the engineer of the conditions to which the completed structure will be exposed. Such control includes the management of the execution process and the consequent direct and related actions, together with the provision of temporary support and bracing. Structural behaviour and the execution process are therefore integrally related.

6.4 Responsibility for Structural Performance during Execution

An important principle in the inclusion of provision for transient design situations in Part 1 and the associated actions during execution in Part 8 refer to the eminent need to consider the structure properly under these conditions, including the implication that appropriately qualified professionals should be appointed, to take responsibility for this phase.

This does however not imply that the design engineer should now take such responsibility because of the inclusion of Part 8 into the Loading Code; on the contrary, the requirements imply the explicit assignment of such responsibilities. The contractual arrangement lies beyond the scope of SANS 10160 and is related to the associated way in which the contract for execution is structured, together with contract fees. In the case of the traditional professional appointment of the designer by the client and tendered/negotiated entering into a contract with the constructor, responsibility for the works is taken by the constructor, including the structural implications of execution. The consultant/designer should therefore ensure that this is properly included in the contract agreement. Part 8 should then be used by the constructor's engineer to make the necessary provision for the structural implications of execution of the structure. In the case of a design-and-build contract the requirements of Part 8 still apply, but responsibility is now internally assigned by the design/build consortium.

Although the designer needs to consider constructability during the design phase, such activity is directed towards the feasibility and economy of the design, using proper structural behaviour as a constraint rather than the primary objective of the exercise. Generally the execution procedure is the prerogative of the constructor, first to obtain a competitive advantage during bidding, and then to minimise cost after the contract has been entered into. Assigning responsibility for structural performance to the designer will therefore be an ineffective arrangement which makes the designer responsible for something over which he/she has limited and indirect control and restraining the constructor in his ability to optimise the execution process. Whatever the contractual arrangement may be, the responsibility for structural performance during execution should therefore rest with the constructor, who is in control of the works at that stage.

7. CONCLUSIONS

Provisions for general actions on buildings in SANS 10160 include the stipulations for the basic actions caused by the self-weight of structural elements and permanently installed non-structural components, imposed loads due to various occupancies and industrial activities in Part 2; thermal loads and their treatment in Part 7; actions and procedures to consider their effects on the incomplete structure in Part 8. These Parts are primarily referenced to the three respective Eurocode Parts, EN 1991-1-1:2002; EN 1991-1-5:2004; EN 1991-1-6:2005.

In SANS 10160-2 the general layout of the respective Eurocode Part is followed, including the formal specification of the nominal value of self-weight as the characteristic value, extension of the tables of weight densities; the classification system for occupancies and industrial imposed loads, with the introduction of new specifications for several classes of imposed loads. Stipulated imposed loads are however determined through independent investigations, including the values used in SABS 0160:1989 and various international structural standards. The requirements of SANS 10160-2 are nevertheless fully compliant with EN 1991-1-1 when the discretion allowed through the scheme of Nationally Determined Parameters is considered. The way in which such discretion as applied in the UK through BS EN 1991-1-1 National Annex which has become available recently is a demonstration of how the effectiveness of the simplified Eurocode requirements can be improved. This BS version can be considered for a future revision of Part 2, and can even be considered locally due to the consistency between the South African and British standards.

SANS 10160-7 on thermal actions is a new South African Standard which is substantially based on the requirements for buildings of the respective Eurocode Standard. Part 7 is however firmly rooted in the South African environment by including the environmental characteristic temperatures from TMH-7.

The new and innovative Eurocode Standard on the actions acting on the structure during its execution is introduced as a South African Standard as Part 8 of SANS 10160. This Part is of such general nature that it could essentially be taken over as is from its Eurocode version, with modifications only for consistency with the other Parts of SANS 10160, particularly Part 1, and omitting the annex on special considerations for bridges.

The main features of the SANS 10160 Parts on general actions on buildings are that load specifications within the scope of SABS 0160:1989 have been updated; new classes of actions have been introduced from Eurocode and adapted to the scope of buildings and South African conditions; whilst full consistency with the Eurocode Parts has been maintained.

8. REFERENCES

ASCE 7-95: American Society of Civil Engineers *Minimum Design Loads for buildings and Other Structures*.

AS 1170.1-1989: Australian Standard, Part 1: *Dead and live loads and load combinations*.

BS 6399: Part 1: 1996: British Standard *Loading for buildings Part 1. Code of practice for dead and imposed loads*.

CIB (1989a) *Actions on structures – Self-weight loads*. CIB Report - Publication 115. International Council for Building Research Studies and Documentation, Rotterdam, The Netherlands.

CIB (1989b) *Actions on structures – Live loads in buildings*. CIB Report - Publication 116. International Council for Building Research Studies and Documentation, Rotterdam, The Netherlands.

Gulvanessian H, Calgaro J-A & Holický M (2002). *Designer's guide to EN 1990 Eurocode: Basis of structural design*. Thomas Telford. 192 pp. ISBN 0 7277 3011 8.

EN 1990:2002. Eurocode – *Basis of structural design*. European Committee for Standardisation, Brussels.

EN 1991-1-5:2004. Eurocode 1 *Actions on structures* Part 5 *General actions – Thermal actions*. European Committee for Standardisation, Brussels.

EN 1991-1-6:2005. Eurocode 1 *Actions on structures* Part 6 *General actions – Actions during execution*. European Committee for Standardisation, Brussels.

EN 1997-1:2004. Eurocode 7 *Geotechnical Design* Part 1 *General rules*. European Committee for Standardisation, Brussels.

ENV 1991-2-1:1995: Ratified European Standard, *Part 2-1: Actions on structures - Densities, Self-weight and Imposed Loads*. European Committee for Standardisation, Brussels.

Hemmert-Halswick A, Sedlacek G & Merzenich G (1988). *Eurocode actions on structures: Background document Part 6: Imposed loads on floors and roofs*.

ISO 2393:1998. General principles on reliability for structures. *International Standard*. International Standards Organisation. Adopted by Standards South Africa as SANS 2394:2004 (Ed 1)

IStructE (2009). *Manual for the design of building structures to Eurocode 1*. The Institution of Structural Engineers, London. ISBN 978-1-906335 (Draft).

JCSS (2001). Joint Committee on Structural Safety. *Probabilistic model code*. JCSS working materials, http://www.jcss.ethz.ch/.

Kemp, A R, Milford, R V and Laurie, J A P (1987). Proposals for a comprehensive limit states formulation for South African structural codes. The Civil Engineer in South Africa, September 351-360.

BS EN 1997-1:2004 NA. *UK National annex to Eurocode 7: Geotechnical design – Part 1. General rules*.

Retief JV, Dunaiski PE & De Villiers PJ (2001). *An Evaluation of Imposed loads for Application to Codified Structural design*. Zingoni A (Ed) Proceedings International Conference SEMC 2001, Cape Town, 2 - 4 April 2001 Elsevier.

Retief JV, & De Villiers PJ (2005). *Reliability model for imposed loads on inaccessible roofs for light industrial steel structures*. Australian Structural Engineering Conference (ASEC 2005), 11-14 Sep 2005, Newcastle, Australia.

SABS 0160-1989. The general procedures and loadings to be adopted in the design of buildings, *Code of Practice*. South African Bureau of Standards, Pretoria.

SANS 10160. Basis of structural design and actions for buildings and industrial structures, *Code of Practice*. South African Bureau of Standards, Pretoria.

Sedlacek G & Gulvanessian H (1996). Eurocode 1 Basis of design and actions on structures Part 2.1: Densities, Self-weight, Imposed Loads. IABSE Colloquium *Basis of design and actions on structures – Background and applications of Eurocode 1*, Delft, IABSE Vol. 4, pp 61-70.

TMH-7:1981C*.ode of practice for the design of highway bridges and culverts in South Africa. NITRR, CSIR, Pretoria.*

Part 3 – Wind Actions

3-1 Review of Codification of Wind-loading for Structural Design

Goliger AM, Retief JV & Dunaiski PE

1 INTRODUCTION

Wind loading constitutes the dominant environmental action on buildings within the South African climate, where snow loading does not occur. Therefore, the requirements to provide for wind actions on buildings form an important and substantial part of the current loading procedures. These are stipulated in SABS 0160-1989 (Amended) South African Standard – *Code of practice for the general procedures and loadings to be adopted in the design of buildings* (renumbered as SANS 10160-1989, referred to briefly as SABS 0160 or the *existing standard*) as the dominant environmental action on buildings within the South African climate. Such standardised design procedures are critical to the design of safe and efficient structures due to the inherently variable nature of wind and the complexity of its effects on structures. Consideration of wind actions therefore plays a prominent role in the revised Standard SANS 10160 *Basis of structural design and actions for buildings and industrial structures* (referred to briefly as SANS 10160).

A brief overview of the historical development of the revision process is presented, consisting of the following in terms of the present provisions of SABS 0160:

- a review of options for the development of revised provisions with particular reference to alternative international and other standards, and ultimately
- an assessment of Eurocode EN 1991 *Actions on structures* Part EN 1991-1-4:2005 *General actions – Wind actions* (subsequently referred to as EN-1-1-4 for brevity).
- This is followed by a summary of the wind engineering aspects that need to be taken into account, including a review of the way in which these aspects are treated in possible alternative reference standards.

A critical assessment of Eurocode EN-1-1-4 of is made to consider the merit of applying it as a reference for the revision of present procedures describing wind actions on buildings. This assessment includes a comparison of the net effects of the application of procedures from EN-1-1-4 for wind actions on representative structures. The application of the results of this review to the formulation of stipulations for wind actions in SANS 10160 is reported in Chapter 3-2.

As one of the prominent loads specified in SABS 0160-1989, wind load specifications were extensively reviewed at the South African National Conference on Loading in 1998 (Goliger *et al* 1998). Subsequently a general survey was made of standards that could serve as reference in order to capture recent advances and modernisation of design procedures (Goliger 2001 *et al*). The main options considered were the previous version of Eurocode ENV 1991-2-4, the American Standard ANSI/ASCE 7-95 and the Australian Standard AS1170.2. A proposed draft South African standard was developed on the basis of Section 3 *Detailed procedure: Static analysis* of the AS 1170.2-1989 (Goliger 2003, 2005). However indications of a radical revision of the Australian Standard, which is of the same vintage as SABS 0160-1989 (Amended), compromised its use as a potentially obsolete reference.

The next development stage was to specifically consider Eurocode EN 1991-1-4 *Actions on structures: General actions – Wind actions* as a possible reference. An important reason for initiating this step was a change in its status from voluntary version ENV 1991-2-4 into the normative Eurocode EN 1991-1-4, as published in 2005.

2 REVIEW OF STANDARDS FOR WIND ACTIONS

The variability of wind with respect to time and space is so pronounced that it is essential to present its characteristics in statistical and probabilistic terms. Such variability in its nature is dependent on the wind generating mechanisms such as synoptic wind storms, thunderstorm activity or tropical cyclonic storms. Environmental influences include terrain roughness and topography even for large distances upstream of the site. Wind loads are generated as a result of interaction between the free-stream wind flow and structures. In a smooth or streamlined flow, Bernoulli's equation applies, and the static pressure is transferred in a direction normal to the surface of the body, which generates surface pressures. In places where flow separation develops such as in the vicinity of sharp corners, Bernoulli's equation is no longer valid due to formation of vortices. The pressure difference between the pressure acting on a structure and a reference static pressure (usually in the approaching wind stream) is required for design purposes. This pressure can either be positive (acting towards the surface) or negative (acting outwards). Furthermore structural response can range from static resistance of wind pressures, dynamic behaviour of the structure, which can even influence wind flow (aero-elasticity), or various forms of resonance even at low wind velocities.

2.1 Standardised procedures for wind action design

In order to cope with such a variety of wind actions on structures for a wide range of structures and their properties, diverse approaches have been adopted in the development of standard procedures for determining design wind actions on buildings by different countries. Such diversity, which goes beyond provision for local conditions, since it also reflects different ways to simplify the complexity of structural wind design, complicates direct comparison of various standard design procedures.

An extensive comparison of the wind load provisions of the current loading standard SABS 0160 was reported by Goliger *et al* (1998) during the *South African National Conference on Loading*. The comparative analysis of international codes has shown that, within the ten years preceding the conference, several international wind loading standards underwent significant transformation and improvement. These standards were able to incorporate modern scientific thinking and recent research data obtained from full-scale and wind tunnel studies carried out across the world. SABS 0160 which was developed in the mid-1980s was however based on the old British Code CP3 of 1952. The comparison by Goliger *et al* (1998) indicated a need for updated wind load stipulations.

The consideration of adopting an international standard has the potential to provide both access to a wider body of technology on which procedures are based and to enhance harmonisation with international design practice. The main properties of such standards for wind actions, including model codes, are reviewed and summarised subsequently. It should be noted that whilst British Standards traditionally served as reference to South African standards, the relevant British Standard BS 6399: Part 2 (1997) is to be superseded by Eurocode.

2.2 International and model standards

The Draft International Standard ISO/DIS 4354-1990, in particular its revised format ISO/DIS 4354-2007 (ISO 2007), provides a harmonised approach to the treatment of wind actions. All the aspects of wind actions important to its probabilistic modelling are reported in a CIB Report (1996). Models and data for probabilistic design against wind actions and calibration for limit states design are provided in the Joint Committee of Structural Safety Model Code (JCSS 2002).

The principles, options and alternatives of wind actions on structures are considered concisely in ISO/DIS 4354-2007 as summarised in Table 1, and the design parameters are presented in Table 2. The main issues regarding different approaches taken by various standards which are resolved by ISO/DIS 4354-2007 are the following:

- **Reference wind speed:** The acceptability of either the peak ($V_{ref,\ peak}$ of 3 second average) or mean ($V_{ref,\ mean}$ of 10 minute average) velocity pressure approach is sanctioned, and their equivalence established in terms of an averaging time factor k_T.
- **Site exposure representation:** Although there is a theoretical base for scaling of k_T with height z as $k_T(z)$ in terms of a logarithmic function, the empirical but convenient power law function is acceptable over a limited range of heights (up to 100 m).
- **Thunderstorm climate:** Design should be based on 3 second averaging and a single 'envelope' profile applied for open sea, country and suburban sites, using synoptic storm turbulence intensities; it is noted that profile data for strong winds are sparse due to the transient nature of thunderstorms.

Table 1 Principle features of standard for wind actions – ISO/DIS 4354-2007

Structures	**Provides for:** buildings, towers, chimneys, bridges and other structures, - as well as their components and appendages
Methodologies	**Two alternatives:** Peak and mean velocity pressure methods applied by different standards - equivalence of two methods endorsed - conversion established by fixed ratios of peak/mean (synoptic) wind velocities **Peak velocity pressure method:** based on maximum *3 second averaging period* wind speed - more generally applied in wind action standards - particularly applicable to thunderstorm climates **Mean velocity pressure method:** based on maximum *10 minute average* wind speed - dynamic response transparently represented in terms of mean and fluctuating components, providing more accurate description of response process - more appropriate for complex structures that require load distributions with height and suitable load combinations; - appropriate for wind tunnel applications
Storm types	**The three main storm types:** - **synoptic winds** of large scale mature conditions with well developed wind speed and turbulence profiles - **thunderstorms** produce strong winds with high turbulence for a short period of time - **tropical cyclones** (hurricanes, typhoons) with three-dimensional vortex structure
Basis of design	**Design standard:** For use in conjunction with ISO 2394 and other International Standards **Limit states** design methods - consider applicable storm for respective limit state **Probability of exceedance** – characteristic, alternatively design value (importance class; cyclones) **Elastic behaviour** assumed – design with plastic behaviour requires special consideration
Wind-sensitive structures	- **Particularly flexible**, slender, tall or light-weight structures - **Unusual geometry** may result in unexpectedly large response - **Supplementary studies** may include wind tunnel tests
Wind actions	**Wind action effects:** Wind actions may produce the following - **excessive forces**, instability in structure, members, elements - **excessive deflection** or distortion of structure or its elements - **repeated dynamic forces** causing fatigue of structural elements - **aeroelastic instability**, aerodynamic forces caused by motion of structure - **excessive dynamic movements** causing concern or discomfort - **interference** effects from existing and future buildings

Table 2 Main parameters for wind actions – ISO/DIS 4354-2007

Reference wind speed (V_{ref})	**Specified value of the wind speed** for the geographical area **Standard exposure** i.e. roughness, height and topography - at 10 m height; - open country terrain **Provide for:** - averaging time – peak (3 second) or mean (10 minute) - probability specification – from wind record statistics
Site exposure factors (C_{exp})	**Variability of the wind speed at the site** of the structure due to: - storm type - height above ground level; roughness of the terrain, including change of roughness - topography **Profile** of wind velocity with height, normalised to reference height - logarithmic profile theoretically based - exponential profile pragmatically fitted across practical range
Pressure, Force Coefficients (C_p, C_F)	**Aerodynamic wind induced effect:** Pressure (C_p) / Force (C_F) - ratio with a reference velocity pressure / and a reference area **Influenced by:** - shape of the structure, exposure of site, relative wind direction - Reynolds number, averaging time. **Internal pressure** for enclosed structures determined by - size and distribution of the openings
Dynamic response (C_{dyn})	**Fluctuating pressures** due to: - **Random wind gusts** acting over all or part of the surface area of the structure - **Wake of the structure** (vortex shedding), producing resultant cross-wind and torsional along-wind forces - **Induced by motion** of structure due to wind

2.3 Major structural design standards

Two major structural design standards which incorporate extensive bodies of research, technology and experience are the American Society of Civil Engineers standard ASCE 7 of which the 1995 version was initially considered, and the 1998 version finally; and the Eurocode standard, of which the voluntary version ENV 1991-2-4:1995 was initially considered, and subsequently the normative version EN 1991-1-4.

ASCE 7: This standard is revised in three year cycles, with provisions for tropical cyclones receiving much attention and development during the revision cycles. The procedures are based on the peak velocity pressure method which has been converted from the previous quantity of the *fastest mile* to the 3 second average reference wind speed. The basic wind speed varies from 40 m/s for inland areas to 67 m/s in coastal areas exposed to tropical cyclones. Provision is made for buildings of different importance classes and heights; static and dynamic structural response (excluding vortex shedding); pressure coefficients for a range of standard building geometries; a simplified procedure for buildings of limited size and shape (roof slopes < 10°, height less than 9 m) is provided.

EN 1991-1-4: The draft of ENV 1991-2-4:1995 was started from ISO/DIS 4354-1990, to provide for turbulent wind acting over the structure, causing static and fluctuating pressures, including forces induced by the motion of the structure (Ruscheweyh 1996). The scope of structures comprises buildings and civil engineering structures such as bridges, towers, masts and chimneys, excluding structures of unusual complexity, buildings higher than 200 m or bridges with a span of more than 200 m. The reference wind velocity is based on the 10 minute average with a 50 year return period. The basis and process for its conversion to EN-1-1-4:2005 is presented by Zimmerli (2001). Zimmerli observes that there is a general increase in the magnitude of wind actions resulting from the Eurocode procedures, in comparison to that from the national codes. This is ascribed to the provision of the peak factor and marginally higher forces due to pressure coefficients (Geurts 2001). A notable feature of the

Eurocode procedures is the extended set of pressure coefficient values, providing for a variety of buildings forms.

3 ASSESSMENT OF EUROCODE EN 1991-1-4:2005 *WIND ACTIONS*

The general attributes of Eurocode are that the advances in structural design practice are captured, with application across an extensive range of structures and design situations, reaching harmonisation between diverse traditions and practice of member countries. These attributes apply particularly to EN 1991-1-4, where advanced procedures are applied across a wide range of conditions, provision is made for an extensive scope of application, and a high degree of harmonisation is achieved amongst a diverse range of practices across Europe. A thorough assessment of its merit to serve as a reference to the revision of the provisions for wind actions in SANS 10160 was therefore imperative, as reported by Goliger (2005).

3.1 General principles, scope and advances

The general principles and features of EN-1-1-4 are summarised in Table 3, indicating the extensive range of structures and structural response for which advanced and intricate procedures have been captured. This leads to more complex procedures as compared to those of the existing national standards, particularly in the case of small and simple structures (Vrouwenvelder *et al* 2005).

Table 3 General principles and features of EN 1991-1-4:2005

Scope: - Structures	**Buildings and civil engineering structures** up to 200 m high - Whole or parts, e.g. components, cladding and their fixings - Bridges, single & multi span, up to 200 m span; pedestrian bridges up to 20 m span - Towers, masts Not *fully* covered, requiring specialist advice and wind tunnel tests: - Lattice towers, guyed masts – treated in EN 1993-7-1 - Tall buildings with central core, susceptible to torsional vibrations - Cable stayed, suspension, arched bridges, multiple/curved decks; Offshore structures
Scope: - Actions	**Structural response:** Provides for various modes - Static behaviour of structures, including a large range of shapes and geometries - Dynamic behaviour, including fatigue and serviceability displacement & acceleration • Along wind structural dynamic behaviour, including criteria for its possible occurrence • Cross-wind resonance, vortex shedding and some aero-elastic instabilities - Not provided for: Torsional & transverse vibrations, > 1 fundamental mode shape
Modelling of wind actions	**Nature:** Fluctuating wind actions on structures - act directly as pressures on the external surfaces - due to porosity, also act indirectly on internal surfaces; resulting in forces normal to surface - friction forces acting tangentially to the surface **Characteristic values:** Basic values of wind velocity or velocity pressure, coefficients or models to derive annual probability of exceedance of 0,02 or mean return period of 50 years
Basic wind velocity ($v_{b,0}$)	**Averaging time:** 10 minutes (mean velocity pressure method) **Reference site:** Open country with low vegetation (Terrain Category II) @ 10 m height **Probability of exceedance:** Annually 0,02 (return period 50 yrs)
	Regional values treated as Nationally Determined Parameter - Wind map of Europe: Values vary between 20 m/s (Baltic countries) and 36 m/s (Greece)
General assessment (Vrouwenvelder 2005)	**Comparison with national standards** (+ Improvement; - Disadvantage) + New extensive set of pressure and force coefficients + More accurate description of wind loads + Suitable for lightweight structures + Additional investigations of experts is in many cases avoidable - Application demands more skills than simple standards - Difficult to identify the relevant information in the case of simple structures

3.2 Regional and site effects

Procedures are summarised in Table 4 for basic and peak wind velocities, including exposure due to terrain roughness and elevation; effects of orography, adjacent high buildings and closely spaced buildings for urban sites.

Table 4 EN 1991-1-4:2005 procedures for basic wind velocities and site effects

Pressure	**Basic velocity pressure:** Conversion from basic velocity	$q_b = \rho/2 \bullet (v_b)^2 \quad (\rho = 1,25\ \text{kg/m}^3)$
Velocity at reference height	**Basic velocity** (v_b): Directional & seasonal effect	$v_b = c_{dir} \bullet c_{season} \bullet v_{b,0}$
	Mean velocity at structural reference height (z)	$v_m(z) = c_r(z) \bullet c_o(z) \bullet v_b$
	Roughness factor $c_r(z)$ – Logarithmic profile	$c_r(z) = k_r \bullet \ln(z/z_0)$
	Terrain factor (k_r) in terms of roughness length (z_0) Scaled to Terrain Category II for other four categories	$k_r = 0,19 \bullet (z_0/z_{0,II})^{0,07}$ for $z \geq z_{min}$

	TC	Characteristics of terrain	z_0 [m]	z_{min} [m]
Terrain Categories	0 / I	Sea or coastal / Lakes, no obstacles	0,003/0,01	1
	II	Low vegetation; isolated obstacles, spaced ≥ 20 times height	0,05	2
	III	Regular vegetation; forests; suburbs; villages	0,3	5
	IV	≥ 15% of area covered with buildings ≥ 15 m high	1,0	10

Peak velocity pressure	**Variation** of peak gust velocity derived from mean	$q_{peak}(z) = q_b \bullet [c_r(z)]^2 \bullet G^2$
	Gust factor corresponding to peak factor of 3,5 (consistent with stipulated pressure coefficients) for orography factor $c_o(z) = 1,0$)	$G^2(z) = 1 + 7/\ln(z/z_0)$
Orography	**Orography factor** ($c_o(z)$ dependent on site topography; provides for isolated - Cliffs and escarpments; isolated hills and ridges - Upwind and downwind of feature	

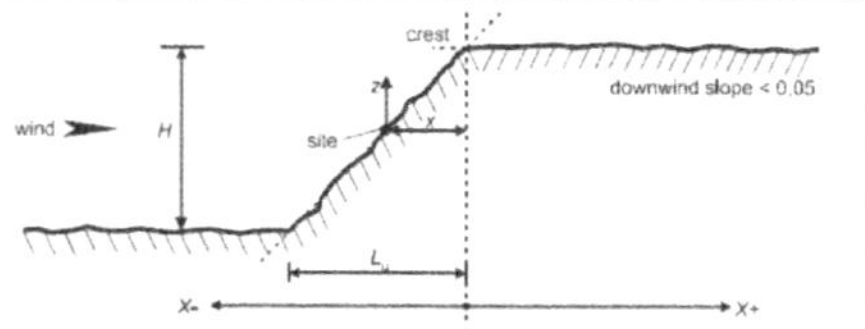

Cliffs and escarpments

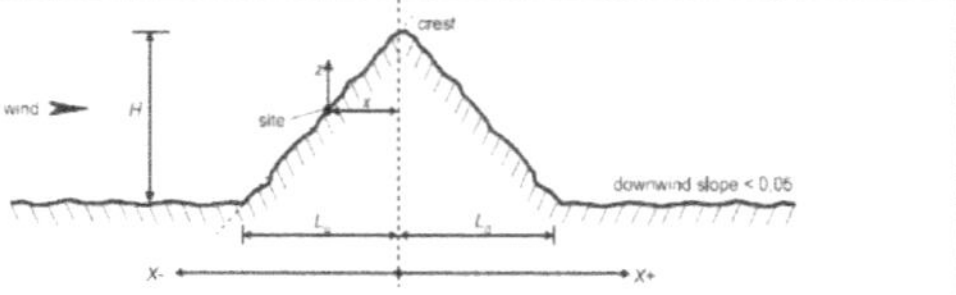

Hills and ridges

Neighbouring structures	Buildings adjacent to high rise building - Effective height z_n applied for design of neighbouring structures - Dependent on ratios heights, dimensions and distances	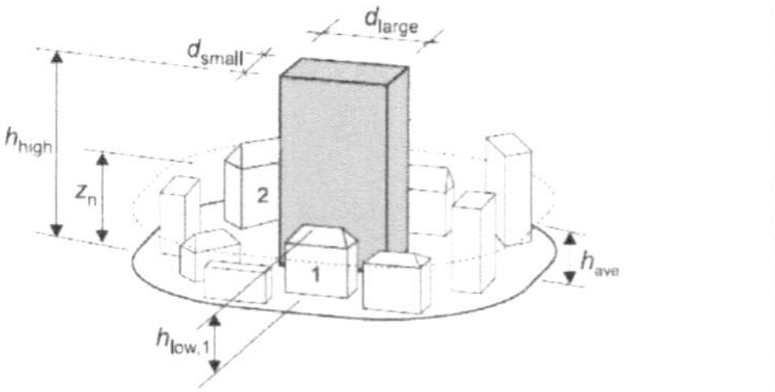
Closely spaced buildings	For Terrain IV with closely spaced buildings and other obstructions: - Effective ground level raised by h_{dis} - May take h_{ave} = 15 m in absence of more accurate information	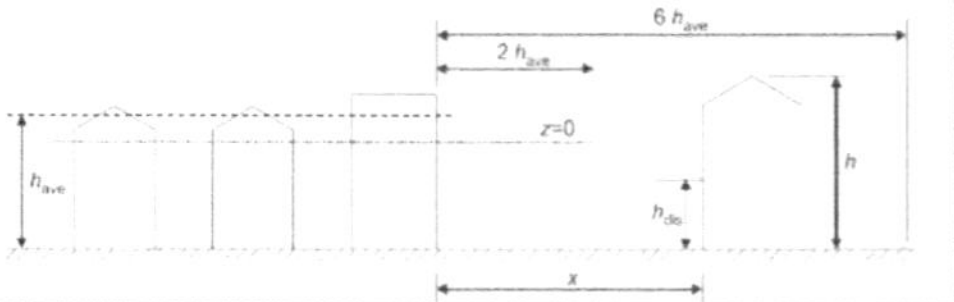

3.3 Structural response

Procedures to represent structural response are summarised in Table 5. In addition to alternative procedures for structural response, extensive information on structural dynamic effects are presented, including guidelines for situations where they do not have to be considered, dynamic characteristics for a range of standard structures such as framed steel and concrete high-rise buildings or steel and concrete chimneys.

Table 5 EN 1991-1-4:2005 procedures for structural response to strong winds

Aerodynamic coefficients	**Pressure coefficients:** External and internal coefficients for building walls, roofs of various shapes **Net pressure coefficients:** Resultant effects for canopy roofs; free-standing walls, fences & parapets **Friction coefficients:** Forces resulting from friction of wind parallel to external surfaces **Force coefficients:** Signboards; elements; cylinders, spheres; lattice structures & scaffoldings; flags **Bridges – force coefficients:** Bridge decks (constant cross section) – x, y & z directions; bridge piers.	
Pressure on external surfaces	**Pressure coefficient** (c_{pe}) converts peak velocity pressure at reference height (z_e) to surface pressure	$w_e = q_{peak}(z_e) \cdot c_{pe}$
Area size effect	**Overall:** Large areas (> 10 m^2) apply $c_{pe,10}$ **In-between:** Logarithmic interpolation	**Local:** Small areas (< 1 m^2) apply $c_{pe,1}$; $c_{pe} = c_{pe,1} - (c_{pe,1} - c_{pe,10}) \cdot \log_{10}(A)$
Internal pressure	**Internal pressure coefficient** (c_{pi}) taken as (0,75c_{pe}; 0,90c_{pe}) for dominant face openings (2; 3) times for remaining faces	$w_i = q_{peak}(z_e) \cdot c_{pi}$
Structural factor (c_sc_d)	**Application:** Structural factor c_sc_d, applied together or separately to provide for size effect (non-simultaneity) of peak wind pressures (c_s) and effects of vibrations of the structure due to turbulence (c_d). **Quasi-static** (c_sc_d = 1): Building height < 15 m; roof elements with $f_{natural}$ > 5 Hz; framed buildings with depth d, height h < 100 m & $h/d_{in\text{-}wind}$ > 4; chimneys with circular section h < 60 m & 6,5 diameter. **Procedures:** Alternative procedures provided - Procedure for limited shapes of structures within normative standard - Alternative procedures (1 & 2) in annex for wind turbulence; structural factor (c_sc_d); number of loads for dynamic response; displacements and accelerations for serviceability - Values of c_sc_d for a range of structures: • multi-storey steel & concrete buildings; • steel chimneys with & without liners; concrete chimneys without liners	

3.4 National adjustment through Nationally Determined Parameters

In addition to the obvious provision made for selecting appropriate values to represent local strong wind conditions by Eurocode Member States, a list of as many as 53 items are identified as Nationally Determined Parameters (NDP), where alternative procedures and values are allowed to be selected nationally. These NDP values include critical basic parameters such site exposure, transition between terrain categories and orography, turbulence parameters, pressure coefficients, dynamic response procedures. With a few exceptions where alternatives are provided, the procedures are based on recommended NDP values. EN-1-1-4 therefore represents a high degree of consensus, but still allowing a wide margin of adjustment to national conditions, requirements and preference.

4 APPLICATION OF EN 1991-1-4:2005 AS REFERENCE

Important issues to be taken into account in assessing EN 1991-1-4:2005 as reference to the revised provisions for wind actions in SANS 10160 include:

- **Scope of structures and wind actions:** Whereas SANS 10160 provide for buildings and similar structures, EN-1-1-4 provides for buildings and a comprehensive range of civil engineering structures, including provision for an extended range of modes of structural response.
- **Climate of strong wind conditions:** The dominance of thunderstorms in inland areas of South Africa differs significantly from the climatic conditions of Europe. Furthermore constraints on resources for the development of SANS 10160 wind action procedures necessitated maximum utilisation of the existing representation of the national strong wind conditions.
- **Reliability levels applied in procedures:** As indicated in Chapter 1-1 the general level of reliability on which SABS 0160-1989 has been based is to be maintained in its revision, with changes only applied to improve consistency of reliability across the range of design conditions, or to rectify clear deficiencies. Differences in target levels of reliability between the revised and reference procedures therefore need to be taken into account.

The implications of applying EN-1-1-4 stipulations to revised SANS 10160 procedures for wind actions are considered against this background in terms of the main steps of the process. Table 6 provides a summary of the main considerations of applying the procedures from EN 1991-1-4 to South African conditions, also taking account of issues from ASCE-7 and ISO/DIS 4354-2007.

4.1 South African strong wind climate

Since the European wind climate is dominated by synoptic winter storms (Drayton *et al* 1999), EN-1-1-4 is primarily based on such conditions and the calculation process is based on basic wind speed derived from an averaging period of 10 minutes. South Africa is however a country of a diverse climate with more than 20 climatic zones (Kruger, 2002). Despite this diversity, one can identify two distinct types of strong wind events, namely of frontal and convective origin. Frontal winds are generated by low-pressure frontal systems off the coast of the country. Strong inland winds originate typically as a result of convective activity in which a significant upward movement of hot air cools-off, saturates and forms thunderstorm cells. Thunderstorms can also produce other types of extreme wind events such as tornadoes or downbursts.

A division of the country into regions of dominant types of strong wind events has been proposed by Goliger and Retief (2002). A graphical representation of wind-climatic zones, developed in collaboration with the SA Weather Service, is presented in Figure 1. The process of developing the zones was aimed at those wind characteristics which affect the lowest regions of the boundary layer (i.e. the built environment) as opposed to those related to the climate or upper level mechanisms of the air-mass movement. The occurrence of tropical cyclones over the north-eastern part of the country is not considered in the model as their inland penetration and influence in terms of the wind speeds, which are generated, is not believed to be sufficiently significant.

The calculation procedures of SANS 0160 are based on the 3-sec gust wind speed in terms of wind speed (at a height of 10 m above the ground-level in terrain category 2) based on the analysis undertaken by Milford (1987), for which long-term data from 15 weather stations were used.

The situation where alternative averaging periods of 10 minutes vs. 3 seconds needs to be considered is seemingly neither unique nor critical, as the non-compatibility of definitions of averaging times represents a universal problem while evaluating and comparing various design codes across the world. For mature winds, a conversion of the magnitude of wind speeds to various averaging periods constitutes standard textbook information (Simiu and Scanlan, 1978). Such conversion is also incorporated in ISO/DIS 4354-2007.

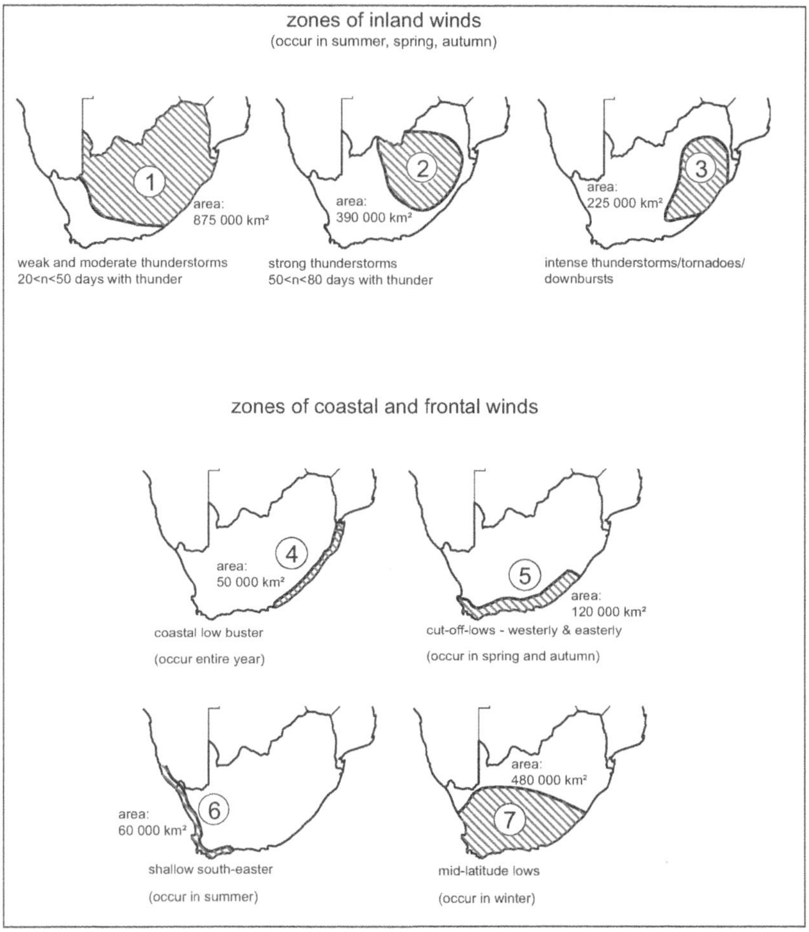

Figure 1 **Zones of strong winds in South Africa**

4.2 Reference wind speed as derived from the strong wind climate

SABS 0160 presents two wind maps for 3 second and 1 hour averages for South Africa, as obtained from Milford (1987). The 3 second value of 40 m/s can be taken as representative of most of the country; with the 1 hour average values of 25 m/s and 20 m/s representative of coastal and inland areas respectively. The different characteristics of coastal and inland winds are demonstrated by the respective gust factors of 1,6 for coastal regions and 2,0 for inland regions as obtained from the ratio of 3 second to 1 hour averages, as shown in Table 6.

The respective velocities are converted to 10 minute average values by using conversion factors k_T as tabulated in ISO/DIS 4354:2007. Also shown are the calculations of the respective basic velocity pressures according to EN-1-1-4 and SABS 0160. Two issues arise from such a conversion:

- **Magnitude of free-stream velocity pressure:** The reference wind action increases by a factor of 1,13 when calculated from the conversion of the 3 second value.
- **Thunderstorm conditions:** Different values of the basic wind speed for inland areas are obtained when the 10 minute average value is obtained from the 3 s and 1 h values

respectively, resulting in two values of basic wind velocity pressure of 1,13 and 0,68 times that of the present procedure.

The 10-minute or hourly mean wind speeds are suitable for application to climates dominated by mature storms e.g. for Europe and the coastal regions of South Africa. This is however not the case for areas dominated by convective weather systems such as for large areas of South Africa or Australia. This issue is also emphasised in the proposed ISO/DIS 4354-2007 standard in which, for climates dominated by thunderstorms, an averaging time of 3 sec or less is recommended for a meaningful determination of extreme value design wind speeds.

Representation of the South African strong wind climate, and particularly for the large inland areas, with prevailing thunderstorm winds, is therefore a critical issue in the possible application of EN-1-1-4 as reference.

4.3 Site exposure

The characteristics of the free-stream wind flow approaching a site are influenced by the roughness of the upwind terrain which affects mainly the magnitude of wind speed and its rate of change (typically described in terms of the intensity of turbulence). The principle of site exposure is generally introduced to represent the influence of terrain roughness and topography (orography) on profiles of wind velocity and variability (gustiness) with height in EN 1991-1-4:2005, ASCE 7-98 and SABS 0160-1989. To some extent, the site exposure also depends on the structure/character of strong winds, and is consequently treated separately for: synoptic, tropical cyclone and thunderstorm winds in the ISO/DIS 4354-2007.

Terrain roughness can be treated in terms of an exposure factor $c_r(z)$, which is a multiplier of the basic wind speed for the site, consisting of the following components:

- **Terrain characterisation:** Classification of a site in terms of its roughness properties, adjusted from the reference site of open country to which the basic wind speed applies:
  - Expressed in terms of the roughness length or characterised by $c_{r,\,class}(z = 10\text{ m})$;
  - Description of roughness properties of the site as given in the procedure.
- **Changes of wind characteristics with height:** The vertical profile of $c_r(z)$ as a function of height for a given roughness, providing for changes in the average speed and gust effects with height:
  - In terms of a logarithmic function of height based on theoretical considerations,
  - Alternatively an exponential function provides a convenient empirical fit over a limited range.
- **Wind profile at low elevations:** A minimum height below which the exposure factor is to be taken as constant.
- **Transition distances:** Local site roughness is dependent on upstream terrain conditions which are expressed by transition distances for site categories as a function of height.

A comparison of EN-1-1-4 characterisation of site exposure with the present provisions of SABS 0160 and related requirements of ISO/DIS 4354-2007 and ASCE 7-98 is summarised in Table 7. Since SABS 0160-1989 treats area effects in terms of different terrain exposure for classes A (local effects), B and C (large) structures, the exposure of class B structures is used as representative of SABS procedures.

4.3.1 Terrain categories

Due to the roughness of terrain, wind flow becomes retarded, reaching stagnation at ground level. This phenomenon is reflected in the division into terrain categories. A distribution of terrain categories across the practical range of roughness values is provided by considering the site exposure factor $c_r(z)$ for different terrain categories at the reference height $z = 10$ m as indicated in Table 7.

Table 6 Application of EN procedures to South African conditions for basic wind velocity pressure

<table>
<tr><th>TOPIC</th><th>EN 1991-1-4:2005</th><th colspan="3">Application to SANS 10160 and existing SABS 0160-1989 procedures</th><th>Comments and considerations from ISO 4354 & ASCE 7</th></tr>
<tr><td>Strong wind climate</td><td>Synoptic strong wind primarily considered
- Represents European wind climate</td><td colspan="3">Synoptic and thunderstorm strong wind climate
- Reflected in design gust and hourly mean wind maps</td><td>- ISO treats synoptic, tropical cyclone and thunderstorm strong winds separately.
- ASCE 7 includes all three mechanisms into one map.</td></tr>
<tr><td rowspan="9">Basic wind speed
- Regional wind speed
- Annual p-exceedance
- Open terrain (II / 2)
- Height 10 m</td><td rowspan="9">Procedure based on 10 minute average maximum
- Useful for representation of structural dynamic response, particularly for synoptic winds
- Map showing values for Europe of 20 – 30+ m/s; with large areas ~ 26 m/s</td><td colspan="3">SABS 0160-1989</td><td rowspan="9">ISO provides alternatively for
- Peak 3 second average - recommended for thunderstorms
- Mean 10 minute average – particularly suitable for dynamic effects

ASCE 7 utilises 3 s peak values
- Converted from fastest mile specification used previously</td></tr>
<tr><td>Wind characteristics</td><td>Coastal (m/s)</td><td>Inland (m/s)</td></tr>
<tr><td>3 second</td><td>40</td><td>40</td></tr>
<tr><td>1 hour</td><td>25</td><td>20</td></tr>
<tr><td>Gust = 3s/1h</td><td>1,6</td><td>2,0</td></tr>
<tr><td colspan="3">Conversion to 10 minute mean per ISO
(3600 – 1,00); (600 – 1,05); (3 – 1,53)</td></tr>
<tr><td>From 3 s</td><td>27,5</td><td>27,5</td></tr>
<tr><td>From 1 h</td><td>26,3</td><td>21,0</td></tr>
<tr><td>Assumed:</td><td>27</td><td>27 / 21</td></tr>
<tr><td rowspan="8">Basic wind velocity pressure</td><td rowspan="8">Basic wind velocity converted to pressure; providing for gust effects from mean value:

$w_e = \rho/2 \cdot G^2 \cdot [c_{r,II}(z = 10\text{ m})]^2 \cdot (v_b)^2$</td><td colspan="3">Present situation: 3 s; 40 m/s; Class B</td><td rowspan="8"></td></tr>
<tr><td>Procedure</td><td colspan="2">Basic pressure (N/m^2)</td></tr>
<tr><td>SABS 0160</td><td colspan="2">941</td></tr>
<tr><td>EN – 27 m/s</td><td colspan="2">1065</td></tr>
<tr><td>EN – 21 m/s</td><td>-</td><td>644</td></tr>
<tr><td>Comparison</td><td colspan="2">Ratio New/Present</td></tr>
<tr><td>28 m/s</td><td colspan="2">1,13</td></tr>
<tr><td>22 m/s</td><td>-</td><td>0,68</td></tr>
</table>

Table 7 Comparison of EN, SABS, ISO and ASCE procedures for site exposure characterisation

Site exposure	EN 1991-1-4:2005	Comparisons	ISO/ASCE
- **Terrain Categories**	Five categories 0, I – IV, with decreasing roughness, (with TC 0 exposed to open sea – only relevant for serviceability at less than max design wind) - Roughness length according to description of terrain - Roughness lengths not quite evenly distributed {0,003; 0,01; 0,05; 0,3; 1}.	**Comparison of terrain velocity exposure factor (10 m)** (see table below)	**ISO** - Distribution of categories in terms of roughness length evenly distributed {0,003; 0,03; 0,3; 3} - Separate single profile for thunderstorm wind (Categories 1 – 3) **ASCE 7** – Four terrain exposure categories A – D with decreasing roughness - Reference category C – open terrain
Description of site - In order of increasing roughness	**EN 1991-1-4:2005**	**SABS 0160-1989**	**ASCE 7**
	0 Sea or coastal area exposed to the open sea (serviceability wind actions). **I** Lakes or flat and horizontal area with negligible vegetation and without obstacles (ultimate limit state, for exposure to open water)	**1.** Exposed smooth terrain with virtually no obstructions and in which the height of any obstruction is less than 1,5 m . - Open sea coasts, lake shores and flat, treeless plains with little vegetation other than short grass.	**D.** Flat, unobstructed areas exposed to wind flowing over open water (excluding shorelines in hurricane prone regions) for a distance of at least 1 mi (1.61 km).
	II Area with low vegetation such as grass and isolated obstacles (trees, buildings) with separations of at least 20 obstacle heights	**2.** Open terrain with widely spaced obstructions (more than 100 m apart) having heights and plan dimensions generally between 1,5 m and 10 m . - Large airfields, open parklands, farmlands and undeveloped outskirts of towns and suburbs, few trees.	**C.** Open terrain with scattered obstructions having heights generally less than 30 ft (9.1 m). This category includes flat open country, grasslands and shorelines in hurricane prone regions.
	III Area with regular cover of vegetation or buildings or isolated obstacles with separations of maximum 20 obstacle heights (such as villages, suburban terrain, permanent forest)	**3.** Terrain having numerous closely spaced obstructions generally having the size of domestic houses. - Wooded areas and suburbs, towns and industrial areas, fully or substantially developed	**B.** Urban and suburban areas, wooded areas, or other terrain with numerous closely spaced obstructions having the size of single family dwellings or larger.
	IV Area in which at least 15 % of the surface is covered with buildings and their average height exceeds 15 m.	**4.** Terrain with numerous large, tall, closely-spaced obstructions. - Large city centres.	**A.** Large city centers with at least 50% of the buildings having a height in excess of 70 ft (21.3 m).

Comparison of terrain velocity exposure factor (10 m)

Terrain	1	2	3	4
SABS (B)	1,08	0,98	0,71	0,62
EN	1,08	1	0,85	0,71
ASCE (interpolated)	1,18	1	0,72	0,44
ISO	1,13	1	0,81	0,7
ISO thunderstorm		1		-

Minimum height

EN	I	II	III	IV
z_{min}	1 m	2 m	5 m	10 m

SABS	1	2	3	4
z_{min}	5 m	5 m	7,5 m	20 m

ISO	3 m	3 m	10 m	20 m
ASCE	2,1 m	4,5 m	9 m	18 m

4.3.2 Roughness lengths

The comparison in Table 7 provides an indication that site categories provided by SABS 0160-1989 are not distributed evenly across terrain roughness values as compared to EN-1-1-4 and ISO/DIS 4354-2007. This conclusion is graphically illustrated by comparing an effective roughness length calculated for the SABS 0160-1989 values according to the EN-1-1-4 function (SABS category 2 normalised to EN category II), as compared to the EN and ISO values, shown in Figure 2.

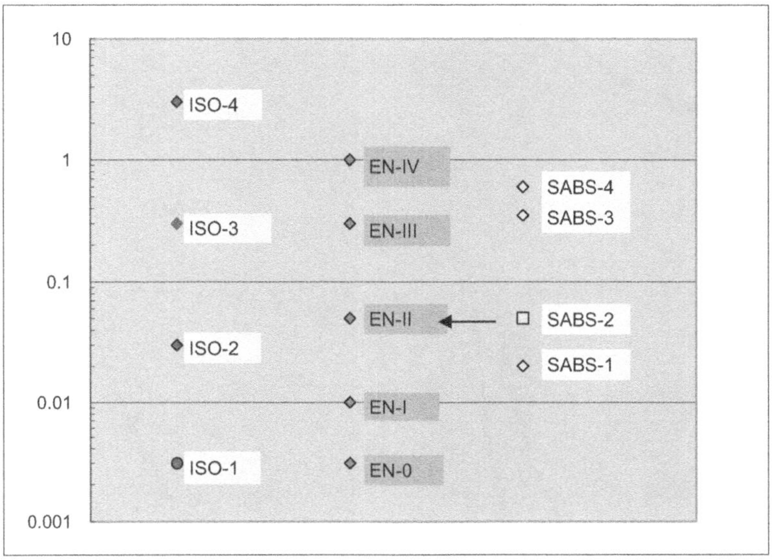

Figure 2 Roughness lengths z_0 (m) for terrain categories (effective values for SABS)

Although there is general correspondence between the respective descriptions from which a terrain category is selected in the design process, as tabulated in Table 6, the careful formulation of this classification system is important, due to its significant influence on the resulting wind actions.

4.3.3 Vertical profiles

The rate of the increase of wind speed with elevation for different terrain categories is reflected by vertical wind velocity profiles as expressed by the related exposure factor $c_r(z)$. The exposure factors at reference height (z = 10m), for the alternative site roughness values, as presented in Table 4, provide an indication that EN-1-1-4 wind velocities are higher for terrain categories TC III & IV as compared to SABS 0160-1989 TC 3 & 4.

This tendency is confirmed when the wind velocity profiles are compared for the respective standards as shown in Figure 3: The EN-1-1-4 profiles (synoptic wind), SABS 0160 profiles (used for both synoptic and thunderstorm wind), ISO/DIS 4354-2007 profiles (synoptic wind) and ASCE 7-98 profiles (synoptic, tropical cyclones and thunderstorms combined) are presented in Figure 3.

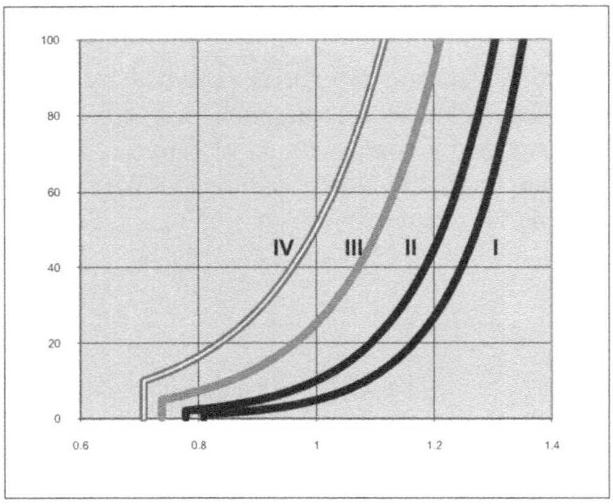

EN 1991-1-4:2005 (normalized)

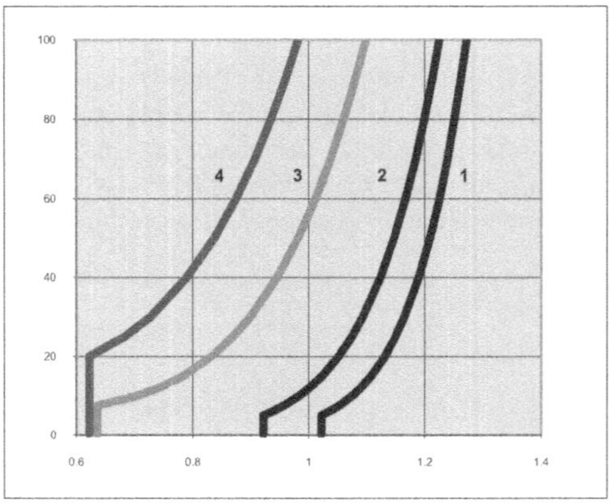

SABS 0160-1989

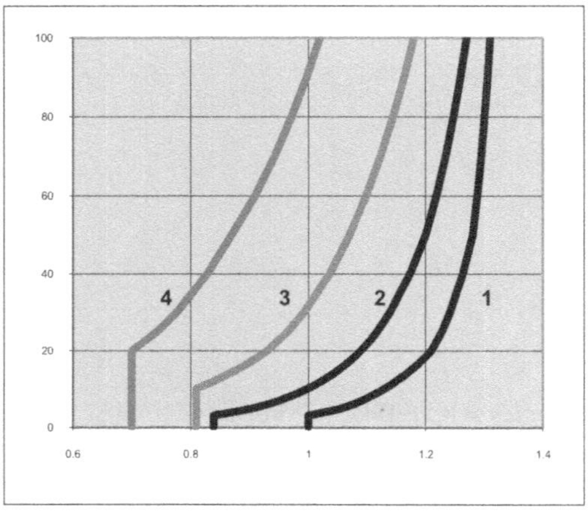

ISO/CD 4354-2007

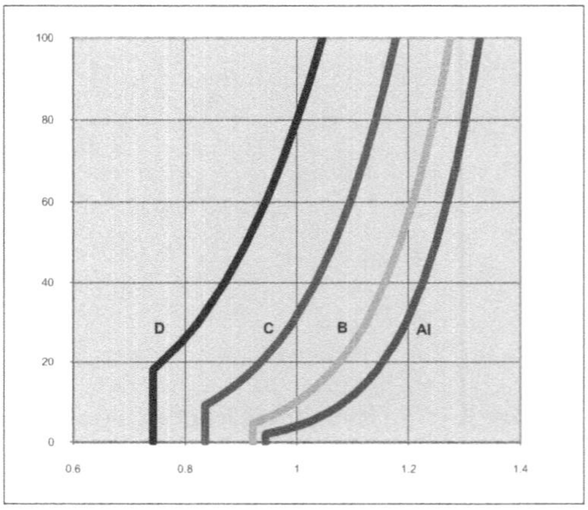

ASCE 7-98

Figure 3 Terrain exposure factors (in terms of velocity) for various standards

Direct comparisons presented in Figure 4 for different terrain categories indicate that for the reference site of open terrain (TC 2/II) the SABS 0160 profile provides lower velocities (~ 7% lower than the EN-1-1-4 values, resulting in ~ 14% lower pressures). For a suburban terrain the SABS 0160 profile is shifted to even lower values, compared to EN-1-1-4 values, with EN values higher by a factor of 1,27 at 7,5 m to 1,10 at 100 m. Similar results are obtained for urban sites, with EN values higher by a factor of 1,14 at 10 m and 100 m, peaking to a factor of 1,34 at 20 m.

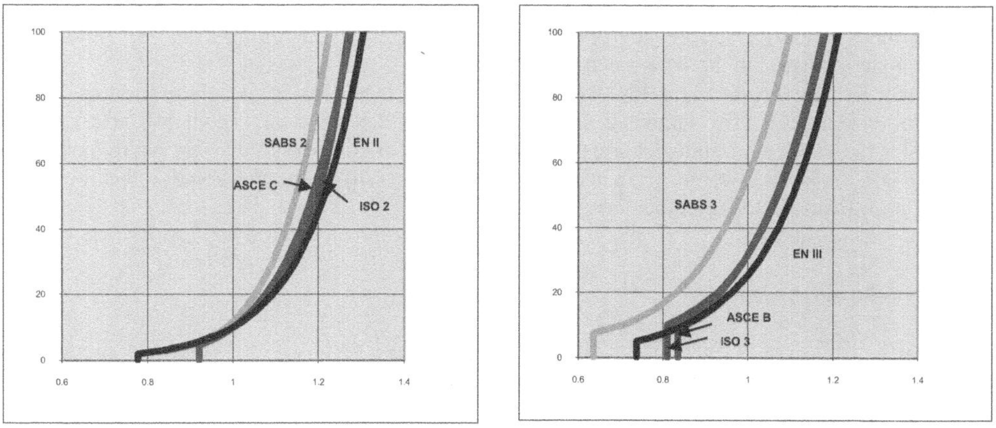

Reference terrain (open country) Suburban terrain

Figure 4 Comparison of terrain exposure factor for open and suburban sites

As shown in Figure 5 the profile of SABS TC 3 compares, however, particularly well with EN TC IV, with both profiles being straddled by ISO TC 3 (suburban) and TC 4 (urban). This observation should be taken into account when considering that South African city centres are not sufficiently large to justify the development of a profile corresponding to TC 4 as described in Table 4 for SABS 0160.

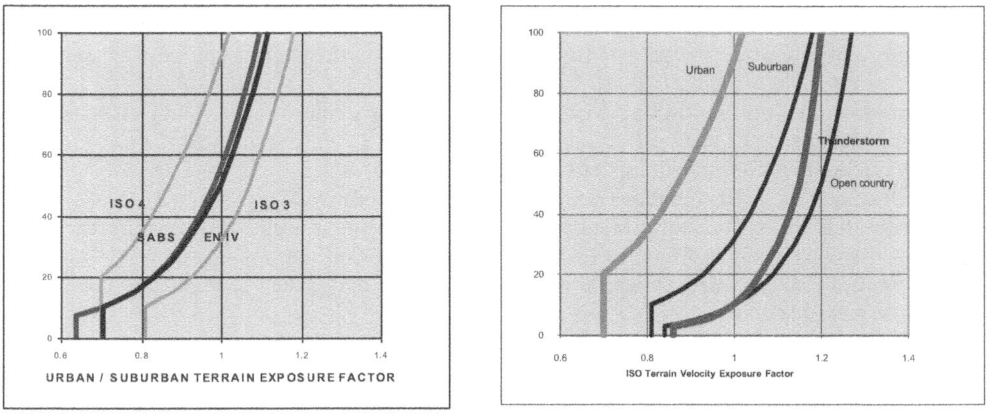

Figure 5 Comparison of alternative terrain exposure factor schemes

Also shown in Figure 5 is the treatment of thunderstorm strong winds by ISO/DIS 4354-2007 where a separate single profile provides for open sea, open country to suburban terrains, irrespective of their roughness. Although the exposure factor profile is similar to that for TC 2 for synoptic strong winds at low heights, it tends toward TC 3 at an elevation of 200 m, reducing to a factor of ~ 1,0 at 500 m and beyond.

4.3.4 Minimum height (z_{min})

At low elevations the wind distribution is significantly influenced by the configuration of local obstructions and the description of the profile becomes artificial and irrelevant. Surrounding obstructions can either accelerate or decelerate wind flow. In view of the level of uncertainties most of the design codes introduce various conservative sets of stipulations regarding a 'cut-off', elevations below which no reduction in the windspeed is permitted.

The minimum height z_{min} below which the profile is kept constant is summarised in Table 7 for the various standards. The tabulated values specified by EN-1-1-4 are about half that of other standards, for the various terrain categories. This implies that significantly lower exposure factors are allowed by EN-1-1-4 for lower structures. SABS 0160 is generally more conservative than both ISO and ASCE specifications for smoother terrains.

4.4 Structural response

SABS 0160 does not include the dynamic response of structures in its scope. In an appendix commentary and guidance are given on along-wind response; quantitative procedures are provided for assessment of vortex shedding, including expressions for the hourly mean speed profiles for respective terrain categories and conversion of the hourly mean wind speed to other averaging times.

4.4.1 Comparison of procedures

Provisions for structural dynamic effects of wind actions can therefore not be directly compared between EN-1-1-4 and SABS 0160. The following matters should however be assessed in the process:

- **Scope of structures and associated dynamic response:** The application of the comprehensive suite of structural dynamic procedures presented in EN-1-1-4 and the scope of structures to be provided for in SANS 10160:
  - SANS 10160 will consider only buildings and similar industrial structures, as compared to the comprehensive range of structures provided for by EN-1-1-4, particularly including towers, masts, chimneys and bridges.
  - Provision is made for an extensive range of structural dynamic response mechanisms in EN-1-1-4
- **Strong wind representation:** Matching EN-1-1-4 procedures with South African wind conditions need to be resolved:
  - Presentation of mean wind speed values for South Africa
  - Provision for wind conditions due to strong thunderstorms.
- **Parametric limits for static response:** EN-1-1-4 presents a useful set of parametric limits within which structural dynamic effects need not to be considered:
  - Framed buildings with structural walls, less than 100 m high and a height less than 4 times the in-wind depth;
  - All buildings with a height less than 15 m;
  - Façade and roof elements with a natural frequency > 5 Hz; glazing spans < 3 m usually lead to natural frequencies > 5 Hz;
  - Chimneys with circular cross-sections, height < 60 m and 6,5 times the diameter.

4.4.2 Level of application

It should be noted that although the general objectives for Eurocode are stated to provide for *common design rules for everyday use* (Foreword to Eurocode), the provisions for structural dynamic response to wind loads are of a specialist nature which requires sound understanding of wind engineering to be used properly. The provisions of EN-1-1-4 for the dynamic effects of wind actions consequently go well beyond the scope of application of SABS 0160.

Provision for structural dynamic wind actions in the revised SANS 10160 procedures could be based on either the incorporation of EN-1-1-4 procedures, or by maintaining sufficient compatibility and information so that the design can be based on the EN procedures where they apply, or various combinations of the two options.

4.5 Pressure coefficients

An elaborate set of coefficients is provided in EN-1-1-4, from which pressures and forces for structures of various configurations can be calculated. These coefficients were derived from the compilation of the provisions of different national standards, including consideration of the references on which these stipulations are based (Geurts *et al* 2001). Aspects considered include exceedance probabilities of the coefficients modelled with extreme value analysis, effects of averaging time and tributary area, properties of incoming flow in terms of wind tunnel testing and turbulence characteristics of the wind flow, and related calibration to the EN procedures.

Area effects are taken into account in EN-1-1-4 in terms of different coefficients for small areas (< 1 m^2) and large areas (> 10 m^2) with logarithmic interpolation in between. This procedure is similar to that used in ASCE 7-98. A different approach is taken in SABS 0160-1989, where different building classes (A, B & C) are stipulated, which are related to different averaging times of 3 s, 5 s & 10 s, as expressed in terms of exposure factors for the respective building classes. Different local pressure coefficients are stipulated for various building classes and geometrical shapes.

The various sources of pressure coefficients applied in EN-1-1-4 result in a lack of consistency in the way in which the information is presented, particularly because they use a mix of tabulated and graphical presentation. In the case of SABS 0160, much better consistency is achieved since the stipulations derive from the well established British wind loading code from the early 50's. However the wide scope of configurations for which pressure coefficients are provided for in EN-1-1-4 facilitates the convenient and reliable selection of appropriate information for a specific structure, which completely compensates for the inconsistencies in the format of presentation.

5 COMPARATIVE CALCULATIONS

A set of comparative calculations was performed in order to explore and assess the implications of applying the EN-1-1-4 procedures with existing design based on SABS 0160-1989. Design values for small (1 m^2) and intermediate (5 m^2) areas were determined for different positions on a typical large warehouse of 40 m x 100 m, eaves height of 15 m and roof slope of 5°, 15° and 30°, and integral forces on the walls of a medium rise building of 60 m high with a base of 40 m x 30 m; located in a coastal area.

A directed comparison of pressures for small and intermediate areas is presented in Figure 6 for various positions on the roof of the building; by comparing EN and SABS terrain categories {I and 1} for the most exposed site, {II and 2} as the reference site of open country and {IV and 4} as sites with highest terrain roughness.

It is clear that EN procedures generally provide much higher pressures than the existing SABS procedures; particularly for the comparison between sites IV and 4. Furthermore this effect is more pronounced for small areas, where the EN values are on average 2,5 times higher than those obtained from the SABS; as compared to a value of 1,9 for intermediate areas.

The EN-1-1-4 roof loads were then adjusted to values which would be similar to the application of SABS 0160 procedures for site effects by a simple normalisation process. The results provide an approximate indication of the effect of EN -1-1-4 pressure coefficients. Such an effective comparison of the pressure coefficients of EN-1-1-4 and SABS 0160 is presented in Figure 7. The comparison indicates that EN-1-1-4 values for small areas are about 1,75 that of SABS 0160, and for intermediate areas the ratio is 1,3.

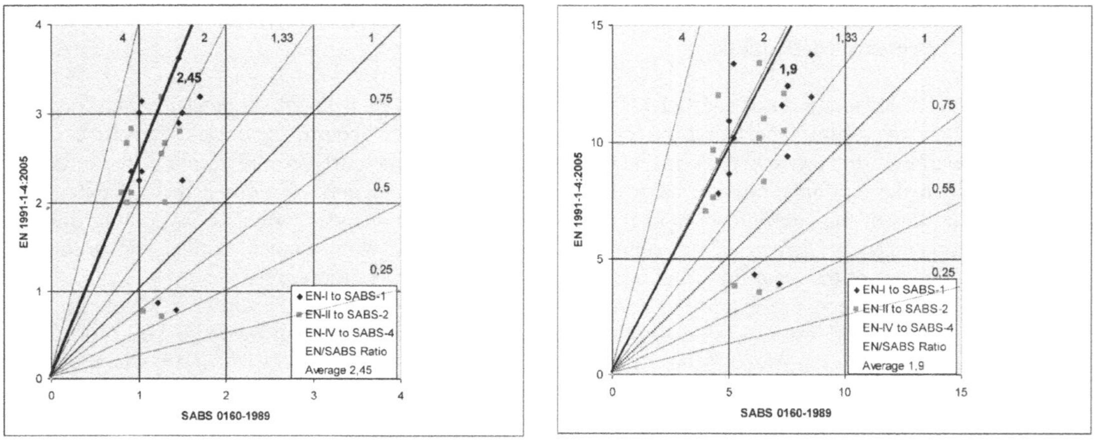

Small areas (1 m^2) Intermediate areas (5 m^2)

Figure 6 Comparison of roof loads according to EN and SABS procedures

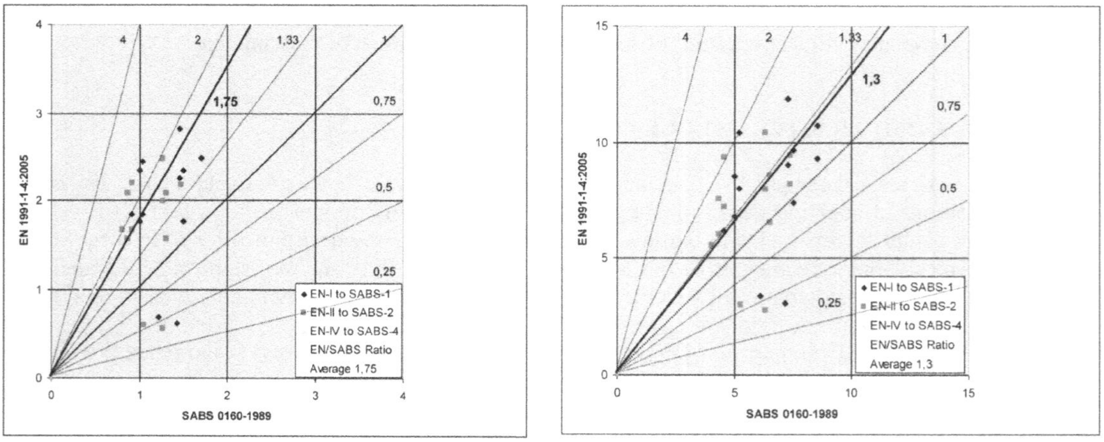

Small areas (1 m^2) Intermediate areas (5 m^2)

Figure 7 Comparison of EN and SABS: Effects of roof pressure coefficients

The EN stipulation of the *roof ridge* as reference height for wind load on walls also results in higher wind loads, particularly for steep roofs; as opposed to the SABS stipulation of eaves height as reference. The SABS procedure is similar to that of ASCE-7, which applies to eaves height if the roof slope is $\leq 10°$.

For the medium rise building the magnitude of pressure coefficients over the windward walls, as stipulated in both codes, was found to be similar. However, differences are apparent in the application method of pressure distribution over the elevation, as presented schematically in Figure 8. Two methods of application are allowed in SABS 0160 (conservative; accurate), whereas EN-1-1-4 prescribes a distribution into zones, defined in terms of the aspect ratio of the windward wall (i.e. the relationship between the height and breadth).

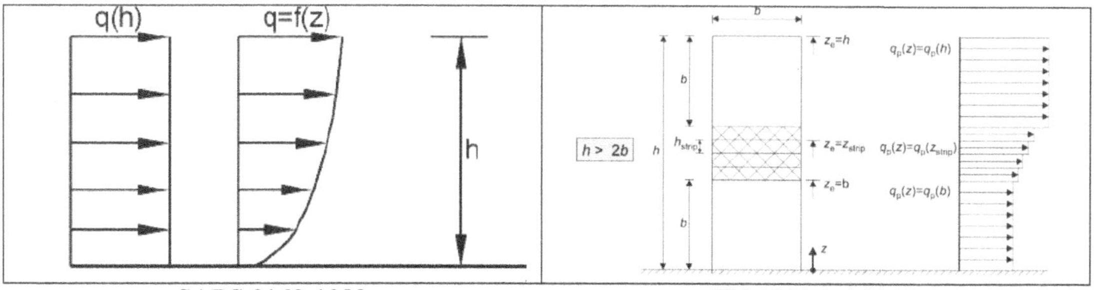

SABS 0160-1989 EN 1991-1-4:2005

Figure 8 Representation of pressure distribution on walls

Significant differences in overall horizontal loads were found however, which in turn determine the overturning moments. This can largely be attributed to the differences in the magnitude of negative pressures over the leeward wall, as stipulated in the respective codes. Depending on the aspect ratio, EN-1-1-4 specifies values of between –0.3 and –0.7, whereas SABS 0160 values are between –0.2 and –0.4, resulting in a substantial increase in horizontal leeward forces according to the EN procedures.

From the comparisons it is concluded that EN-1-1-4 procedures result in significantly higher wind loads as compared to SABS 0160. In approximate terms the basic velocity pressure is increased by a factor of ~ 1,1; terrain velocity exposure effects are increased by a factor of up to 1,2; the effects of pressure coefficients vary, with an increase in loads over small areas by as much as a factor of 1,75!

6 CONCLUSIONS

Recent advances made in standards for determining wind actions in the design of structures confirm the need to revise the provisions of SABS 0160-1989 (Revised). The Eurocode Standard EN-1-1-4 captures the advances in terms of the scope of structures and wind action mechanisms in a comprehensive set of procedures. It contains a large pool of wind engineering knowledge on which consensus has been achieved amongst a large number of countries / wind engineering experts, and is consistent with the higher level international draft standard ISO/DIS 4354-2007.

A technical analysis of EN-1-1-4 confirmed its merit to serve as reference in the formulation of wind action procedures of the revised standard SANS 10160. A number of critical issues that need to be resolved have been identified, including:

- Scope of structures and structural response provided for in SANS 10160;
- The use of EN 1991-1-4 beyond this scope, particularly for dynamic structural response;
- Regional wind speed: adjustment of map, within the constraints of resources to do substantial updating;
- Representation of free-field wind velocity: conversion to 10 minute mean values used in the EN-1-1-4 procedures;
- Procedures for conversion to free-field wind velocity to pressure;
- Terrain characteristics and wind profile;
- Pressure coefficients for the range of structural configurations and geometry;
- Conversion of SA climate to Eurocode procedures for dynamic response
- Level of reliability implied by changes in the provisions for wind actions.

The way in which the difficulties of applying EN 1991-1-4:2005 procedures to South African conditions and practice have been resolved in formulation of the revised specifications for wind actions in the revised standard SANS 10160 will be discussed subsequently in Chapter 3-2.

7 REFERENCES

ASCE 7-98 1998. *ASCE Standard: Minimum design loads for buildings and other structures.* American Society of Civil Engineers.

AS 1170.2-1989. *Australian Standard: SAA Loading code, Part 2: Wind loads.* Standards Australia.

BS 6399: Part 2: 1997. British Standard: Part 2, Code of practice for wind loads, BSI.

CIB (1996). *Actions on structures*: Wind actions. CIB Report Publication 193. Conseil International du Bâtiment pour la Recherche l'Etude et la Documentation (CIB), Rotterdam.

Drayton M J, Page B A & Muir-Wood R (1999). Building a stochastic model for windstorm loss in north-west Europe. Proceedings of 10th International conference : Wind Engineering into 21st Century, Copenhagen, Denmark June 1999, Vol 2, pp. 1395-1440.

EN 1990:2002 *Eurocode – Basis of structural design*. European Committee for Standardisation (CEN), Brussels.

EN 1991-1-4:2005 Eurocode 1 *Actions on structures*, Part 1-4 *General actions – Wind actions*. European Committee for Standardisation (CEN), Brussels.

Geurts C, Blackmore P, Hansen SO *et al* (2001). Transparency of pressure and force coefficients. WISSE J *et al* (Ed). Third European & African Conference on Wind Engineering. Eindhoven, The Netherlands.

Goliger AM, Niemann H-J & Milford RV (1998). *Assessment of wind-load specifications of the SABS 0160-1989*. South African National Conference on Loading. SAICE.

Goliger AM, Milford RV & Mahachi J (2001). *South African wind loadings: where to go*. Zingoni A (Ed) Structural Engineering Mechanics and Computation, Cape Town, SA.

Goliger AM & Retief J V (2002). *Identification of zones of strong wind events in South Africa.* Journal of Wind Engineering and Industrial Aerodynamics, No 90 (2002), pp. 1127-1235.

Goliger AM (2003). *Proposed wind loading stipulations for a new revision of the South African Loading Code*. Draft document submitted to SAICE WG on Loading Code, unpublished, Division of Building Technology, CSIR.

Goliger AM (2005). *South African wind loading specifications: the Euro way?* Naprsteck J & Fischer C (Ed) 4th European & African Conference on Wind Engineering, Prague.

ISO/DIS 4354-1990. International Standards Organisation. Draft International Standard *Wind actions on structures.*

ISO/DIS 4354-2007. International Standards Organisation. Committee Draft: *Wind actions on structures*. TC 98 SC 3 WG8.

JCSS (2001). Joint Committee on Structural Safety. *Probabilistic model code.* JCSS working materials, http://www.jcss.ethz.ch/.

Kruger AC (2002). Climatic regions of South Africa. WS44, Climate of South Africa. South African Weather Service, Pretoria.

Milford RV (1987). Annual maximum wind speeds for South Africa. *The Civil Engineer in South Africa*, January 1987.

Ruscheweyh HP (1996). *EC 1: Wind actions.* IABSE Colloquium – Background and application of Eurocode 1. Delft.

Simiu E & Scanlan R (1978). Wind effects on structures: An introduction to wind engineering. John Wiley& Sons Inc., New York / Toronto.

Vrouwenvelder T, Steenbergen H, Kraus O & Butz C (2005). Wind actions, Chapter III in Implementation of Eurocode, Handbook 3 – Actions effects for buildings. Leonardo da Vinci Project. http://www.ctn-eurocodes.cz//leonardo/ .

Zimmerli B (2001). Process of drafting the Eurocode on wind action. WISSE J *et al* (Ed). Third European & African Conference on Wind Engineering. Eindhoven, The Netherlands.

3-2 Revised Wind-loading Design Procedures for SANS 10160

Goliger AM, Retief JV, Dunaiski PE & Kruger AC

1 INTRODUCTION

The importance of wind loads as the dominant environmental load to be considered in the design of structures in South Africa has led to a critical assessment of the status of wind load provisions (Goliger *et al* 1998) during the review of the present South African Loading Code SABS 0160-1989 *The general procedures and loadings to be adopted in the design of buildings* (Day & Kemp 1999). The revised procedures for wind actions incorporated as Part 3 *Wind actions* consequently also form a critical component of the proposed revised standard SANS 10160 *Basis of structural design and actions for buildings and industrial structures*.

The preceding Chapter concluded that the Eurocode Part on wind actions EN 1991-1-4:2005 *Actions on structures: General actions – Wind actions* (referred to subsequently as EN-1-1-4 for brevity) could serve as reference in the formulation of appropriate wind action procedures. However a number of critical issues and complications in the application of EN-1-1-4 to South African conditions and practice are also identified in Chapter 3-1. This chapter presents the considerations taken into account in resolving these issues to formulate the proposed procedures incorporated into SANS 10160 Part 3.

2 GENERAL CONSIDERATIONS

2.1 Features of EN 1991-1-4:2005 Procedures

The biggest concerns regarding the applicability of the EN-1-1-4 stipulations to the South African context are related to the following:

- **Climatic conditions of Europe:** The dominance of synoptic winds in Europe results in the use of procedures based on 10-minute mean wind speed values to represent the strong wind conditions.
- **Scope of structures and design conditions:** The formulation of EN 1991-1-4 had the specific objective of allowing for the design of a comprehensive range of buildings and civil engineering structures (including bridges, towers, masts etc.) and structural response (various modes of wind structure interaction).
- **Complexity of procedures:** The wide scope and advanced models applied in EN-1-1-4 result in an extensive and complex set of stipulations, resulting also in an increase of calculation effort.
- **Magnitude of wind loads:** The wind loads resulting from EN-1-1-4 procedures are reported to be generally substantially higher than present practice of the Eurocode Member States, with indications that the wind loads are also higher than the values resulting from the present South African code SABS 0160:1989.

2.2 South African Conditions

In contrast to the scope of application, procedures and conditions relevant to EN 1991-1-4, the needs for a South African standard are related to the following environmental conditions and preference of practice:

- The climatic differences between Europe and South Africa, where inland regions are dominated by strong winds originating from convective activity.
- The characteristics and composition of the built-environment in South Africa in which, due to several factors (e.g. accessibility and cost of land, social / traditional aspects or architectural trends) most of the built assets constitute low-rise, in particular single storey, structures.
- The overall less rigorous standard of technical procedures and requirements within the SA construction industry i.e. a questionable rationale behind introducing high levels of sophistication in structural calculations, the output of which will be combined with relatively lenient standards of material, labour, supervision and approval processes in South Africa.

These concerns were kept in mind during the process of developing the wind-loading stipulations. The modifications applied to the EN-1-1-4 procedures are presented in the following sections.

The proposed changes to the present South African wind loading stipulations as given in SABS 0160:1989 will have far-reaching implications as they will change the design paradigm, and in many aspects break the historical links with the old CP3 code (CP3 1952) and will consequently increase the amount of design effort. Some of these changes are necessary in order to implement a modern engineering approach which is based on wide-ranging research in wind engineering. This is particularly relevant to the data on pressure coefficients. The proposed changes will increase the clarity and logic of the design process, while maintaining, where relevant, the current levels of reliability.

2.3 Guidelines for Referencing SA Procedures to EN 1991-1-4

A set of guidelines was formulated according to which EN-1-1-4 procedures could be adapted optimally to South African conditions for SANS 10160 Part 3, consisting of the following:

- **Scope of structures and structural response:** The general SANS 10160 scope of structures was restricted further in Part 3 to limit the procedures to situations where quasi-static structural response could be applied. The objective of this measure is primarily to simplify the procedures, yet improve their rational basis.
- **Moderation of high wind loads:** The high wind loads predicted by EN-1-1-4 procedures were moderated by effectively maintaining present levels of free-stream wind speed pressures. However increases of wind loads which result from the EN-1-1-4 pressure and force coefficients are accepted to be the result of improved and updated models and information.
- **Representation of convective strong winds:** The EN-1-1-4 procedure for conversion of the basic regional reference wind speed into pressure was modified to effectively maintain its present magnitude for inland regions where strong winds result predominantly from thunderstorm activities.
- **Wind map of basic regional wind speed:** The wind map is modified to express the basic regional wind speed in terms of maximum 10-minute mean values, in accordance with EN-1-1-4 procedures. Only a limited update of the map could be afforded due to limitations of resources and time.
- **Terrain exposure:** The present terrain exposure characteristics were generally maintained. The need to apply some modifications was identified. The reasons include the need to achieve a more realistic and convenient set of terrain profiles.
- **Compatibility with advanced EN-1-1-4 procedures:** Sufficient compatibility was to be maintained with EN-1-1-4 in order to be able to apply its procedures for structural dynamic response and wind-structure interaction under South African conditions.

2.4 Scope of Application

The investigation reported in Chapter 3-1 has lead to the following submission regarding the scope of applicability of the proposed code in order to limit the procedures to quasi-static structural response, specified in EN-1-1-4 for conditions under which quasi-static response could be assumed:

- Buildings and structures with an overall height of up to 100 metres
- Elements of buildings and structures having natural frequency greater than 5 Hz
- Chimneys with a circular cross-section, height less than 60 metres and height to diameter ratio of less than 6,5.

Furthermore, the wind loading stipulations will not cover:

- Structures and buildings higher than 100 metres
- Dynamic effects and design of dynamically sensitive structures
- Off-shore structures
- Bridge structures
- Building and structures of unusual shape
- Structures or their components, which are not fixed permanently, but are designed to accommodate movement (e.g. revolving dishes, movable roofs)
- High-risk structures (containing nuclear or biological material)
- Transmission line towers

Finally, a statement will be made in which it is pointed out that the code does not give guidance on wind loads due to high intensity winds like tornadoes or downbursts.

The exclusion of structures higher than 100m was considered in terms of the character of the built environment and architectural trends in South Africa. Furthermore, design of structures susceptible to dynamic excitations (e.g. tall industrial chimneys), typically involve expert consultants and access to specialist input and literature. The various dynamically-related stipulations, which are not relevant to vast majority of designs undertaken in South Africa, could therefore be excluded. This measure should result in a less complex and more user friendly document.

The exclusion of the off-shore and high-risk structures, and structures with unusual shapes, as well as the effects of high-intensity winds is in line with the modern approach of international design specifications. Modern standards assume that structures of such a nature would receive full expert attention, including wind-tunnel testing; while the occurrence of high intensity winds falls outside the acceptable levels considered in the design (i.e. with the probability of occurrence significantly smaller than 1 in 50 years for a specific site.)

3 WIND RELATED INPUT DATA

3.1 Climatologic Input Data

Due to the latitudinal position of South Africa, its weather is influenced both by the weather systems in the mid-latitudes, which is dominated by eastward moving depressions, as well as weather systems originating from the tropics. The country's topography and the fact that it is partly surrounded by oceans further contribute to its diverse range of climate types. Many different climate zones can be identified in South Africa, ranging from desert-like climates in the west to subtropical climates in the east, as presented in Figure 1 (Kruger 2004).

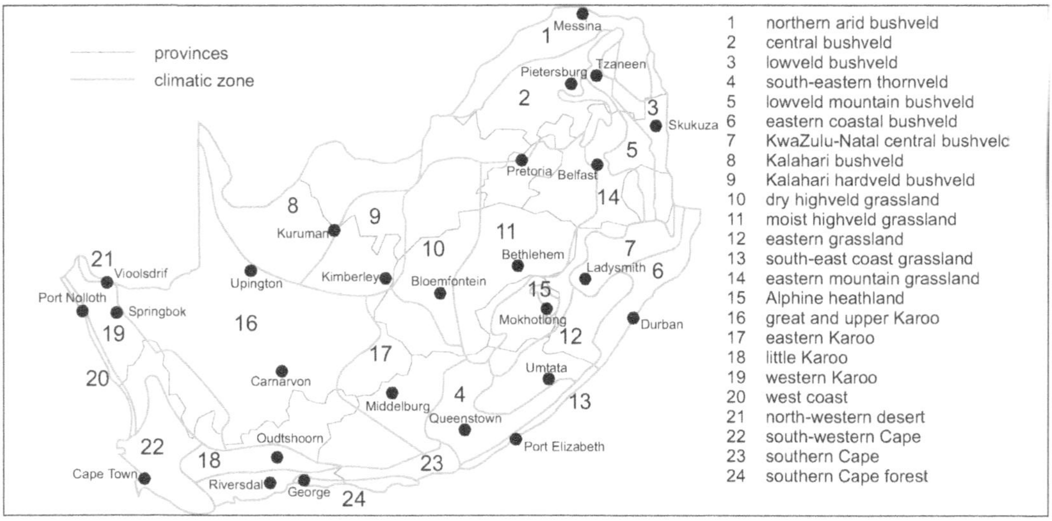

Figure 1 Climatic zones of South Africa

Having such a diverse climate, the wind climatology of South Africa is, as a consequence, relatively complex. Therefore also the origin, mechanism, seasonal occurrence and characteristics of strong and extreme winds are complicated.

However, high wind speeds broadly have two origins. In the interior, high wind gusts usually occur in the surface outflows of well-developed thunderstorms. At the coast and adjacent interior, frontal systems usually play the major role, where strong winds can be associated with the occurrence of berg winds, pre-frontal thunderstorms, or with the frontal bands themselves. Figure 2 presents a schematic layout of two zones in South Africa, with strong winds mainly associated with thunderstorms or frontal activity respectively (Goliger and Retief, 2002). The KwaZulu-Natal coastline represents the most important part of overlapping zones, where significant development occurs.

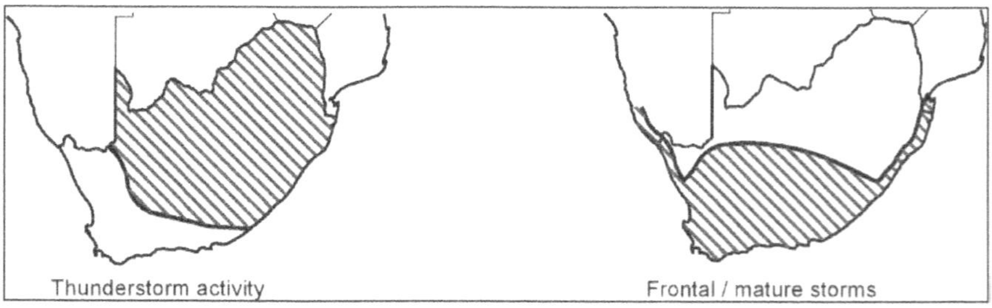

Figure 2 Geographical regions dominated by thunderstorm and frontal winds

3.2 Regional Basic Wind Speed

The main objective of the wind loading design of structures is to assure that (within a certain level of reliability) they will withstand the worst possible wind loads during their lifetimes. Therefore the prediction of the design wind speeds, which are derived from the regional climatic data, is the most important starting point of the design chain.

The present code SABS 0160-1989 provides regional wind maps developed by Milford (1987), based on an extreme-value analysis of long-term records from the 14 South African Weather Service (SAWS) recording stations which were available at the time, as depicted in Figure 3 (a). The use of such a small number of recording stations would, by the nature of the climatic diversity of South Africa, lead to a limited representation of the wind climate.

3.2.1 Update of regional basic wind velocity data

Since the early 1990's, SAWS started to implement automatic weather station (AWS) technology on a large scale, as it was more cost-effective to implement in comparison to conventional weather stations, and also because it became possible to set up recording stations in more remote areas. The AWS measure wind speed and wind direction, as well as maximum wind gust and direction, within a five-minute time interval. At the beginning of 2007, a total of 172 AWS were operational, as depicted in Figure 3 (b). At least 125 AWS of this network have been in operation for 10 years. Therefore, using modern statistical extrapolation models for analysis of full-scale data more than 70% of currently operational wind recording station records can be used to calculate or estimate long-term wind statistics.

It is argued that in the light of the vast increase in the availability of measured data, as well as the accessibility of other wind data sources, such as high-resolution data from weather forecasting models in operation at SAWS, a revision of the South African wind climate models and statistics has become imperative. Such an update could however not be made in time for inclusion in Part 3 of SANS 10160.

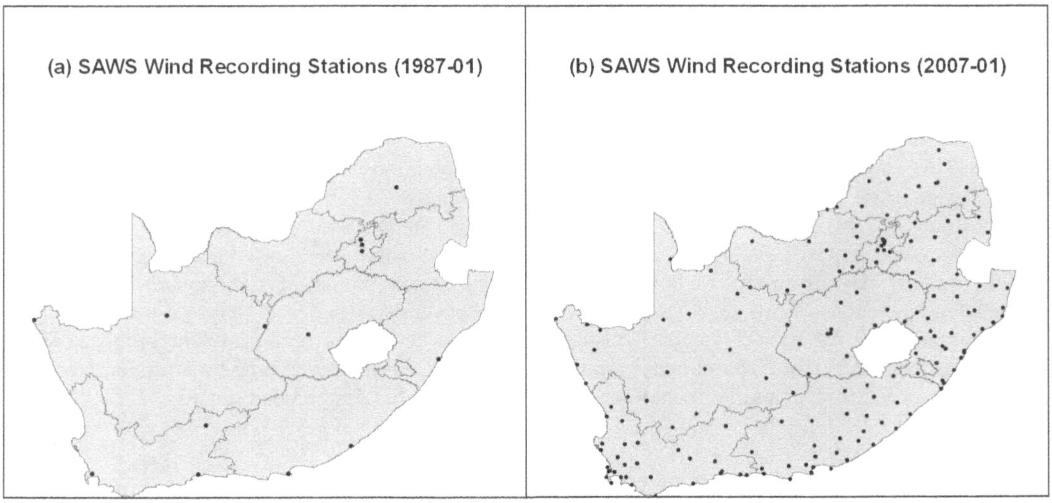

Figure 3 Positions of South African Weather Service wind recording stations

3.2.2 Representation of basic mean velocity

One of the fundamental and preset principles of Eurocode, with far reaching consequences, is the stipulation of a 10-min wind speed averaging period in respect to the basic mean velocity v_b. This requires that the South African strong wind climate needs to adopt the 10-min averaging time, which will enable proper alignment of SANS 10160 to EN 1991-1-4.

For mature winds, a conversion of the magnitude of wind speeds to correspond to various averaging periods constitutes standard textbook information (Simiu and Scanlan, 1978). In fact, it can be derived that a multiplier of 1.1 should be used to convert an hourly mean wind speed to a 10-minute mean, and a multiplier of about 0.7 to convert a 3-second peak gust wind speed to 10-minute mean wind speed.

However, in South Africa a problem exists with converting the wind speeds in various areas of the country, as demonstrated by Table 1, in which a comparison is made (based on the SABS 0160-1989 wind maps) between the coastal and inland areas of the country.

Table 1 indicates large differences in the Gust Factor, and thus the complexity of the South African wind climate, in which the strong coastal winds originate in mature frontal storms, while strong inland winds develop as a result of convective activity (see Figure 2).

Table 1 Comparison of wind speeds

Area	Regional maximum basic wind velocity (m/s)		Gust Factor
	3-sec gust	Hourly mean	
Coast	40	25	1.6
Inland	40	20	2.0

As discussed above, for coastal areas dominated by mature storms, conversion factors of 1.1 or 0.7 can be used to determine the magnitude of the basic wind speed in terms of the 10-minute averaging period from the 3s and 1h averaging times respectively. Either way, a value of 28 m/s is obtained (i.e. 1,1 x 25 = 27,5 or 0,7 x 40 = 28).

This conversion, however, cannot be applied to inland areas, as two different values of the basic design wind speeds would emerge (1,1 x 20 = 22 m/s or 0,7 x 40 = 28 m/s). This situation demonstrates the inability of design wind data, expressed in terms of large averaging periods, to adequately represent wind climates dominated by extreme winds of short duration generated in thunderstorms. The 10-minute or hourly mean wind speeds are suitable for application to climates dominated by mature storms (e.g. Europe) and not by convective weather systems as in the case of Australia or South Africa. This issue is also emphasised in the recent versions of the proposed ISO DIS 4354 standard in which, for climates dominated by thunderstorms, averaging times of 3 sec or less are recommended for a meaningful determination of extreme value design wind speeds.

3.2.3 Regional map of basic wind velocity

For the purpose of developing the new revision of the code a regional map of wind velocity based on the 10 min averaging period is required. A re-analysis of the full-scale data was not a feasible option due to the inherent difficulty of the proper representation of wind speed in short duration windstorms originating in thunderstorms. It was decided to adopt the *actual* fundamental basic wind speed of 28 m/s (i.e. 0,7 x 40) for areas dominated by frontal winds and an *effective* wind speed of 28 m/s for areas dominated by thunderstorm activities. Such a regional wind map would be suitable for quasi-static wind loads across the country.

As a result, for most of the country a basic wind speed $v_{b,0}$ = 28 m/s was stipulated. For parts of the Karoo (Calvinia – Brandvlei – Victoria-West) values of 32 m/s (0,7 x 45 = 31,5) and around Beaufort West 36 m/s (0,7 x 50 rounded up) were adopted. Mapping of the 32 m/s and 36 m/s isopleths was also revised.

The revised layout, as presented in Figure 4, differs from the one included in the current wind speed map of the 3-sec gust wind speed. This modification was derived by taking topographical characteristics of the southern and south-western Karoo into account, as well as directional prevalence of winter storms and the wide range of climatic regions in South Africa (Kruger 2004).

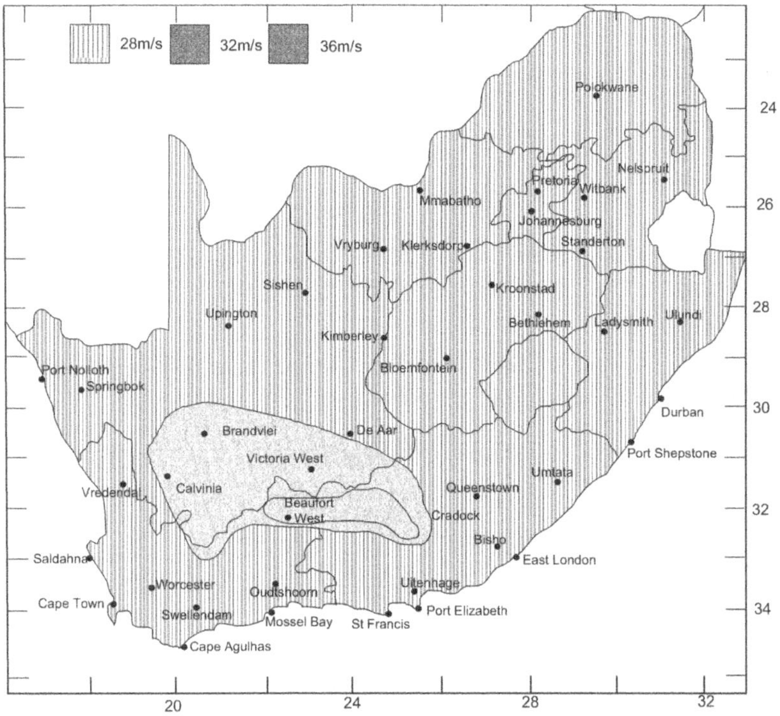

Figure 4 Proposed map of fundamental value of the basic wind speed $v_{b,0}$

3.3 Reference Wind Velocity

For historical reasons, and due to the limitations of wind-tunnel instrumentation, all international design procedures are based on the principle of combining the loading data derived in terms of the mean pressure and force coefficients, with the wind velocities corresponding to short-duration gusts (typically 2-3 second). Therefore, all standards based on the regional mean wind velocities at some stage include a conversion to peak wind velocity or peak wind pressure.

In the Eurocode approach, peak wind velocity pressure is determined as a function of the mean and short-term wind velocity fluctuations, according to a recommended formula given below:

$$q_p(z) = [1+7I_v(z)]\frac{1}{2}\rho\cdot[v_m(z)]^2 \qquad (1)$$

Where $I_v(z)$ is the intensity of turbulence of the flow at an elevation z, with a recommended expression based on the classical inverted natural logarithmic formula. The recommended value for the average air density ρ is 1,25 kg/m^3 but adjustment to provide for its dependence on altitude above sea level, temperature and barometric pressure is allowed.

In essence, the formulation given in Equation (1) provides for the conversion of the mean pressure $\frac{1}{2}\rho \cdot v_m^2(z)$, corresponding to the mean wind velocity component of the flow, to peak pressure by applying a Peak Factor equal to $\left[1+7I_v(z)\right]$.

It is proposed that in the new version of the code, this conversion should be applied to the wind velocity level rather than converting the mean pressures to peak pressures. Such a conversion is more rational from the point of view of physics. Furthermore it constitutes a simplification of the approach adopted in Eurocode. Finally it enables the omission of the entire concept and formulation of wind turbulence.

The following formula is proposed to describe the peak wind velocity at a height z above the ground level:

$$v_p(z) \quad = \quad c_r(z) \cdot c_0(z) \cdot v_{b,peak} \qquad (2)$$

Where:

$v_{b,peak} = 1{,}4\ v_b$, which in effect, for most parts of the country, converts the basic wind speed v_b back to 3-sec gust wind speeds stipulated in the current version of the SANS 10160 (e.g. 1,4 x 28 m/s = 39,2 m/s).

$c_r(z)$ is the terrain roughness factor which describes the changes in wind profile with elevation and will be discussed in the following section, and

$c_0(z)$ is the orography factor, which accounts for the increase in wind speed over isolated hills or escarpments.

3.4 Air Density

South Africa is a country with a large topographical diversity and the development of the built environment is taking place at sites with a variety of altitudes above the sea level. Therefore the approach included in the current SABS 0160:1989 code should be maintained and a table relating the air density and site altitude should be included.

3.5 Terrain categories and wind profiles

Following the background discussions and comparative analysis of various international wind loading specifications presented in Chapter 3-1, a decision was taken to adopt and propose a unique set of boundary profiles which will largely be based on those stipulated in SABS 0160. This is in accordance with options of nationally determined parameters allowed by Eurocode. The decision was also based on the merit of applying the widely accepted and proven power-law format, currently used by SABS 0160.

3.5.1 Basis for boundary layer profiles for terrain categories

A set of reasonable measures was adopted to form the basis for the description of boundary-layer profiles. The measures did not provide for any reliability considerations. The basis for the specification of the system of terrain categories and their associated boundary layer wind velocity profiles consists of the following measures:

- **Number of terrain categories:** The practice of providing four terrain categories to describe the range of terrain roughness conditions is maintained.

- **Stipulated categories:** The categories are referred to as A, B, C and D, starting with the smoothest terrain (A) to the roughest (D). This is done to clearly distinguish the new categories from the current stipulations of SABS 0160 (categories 1 - 4) and Eurocode (categories 0 - IV).

- **Profile parameters:** The profiles of wind velocity with height are determined by maintaining the exponential function, and adopting the parameters corresponding to the 5 sec profiles (i.e. median profiles) stipulated in SABS 0160.

- **City terrain profile:** In line with modern international wind loading norms, the profile category 4 (city centre) is left out as being unrealistic, and nowhere applicable to South African conditions, due to the large distances required for fully developed city-terrain profiles to occur.
  Note 1: Due to the complexity of wind flow within this type of terrain, as well as the lack of adequate development length of the boundary layer, the specification of a city centre wind profile is largely artificial. Within large city centres the wind flow affecting specific sites is mostly determined by the neighbouring structures, which can amplify or reduce the wind loads, if compared with those corresponding to a structure considered in isolation. The only reliable method of investigating the wind loads of buildings located in large city centres is by wind-tunnel testing in which due consideration to modelling of the neighbouring structures and the boundary layer conditions, is given.
  Note 2: Following the Eurocode stipulations, a provision will be made in an annex to the proposed code in which a certain amount of reduction in loading can be considered in a case of closely spaced obstructions which encourage a vertical displacement of the wind profile.

- **Distribution of category profiles:** The terrain categories should be selected to provide an even distribution of profiles with roughness. Although the SABS 0160 profiles are generally followed, the wide gap between categories 2 and 3 should be rectified. This is done by inserting the fourth (new) category (C) between the profiles adopted from SABS 0160 category 2 (new category B) and category 3 (new category D). The values of the parameters are selected to provide the required evenness in the distribution of the profiles.

- **Cut-off height of profiles:** The values for the cut-off height of the profiles (z_c) at low elevations as stipulated in Eurocode are adopted. This will reduce wind loads on structures of low height as compared to the SABS 0160 stipulations.

3.5.2 Stipulated profiles

The parameters of the proposed wind profile at height z (e.g. gradient height z_g, reference plane z_0 and profile exponent α) as expressed in Equation (3) for the terrain roughness factor $c_r(z)$, modified as indicated above are presented in Table 2. The variation of $c_r(z)$ with height is presented in Figure 5.

$$c_r(z) = 1{,}36 \cdot \left(\frac{z - z_0}{z_g - z_0} \right)^{\alpha} \qquad (3)$$

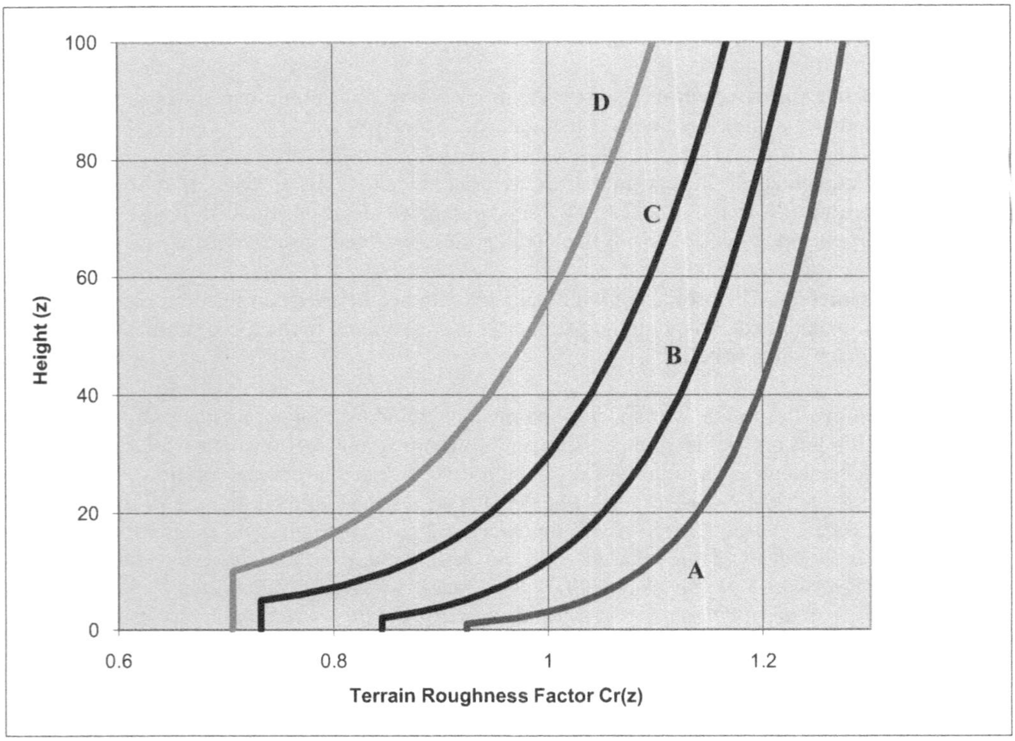

Figure 5 **Variation of the $c_r(z)$ factor with height above ground level for SANS 10160 Categories A - D**

Table 2 **Parameters of wind profile**

Terrain category	z_g	z_0	z_c	α
A	250	0	1	0,070
B	300	0	2	0,095
C	350	3	5	0,120
D	400	5	10	0,150
See text for definition of symbols				

A comparison of the revised profiles with the related SABS 0160 profiles up to a height of 50 m is shown in Figures 6. The comparison indicates slight shifts to lower wind velocities for SANS 10160 Category B as compared to SABS 0160 TC 2 and Category D versus TC 3 respectively. The interpolation of Category C of the wide gap between TC 2 and TC 3 is also clear. Lower velocities for building heights less than 5 m for Category A compared to TC 1, and for less than 4 m for Category B compared to TC2 are also apparent; for Category D wind velocities for buildings between 6 m and 10 m are however increased in the new stipulations.

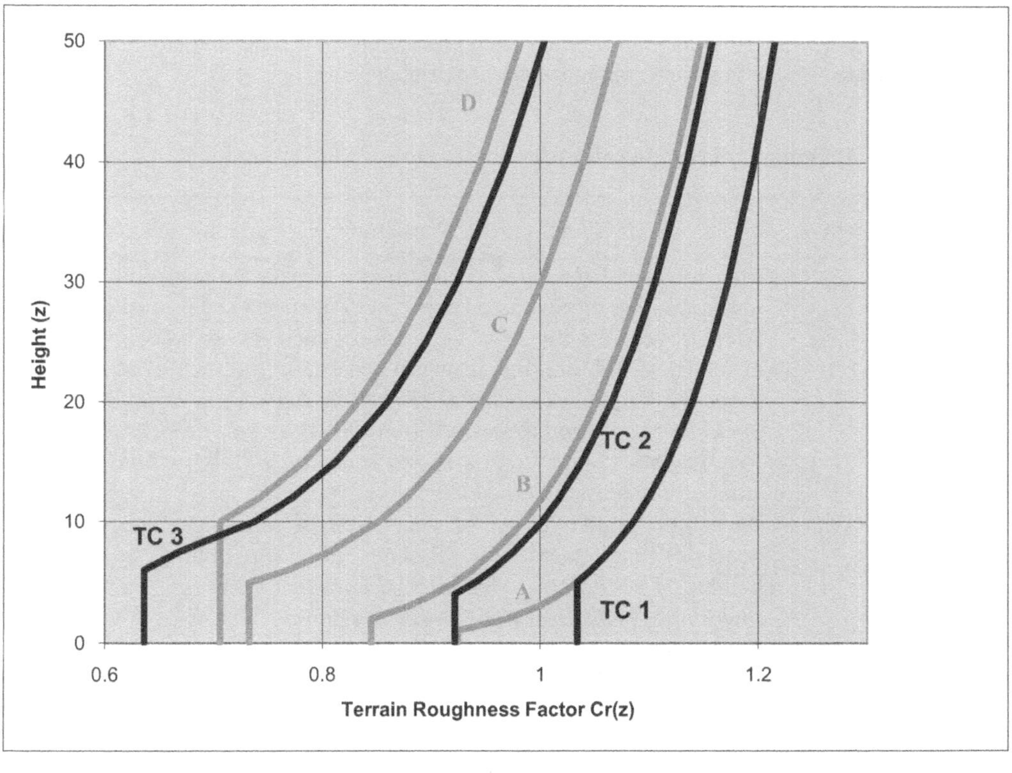

Figure 6 Comparison of revised SANS Categories A – D and SABS 0160 profiles TC 1 – 3

3.5.3 Assessment of profiles

The selection of wind profiles which are less conservative than those of Eurocode and even marginally so in comparison with the present stipulations is a deliberate decision in the light of the much higher wind loads that would result from the direct implementation of Eurocode procedures, as reported and discussed in Chapter 3-1.

Furthermore, as a result of adopting less conservative values of the cut-off elevations (below which the profile remains constant), the set of proposed profiles will markedly reduce the wind loadings at low elevations (below 5 metres) for terrain categories A and B, as compared to TC 1 and TC 2 of SABS 0160. This has significant implications for the design of one- and two-storey structures for exposed sites. For terrain categories C and D the profiles at low elevations will remain fairly similar (slightly more conservative) than current profile TC 3.

A systematic comparison of wind loads for the current code SABS 0160 and the revised procedures presented in SANS 10160 Part 3 for a representative range of structures was done. At this stage it suffices to indicate that the revised procedures generally result in higher wind loads as compared to present practice. This provides sufficient motivation for the moderated stipulation of the wind velocity profiles.

3.5.4 Description of terrain categories

The descriptions of terrain categories were formulated as a combination of those stipulated in Eurocode (with respect to the average height and spacing of obstacles), together with the descriptors of the current SA code. The definition of the terrain categories are given in Table 3. Pictorial information is also provided to assist with the interpretation of the descriptions.

Table 3 Definitions of terrain categories

Terrain category	Description
A	Flat horizontal terrain with negligible vegetation and without any obstacles (e.g. coastal areas exposed to open sea or large lakes)
B	Area covered with low vegetation, such as grass or isolated obstacles (trees, buildings) with a separation distance of at least 20 times the heights of obstacles. *This terrain includes large airfields, open parklands and farmlands and undeveloped outskirts of towns, with isolated trees. This is the category on which the fundamental value of the basic wind velocity,* $v_{b,0}$, *is based.*
C	Area with regular cover of vegetation or buildings with isolated obstacles with separation distances of a maximum 20 obstacle heights. *This category includes suburban terrain and residential townships consisting of one storey structures.*
D	Area in which at least 15% of the surface is covered with buildings with average heights exceeding 15 metres. *This category refers to fully developed industrial suburbs, towns and cities.*

Misinterpretation and miscalculation of terrain categories and corresponding wind profiles constitutes one of the most mistreated procedures in international loading codes, allowing for large amount of leeway in the design process. Designers are often not aware or choose to ignore the fact that the development of full-scale boundary layer profile requires significant distances. As a result, some structures may in fact be subject to 'smoother' profiles (i.e. more onerous loading) than those, which were adopted in their design.

3.6 Changes in Terrain Category

In a situation in which a change of terrain category occurs, the new wind speed profile does not develop immediately to its full height, but gradually with the increasing distance (i.e. fetch) from the place where the change takes place.

The South African code includes a relevant procedure, which stipulates the rules of transition and determination of the resultant wind profile. This procedure requires incorporation of two profiles (*'smoother'* and *'rougher'*) and determination of the representative profile over the transition elevations.

In order to avoid the problems raised above and also to sensitise the designers to the issue of large distances required for the transition of the boundary layer to take place, the Eurocode adopts a more conservative approach and provides two straightforward procedures. These procedures are based on a set of rules in which, for the transition from *'smoother'* to *'rougher'* terrain, a certain minimum distance is required, stipulated in terms of the height of the structure and types of terrain. If the minimum required distance is not met, then the *'smoother'* terrain category should be adopted in the design. The procedure should be applied to all wind directions, with a division into recommended angular sectors of 30 degrees.

It is proposed that the new revision of SANS 10160 should adopt the Eurocode specifications appropriate terrain category without amendment. Due to the conservative approach taken by Eurocode, alternative proven procedures are allowed, such as information derived from proven research, for instance wind-tunnel topographical modelling or full-scale measurements.

3.7 Directional Effects

The current SANS code does not consider directional design, i.e. it does not take into account the directional prevalence of full-scale strong winds in relation to the orientation of the structure. The Eurocode introduces a directional factor c_{dir} with a recommended value of 1,0 but allows the use of other values as nationally determined parameters.

In the proposed version of the SANS code a directional factor is not included. It is envisaged that this might be introduced in the future versions of the code once relevant statistical data for South Africa have been developed.

3.8 Immediate Surroundings

The surroundings of a specific site may have a significant influence on the characteristics of the approaching wind by introducing redirection, acceleration or deceleration of the flow, buffeting of gusts and vortices.

This is most applicable to situations where significant structures (large or tall) are located in the proximity of a site, typically within centres of large cities. The only reliable and feasible way of investigating this issue is by using wind-tunnel technology. Comprehensive full-scale measurements of wind characteristics approaching a site of concern provides an alternative approach, but it is however time consuming and costly.

The Eurocode offers some assistance to the designer in that respect by introducing two generic procedures to take the presence of neighbouring structures into account. Both procedures were presented graphically in Chapter 3-1.

Inclusion of both stipulations is proposed. The procedure for obtaining the vertical displacement of the profile due to the shielding effects due to a city environment should be noted in particular. Upon meeting a certain set of requirements, this stipulation may offer a significant reduction in the magnitude of the wind profile, which will effectively counteract the issue of removal of the option of an urban profile (category 4) as stipulated in SABS 0160.

3.9 Complex topography

The presence of pronounced topographical characteristics of the terrain can significantly modify the characteristics and magnitude of wind affecting the site of concern. EN-1-1-4 adopted a fairly modern approach for calculating the effects of dominant topography for cliffs and escarpments as well as for hills and ridges, as highlighted in the preceding paper. This approach was developed on the basis of extensive wind-tunnel and limited full-scale measurements. The procedure is based on a graphical simplification of the relationship between the geometrical parameters of topography in relation to the position of the site of concern in order to determine the magnitude of an orographic coefficient to be applied to the mean wind speed.

Apart from the graphical procedure outlined in Chapter 3-1, an additional set of exhaustive empirical formulae is given, so as to enable a calculation of the orographic location factor, and subsequently the orographic coefficient. Both procedures should yield similar results. Since the formulae are available in EN-1-1-4 they are omitted from the draft standard due to their redundant nature.

4 PRESSURE AND FORCE COEFFICIENTS

The assessment of the EN-1-1-4 provisions for pressure and force coefficients as reported in Chapter 3-1 led to the conclusion that the extensive set of situations and geometries for which an advanced set of coefficients is given, should be applied without any modification in the revised design procedure. The most important adaptation to be applied is to select the sets of coefficients which are relevant to the scope of SANS 10160 and specifically Part 3.

4.1 Design Data

A review of the relevant literature and initial comparison of the data on pressure and force coefficients indicated clearly that the extent of pressure coefficients stipulated in Eurocode is more elaborate than those specified in SABS 0160. This issue has been raised in the preceding Chapter.

In this context it is important to note that the SABS 0160 information on pressure and force coefficients is based on an outdated set of data introduced to the international loading standards more than 50 years ago (CP3, 1953). It was derived from wind-tunnel studies conducted in the 1940's and 50's, where no due consideration of boundary layer and other scaling parameters was taken into account. Furthermore, the instrumentation which was used in those days had low accuracy and frequency response rate.

The assessment of the set of pressure and force coefficients provided in EN 1991-1-4 reported in Chapter 3-1 concludes that the extensive range of structural geometries and configurations which are included, compensates abundantly for somewhat inconsistent presentation of the information. Incorporation of this modern and advanced information into SANS 10160 is in fact a major benefit of deriving the revised procedures from EN-1-1-4.

4.2 Correlation of Loads

The differences in the concept of applying the correlation of pressures over the loaded surfaces in SABS 0160 and EN-1-1-4 should particularly be noted. The equivalent gust profile approach incorporated in SABS 0160 implies that the maximum load can be generated by a gust profile enveloping the entire structure. The shortcoming of this assumption is that it ignores the effect of the size of the structure i.e. the fact that the larger the loaded area the smaller the correlation of gusts becomes. As compensation, the code introduces a division into classes of structures with different gust wind speed profiles (3, 5 and 10 sec.).

Furthermore, in order to accommodate the development of large magnitude pressures along edges and ridges it introduces the principle of local pressure coefficients applied to small areas in a separate analysis.

The approach of EN-1-1-4, which will be incorporated in the proposed SA code, is more logical and transparent. Being based on updated international loading information it specifies more elaborate and detailed loading zones. Furthermore, to accommodate the correlation of pressures, all pressure coefficients given in the code are given in terms of two values for loaded areas of $1m^2$ and $10m^2$ respectively. The differences in pressures between small and large areas are of the order of 50% and even more, as compared to the range of 4% - 8% differences for different building classes of the present procedures!

4.3 Application of Loads

Two relevant (and not necessarily obvious) issues regarding the differences in application of the loads between SABS 0160 code and the proposed approach need to be mentioned, namely:

- The application of the pressures on the windward walls of buildings
- The application of reference pressures at the top of the wall

The first issue has been presented in the preceding chapter and will not be discussed in detail – in a nutshell the current code provides two methods of application: accurate (based on the integration of the wind profile) and conservative (based on the pressures at the top of the building). The method used in EN-1-1-4, which is proposed also for Part 3, prescribes a distribution of zones in terms of the aspect ratio of the windward wall.

The second issue refers to the way in which the pressure coefficients of walls are being integrated with the free-stream dynamic pressure. SABS 0160 stipulates this in terms of the pressure q_z applicable to the height of the wall, while the EN-1-1-4 refers to the height of the building. For buildings with pitched roofs of a steep slope and the walls parallel to the ridge this may introduce significant differences in the resultant loading. The proposed Eurocode approach is more conservative. Since selection of the reference height is identified as a Nationally Dependent Parameter, maintaining the present practice of using the height of the wall in the new standard would be fully within acceptable Eurocode procedures. Such a procedure is also in agreement with that of ASCE-7:1998.

5 LARGE AND DYNAMICALLY SENSITIVE STRUCTURES

As the majority of structures designed in South Africa are dynamically insensitive, SABS 0160 does not consider dynamic effects apart for guidance on along-wind response and vortex excitations

EN-1-1-4 stipulates a c_sc_d dynamic magnification factor which combines the influence of non-simultaneous occurrences of peaks over the external surfaces of buildings i.e. lack of correlation of pressures (c_s) and dynamic magnification effects (c_d) due to the resonance between the turbulence of the flow and vibrations of the structure.

It is proposed that the presence of the structural factor (c_sc_d) should be maintained in Part 3, being a useful and informative quantity; which may also enable (and encourage) future modifications and further alignment with Eurocode. Furthermore, in view of the limitations of the scope of the code discussed in this paper, the structural factor (c_sc_d) should be adopted as unity. In line with the current code, brief data of an informative nature describing the types of dynamic excitations will be included as an annex.

The procedures presented in the draft standard are however limited to quasi-static structural response. By maintaining compatibility with EN-1-1-4, it would be possible to apply the procedures from that standard to cases where dynamic effects need to be taken into account. The provision is that the relevant South African wind climatic data should be used together with the EN-1-1-4 procedures. Such applications would require some wind engineering expertise and inputs, particularly in the case of structures in inland regions subjected to thunderstorm winds.

6 CONCLUSIONS

The main features of the revised procedures for determining wind actions for structural design, as presented in Part 3 of SANS 10160 consist of the following:

- **Scope of structures:** The SANS 10160 scope of structures is restricted to reduce the complexity whilst providing for most of the design situations encountered in general practice.

- **Compatibility with EN 1991-1-4:** Sufficient compatibility with the Eurocode procedures is maintained to consider the procedures to be sufficiently harmonised to consider them within the tolerances of Eurocode Nationally Determined Parameters; and to allow the use of advanced Eurocode procedures for South African conditions.

- **South African strong wind climate:** The representation of the South African strong wind climate is based on an adaptation of the SABS 0160-1989 specifications, with limited adjustments and improvements to represent the EN-1-1-4 format. The need to update the strong wind climate representation in South Africa is evident.

- **Application of wind loads:** The incorporation of updated and modern procedures for the conversion of the free-stream wind speed into wind loads is considered to be a major improvement on the revised procedures. The higher wind loads that generally result from the stipulations are accepted as deriving from updated stipulations based on improved information. The net effects of the revised stipulations are assessed in the following chapter.

Due to the complex nature of wind actions on structures and the consequent difficulty in providing clear and unambiguous stipulations, their application requires knowledgeable and responsible treatment. A number of steps in the process require particularly careful consideration due to their influence on the outcome of the design:

- **Terrain category:** Due to the wide range of terrain roughness conditions that are represented simply by a set of four terrain categories, this requires the careful and conservative selection of the appropriate category. The large distances required for full development of wind velocity profiles for rougher terrains should particularly be noted!

- **Topography of terrain and surroundings:** Consideration should be given to the application of wind tunnel testing or modelling for important or wind sensitive structures for complex topography or transition of terrain conditions.

- **Structural configuration:** More complex or composite geometry of structures which deviate from the basic and symmetric shapes provided for, would require careful and conservative treatment or wind-tunnel testing.

- **Structural response:** The possibility of dynamic structural response for both integral behaviour and locally for components or subsystems should be kept in mind.

ACKNOWLEDGEMENTS

The authors would like to acknowledge the research support of the CSIR, as well as the benefits of collective guidance of the SAICE Working Group on the Loading Code and in-depth discussions with Dr Rodney Milford on the implications of adopting the wind-speed averaging period.

7 REFERENCES

CP3: Chapter V: Part 2: 1952. Code of basic data for the design of buildings. BSI, 1952.

Day PW & Kemp AR (1999). *South African loading code: Past, present and future*. Civil Engineering 7(4): 11 – 12 (April 1999).

EN 1991-1-4:2005. Eurocode 1: Actions on structures, Part 1-4: General actions – wind actions. CEN, Brussels.

Goliger AM, Niemann H-J and Milford RV, (1998). *Assessment of wind-load specifications of the SABS 0160-1989*. South African National Conference on Loading. SAICE.

Goliger AM and Retief JV (2002). *Identification of zones of strong wind events in South Africa.* Journal of Wind Engineering and Industrial Aerodynamics, Vol. 90 (2002), pp.1227-1235.

Kruger AC (2004). *Climate of South Africa. Climate Regions.* WS45, South African Weather Service, Pretoria.

Milford RV (1987). *Annual maximum wind speeds for South Africa.* The Civil Engineer in South Africa, January 1987.

SABS 0160-1989 (as amended 1990, 1991 & 1993). *South African standard code of practice for the general procedures and loadings to be adopted in the design of buildings*. The South African Bureau of Standards, reprint 1994.

Simiu E and Scanlan R (1978). *Wind effects on structures: an introduction to wind engineering.* John Wiley& Sons Inc., New York / Toronto.

Part 4 – Seismic Actions and General Requirements for Buildings

4-1 Background to the Development of Procedures for Seismic Design

Wium JA

1 INTRODUCTION

This chapter provides the background to Part 4 of the revised Code, SANS 10160. It describes the procedures which were followed during the revision of the sections on seismic loading of SABS 0160:1989 and provides the motivation for specific issues considered during the revision. In this chapter reference is made to the "revised Code" to indicate the revised standard SANS 10160. SABS 0160:1989 is referred to as the "existing Code" or SABS 0160 for brevity.

2 BACKGROUND TO THE REVISION OF SANS 10160

Ever since the publication of the clauses on seismic loading in SABS 0160 designers in the seismic Zone I areas have considered the provisions to be unrealistic and too stringent. Zone I areas are those where structures need to be designed for a nominal peak ground acceleration of 0.1g.

A meeting held in 2003 with a group of designers in the Western Cape region revealed that some designers, although being aware of Code requirements, often apply the rules in such a way to suit economic pressures and requirements from Clients, rather than to fulfil the requirements of the Code. Although designers recognised the need to design for seismic loads, the broader community was not convinced and this often placed pressure on designers to opt for solutions which may not always have complied with the requirements of the Code. This practice often occurred due to the fact that designers did not feel convinced that the Code requirements were reasonable. It was felt that this situation was the result of inadequate consultation and involvement of the broader engineering community when the provisions were compiled, and without adequate information sessions through which the need for seismic provisions in the Code could be explained. A general lack of knowledge and commitment to seismic design of structures was expressed by several practitioners.

A need was thus established in 2003 to re-evaluate the 1989 code provisions (SABS 0160:1989) with three objectives in mind:

- to determine if the SABS 0160 code provisions were realistic.
- to gain the support of the industry by either confirming the code provisions, or by issuing a revision to the existing code.
- to improve the knowledge in the industry about the basic principles of seismic design.

3 GENERAL PROCESS AND THE APPROACH ADOPTED

Having recognized the historical objections to and lack of confidence in the existing Code amongst designers, it was decided in 2004 to establish a local seismic load working group in the Western Cape region, reporting to the Working Group responsible for the revision of the Loading Code. The group consisted of academic staff from the University of Stellenbosch and of representatives from consulting engineering design firms in the region. The chosen process ensured that designers became involved in the process themselves, thereby creating legitimate support for any revisions to the Code. The Western Cape working group had monthly meetings and reported to the Working Group of the SA Institute of Civil Engineering tasked with the revision of the entire loading code SABS 0160.

Once established, the Western Cape working group embarked on a process of revision of the seismic provisions of SABS 0160. Considering the time-frame available to produce a document it was decided to use the existing Code as a basis, and to revise provisions where sufficient information merited such change. On issues for which justification did not exist, or which required additional research input, it was broadly decided that the fall-back option would be to retain the provisions in the existing Code. The objective would be to identify those issues which need more clarification, eventually to be addressed in a longer term research program.

The members of the Western Cape loading code working group had among them experience of seismic design of structures using a number of different codes. These included the Uniform Building Code (UBC:1997), ACI 318-02 (2002), Eurocode 8 (2003) and SABS 0160:1989. Therefore, although the Working Group for the revision of the Loading Code had decided to use the Eurocode (EN-1990:2002, EN-1991) as reference document, it was decided that information for the Part of the Code on seismic loading would be borrowed from different sources. The sequence of paragraphs and the layout of the Part on seismic loads would however follow that of EN 1998-1:2004 : *Design of structures for earthquake resistance – General rules: Seismic actions and rules for buildings* (for brevity referred to as EN 1998-1).

In the following paragraphs the background and reasoning behind the clauses in Part 4 of the revised Code SANS 10160 are presented.

4 SEISMIC DESIGN PHILOSOPHY

International codes are based on the philosophy that structures shall be designed and constructed to withstand design seismic loads by retaining their structural integrity and a residual load bearing capacity after the seismic events without local or global collapse. The design seismic action is generally expressed in terms of the seismic action associated with a 10% probability of exceedance in 50 years or a reference return period of 475 years (EN 1998-1).

Furthermore, the structure shall be designed and constructed to withstand a seismic action having a larger probability of occurrence than the design seismic action, without the occurrence of damage and the associated limitations of use. In the case of the Eurocode (EN 1998-1) the seismic action to be taken into account for the "damage limitation requirement" has a 10% probability of exceedance, in 10 years, or a reference return period of 95 years.

The above philosophy is also adopted for the revision of SABS 0160.

5 SOUTH AFRICA AND SEISMIC DESIGN OF STRUCTURES

The Southern African region is known for its relative seismic stability, where only a small number of medium intensity earthquakes have occurred since the seventeenth century. On the other hand, between 40 and 60 tremors occur monthly, focused primarily in the gold mining areas of Gauteng, North West Province and the Free State. Although the effects of these events are far from being as serious as those caused by the larger earthquakes, extensive damage has occurred in one or two cases (Milford and Wium 1991). For a more comprehensive discussion on the seismicity in the South African region, refer to Kijko *et al* (2003) and Milford and Wium (1991).

SABS 0160 provides a seismic hazard map as developed in 1987 and 1989 based on a peak ground acceleration with a probability of exceedance of 10% in 50 years, also presented here as Figure 1. A more recent update of this map was published in 2003 by the South African Council of Geoscience, shown in Figure 2. The seismic-event catalogue used to publish this map has been extracted from the database of the Council for Geoscience. The original Geoscience catalogue had been compiled from many different sources, covering a period of time from 1620 to December 2000. The catalogue consists of natural, as well as mining induced seismicity (Kijko et. al. 2003).

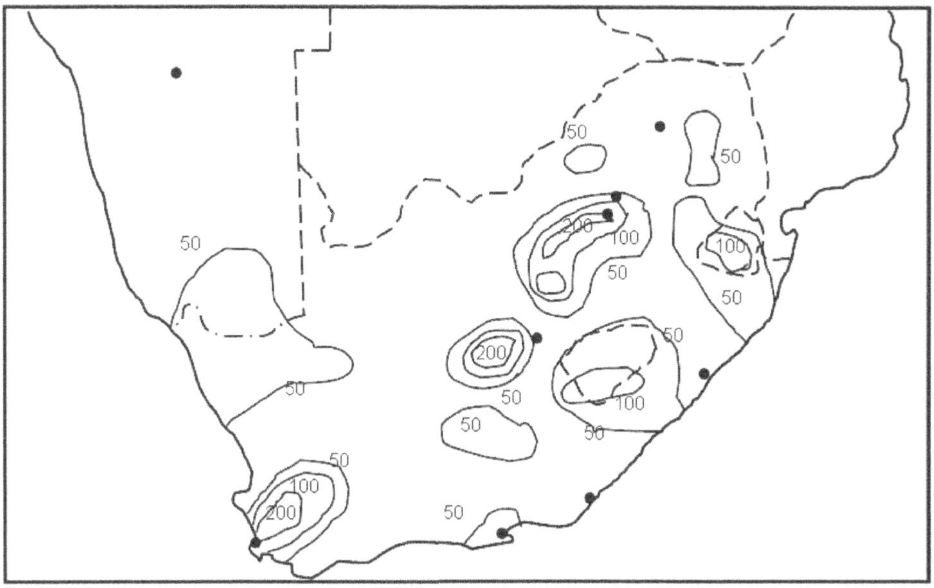

Figure 1 : Seismic hazard map from SABS 0160:1989 showing peak ground acceleration with 10% probability of exceedance in 50 years

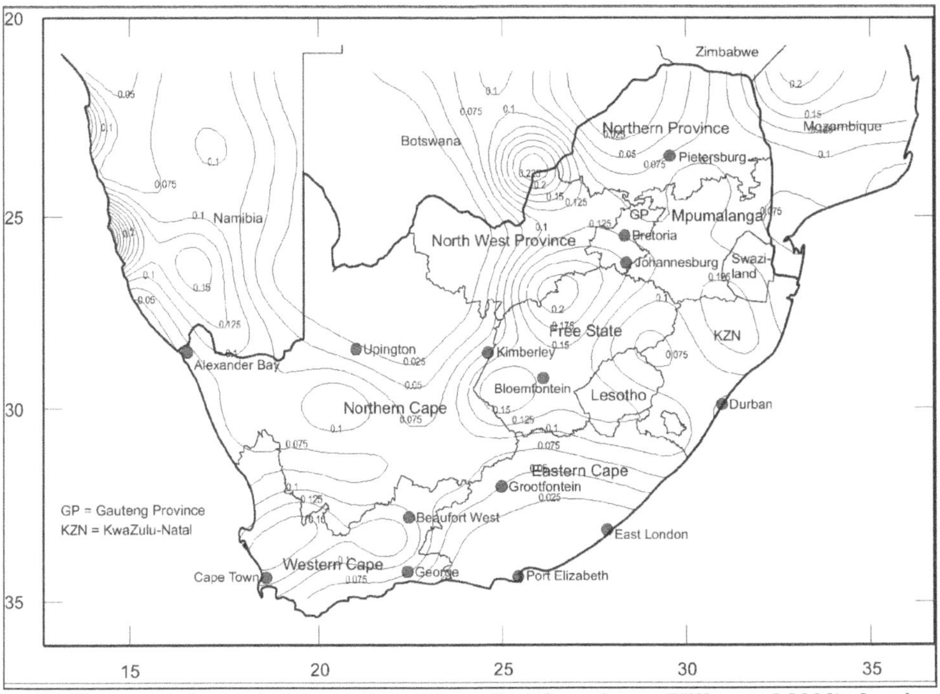

Figure 2 : Contours of seismic hazard from Council of Geoscience (Kijko et al 2003) showing peak ground acceleration with 10% probability of exceedance in 50 years

The seismic active areas in South Africa are broadly divided into two groups, namely those due to natural seismic events, and those due to mining activity. Clearly, regions for mining induced activity are located in the mining areas of the country. It has been shown that mine tremors are not likely to produce any significant structural response for buildings with natural frequencies of vibration less than 2 Hz. Stiff structures such as low-rise load-bearing masonry structures are therefore influenced most by mining tremors (Milford and Wium 1991). Response spectra were developed by Milford and Wium for typical mining induced seismic events. The spectra are compared in Figure 3 to the Type 1 elastic soil response spectrum from EN-1998-1. It can be seen that only structures with a very low first natural period (high natural frequency) will be affected significantly by mining induced activity.

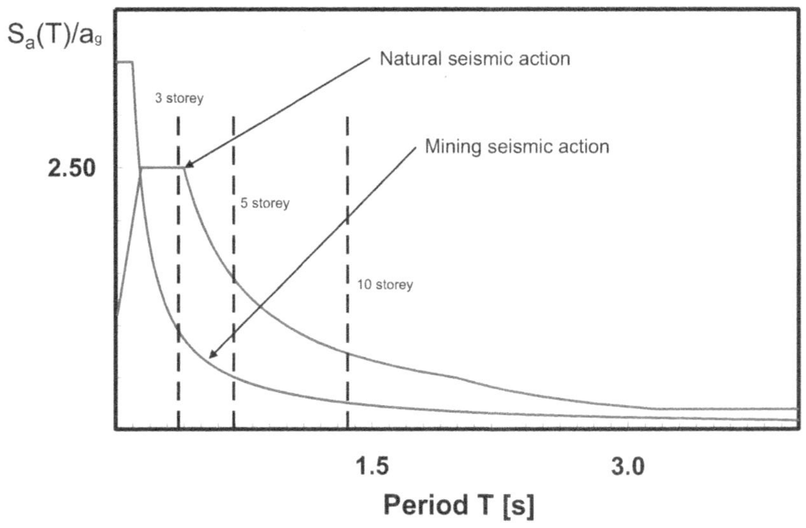

Figure 3 : Comparison of response spectra for mining induced seismic events (Milford and Wium 1991) and natural induced seismic events (EN-1998-1:2004)

Two seismic zones were defined in SABS 0160 to be considered in the design of building structures. The two zones are those areas associated with seismic activity of natural origin, and those associated with mining related activities.

Zone I was defined as those areas with *natural* seismic activity and with a magnitude of peak ground acceleration sufficiently high to warrant structural design to consider seismic action as a design load. Zone I covers a small area of the country and, although there are small patches in less populated areas, the most significant populated area is that in the province of the Western Cape which includes the Cape Town Metropolitan area. A nominal peak ground acceleration value with a probability of exceedance of 10% in 50 years was taken as the ultimate limit state design intensity. Although local peak ground acceleration values of more than 0.2g are shown in Figure 1, the existing Code specified a nominal peak ground acceleration value of 0.1g for Zone 1. The reason for this choice is not clear.

Zone II is defined as those areas related to *mining induced* seismic activity. This area is located in a limited region towards the west of Johannesburg, located on the gold mining area. As described above, the *mining* related seismic activity is of a different nature from that of *natural* seismic activity. The frequency content of mining related seismic events is much higher, and mainly influences structures with a high natural frequency (Milford and Wium 1991). Structures located in this Zone are only required to comply with certain layout requirements and with provisions for non-structural components.

Although the seismic hazard map published in 2003 (Figure 2) shows peak ground accelerations of up to 0.2g and more, the information in the 2003 seismic hazard map does not differ substantially from that in the 1987 and 1989 map (Figure 1). The maximum values in the mining areas are up to 0.2g in the 2003 map, which is similar to the 1989 information. However, the maximum peak ground acceleration values are 0.15g in the Western Cape as opposed to 0.2g shown in the 1989 map. It should be noted that the values shown in Figure 2 have been obtained from the Council of Geoscience. The Council plotted the same information in a coloured format where higher vales are shown (Kijko et. al. 2003). The working group was advised by the Council of Geoscience that the contour plot (Figure 2) is a better representation of the values (Kijko 2008).

The definition of a representative peak ground acceleration value for a structural design code requires that for a specific region a large variety of parameters needs to be considered. These include:

- The determination of the magnitude of an event at a specific site with a probability of occurrence of 10% in 50 years.
- An understanding of the influence of the geology and site soil conditions prevalent in the area.
- A thorough understanding of the risks involved to the infrastructure and to the inhabitants in the region.
- A consensus by stakeholders (scientists, engineers, politicians, communities) on the level of risk considered to be acceptable.
- For an acceptable level of risk to be agreed upon, decision makers need to be aware of the consequences of their decisions. The possible damage to infrastructure and risk to lives need to be quantified in order to make a decision.
- Knowledge of the impact of a decision on the local economic environment, and an acceptance of these effects.

The Working Group for the revision of the seismic loads in SABS 0160 is not in a position to make a decision based on the above information on a value to be used as design peak ground acceleration for any region in the country. They are well aware that a need exists to quantify parameters as set out above, and they support actions which can lead towards a better comprehension of the risks involved.

With this in mind, a decision nevertheless had to be taken for the definition of the peak ground acceleration for Zone I areas. Figure 4 shows the seismic zones defined for the revised Code. The following reasoning gives the background for the zones presented in Figures 4 and for the choice of a peak ground acceleration of 0.1g to be used in these areas. It is suggested that any change to this value needs to be motivated with the above parameters in mind, none of which are currently quantifiably available.

It is well recognized that the seismic activity in large parts of the interior of the country has a mining origin, specifically in the Witwatersrand and the northern Free State. It is also accepted international practice (EN 1998-1) to design structures for seismic loads when the nominal peak ground acceleration values (1:475 years) exceeds a value of 0.1g. It would therefore have been reasonable to choose the magnitude of 0.1g as the value to define the seismic zones in the country. Such a choice is shown in Figure 5 where the areas are shown with nominal peak ground acceleration of more than 0.1g. However, it is recognized that mining related activity has, at most, an influence of 120 km to 150 km from the origin of an event (Kijko 2008). The Zone II area (mining induced) would be limited to the close proximity of mines. From this information it is evident that if the seismic zones are defined on the basis of a peak ground acceleration of more 0.1g, then a very large part of the interior of the country would now subjected to the design of structures for seismic loads, where this has not been the case before. A significantly increased Zone I area (natural induced) would be the result based on these parameters.

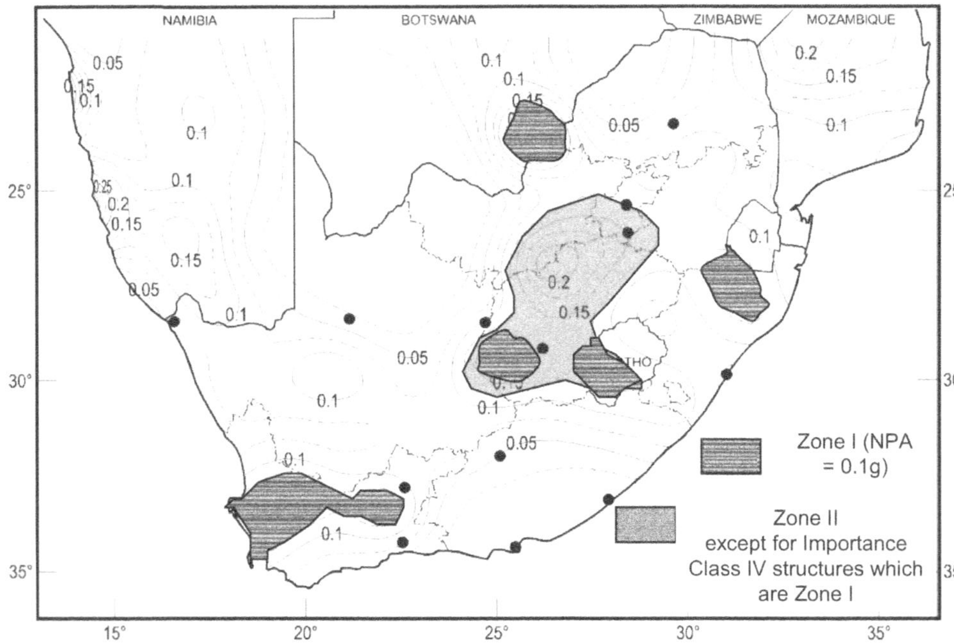

Zones chosen using NPA > 0.125g

Figure 4 : Seismic zones as defined in the revised Code for South Africa.

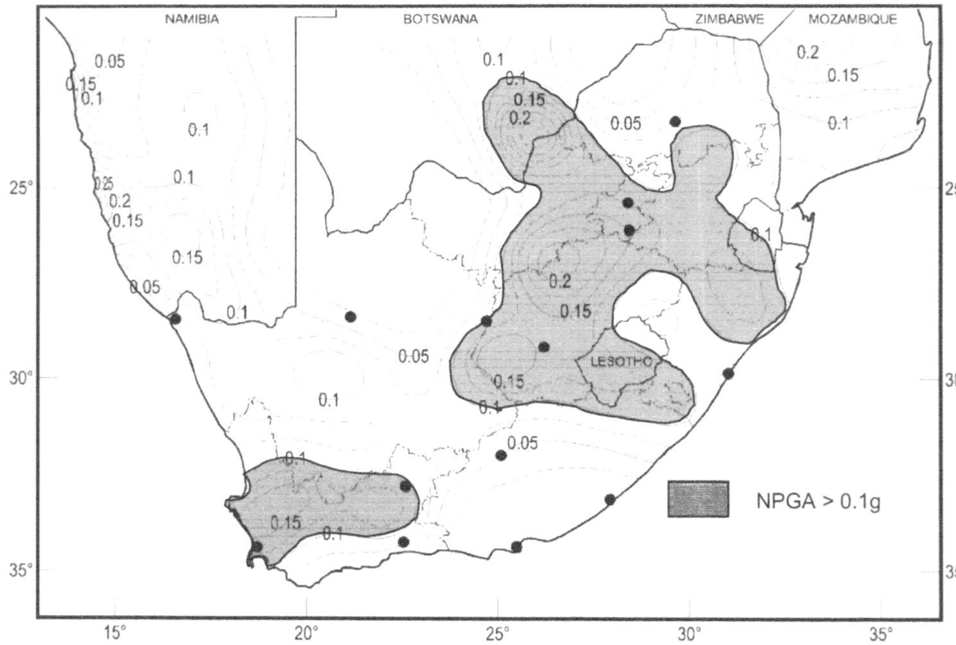

Figure 5 : Zones in South Africa with nominal peak ground acceleration of more than 0.1g for a 10% in 50 years probability.

For this reason, it was decided to choose the magnitude of peak ground acceleration of 0.125g as the value to be used for the definition of seismic zones. These are the zones shown in Figure 4. The only area for which an exception was made, is the Zone I region in and around Lesotho. Recognizing that the seismic activity in this region is due to natural activity, and that this region would not be included in the Figure 4 zones if 0.125g is used as the criterion for defining zones, a Zone I has been chosen in this region to coincide with the Zone I area in the 1989 map. It was furthermore decided to specify that structures of Importance Class IV (see Table 1 below) shall be designed for a peak ground acceleration of 0.1g in the Zone II area.

Considering that the revision of the loading code was initiated from the criticism by designers that the design earthquake was already too conservative, it was deemed inappropriate to increase the magnitude of the nominal peak ground acceleration from 0.1g to 0.15g (maximum values in the Western Cape, refer to Figure 2). For this reason the nominal peak ground acceleration for Zone I areas (natural induced seismic events) was chosen as 0.1g. The specification of the nominal peak ground acceleration was however considered together with the specification of a redundancy factor, as discussed in Paragraph 7.3 below.

6 CONCEPTUAL DESIGN

A report on the Erzincan earthquake in Turkey in 1999 (Earthquake Hazard Centre Newsletter 1999) states that sophistication of calculations and designing for a greater total base shear force do not necessarily lead to improved earthquake resistance. Considering the possible magnitude of seismic events in South Africa, it is more important to instil in designers and developers the principle that applying correct structural concepts and appropriate detailing of structural elements will be more effective than extensive calculations based on a flawed concept.

The revised Code therefore provides the basic principles of conceptual layout for acceptable behaviour under seismic loads. These principles have to a large extent been taken from the existing SABS 0160, supplemented with additional guidelines from EN-1998-1.

7 STRUCTURAL DESIGN AND DESIGN LOADS

Several factors play a role in determining load values for the design of a structure. This paragraph presents the concept of a seismic action acting from any direction, as well as applicable load factors to be used in design, the design methods and a redundancy factor which is used to increase the magnitude of the design seismic action.

7.1 Direction of seismic action

The total design load is a result of seismic action which can occur in any direction, including vertical. The occurrence of a seismic event may not coincide with the principal axes of the structure. This is taken into account by the combination of seismic effect in two perpendicular horizontal directions as in the existing SABS 0160. The requirement is that 100% of the seismic force be applied in any horizontal direction simultaneously with 30% of the seismic forces from a perpendicular direction. This is similar to the requirement in the existing Code.

In general, structures are designed for gravity loads, and the vertical component from an earthquake normally has a negligible influence on the design of the structure for seismic events of light to moderate magnitude. The Code requires consideration of the vertical component only when a seismic event with peak ground acceleration of more than 0.25g is considered (EN 1998-1).

7.2 Load factors

In order to accommodate the magnitude of accelerations and resulting displacements during a seismic event, structures are assumed to behave in a non-linear fashion to absorb the energy from the seismic event. The extent to which non-linear behaviour can take place is a function of the ability of the structural elements to accommodate non-linear deformations in a ductile manner. However, calculation methods commonly available to designers, are based on linear analysis techniques, and it is seldom that full non-linear analyses with material non linearity are performed by designers.

For this reason, the concept of *behaviour factors* has been introduced in design codes (EN 1998-1; the Uniform Building Code UBC:1997; the New Zealand Standard NZS 4203:1992 (referred to as UBS & NZS for brevity). The *behaviour factor* is a measure of the ability of the structural elements to deform in a ductile manner. These factors are discussed in Paragraph 8.3 below. By the use of *behaviour factors*, material linear elastic analyses can now be performed, but it is important to remember that actual deformation will be much higher than those values obtained from the linear elastic analysis. The Code now includes a formula for calculating the non linear displacement of the structure following a linear elastic (material) analysis (UBC 1997).

SABS 0160 requires buildings subject to seismic loading to be designed for the ultimate limit state with a partial load factor of 1.6 on the seismic load effect, including a load factor of 1.2 for dead load effects. Both EN 1990 and NZS require verification of buildings in the serviceability and the ultimate limit states. For the ultimate limit state a seismic event is considered in the same manner as accidental loading with load factors of unity on the seismic load effect.

Based on this approach of considering a seismic event as an accidental condition, the revised Code now uses a load factor of unity in the ultimate limit state for the seismic condition, including both seismic loads and permanent loads. The load combinations applied to seismic and accidental design situations are discussed more extensively in Chapter 1-1 and Chapter 1-2.

For the verification of the structure in the serviceability limit state (deflection verification), it is recognized that a seismic event with a smaller return period should be used (Paragraph 4 above). By assuming the response spectra for the reduced return period to have a similar shape to that for the design (ultimate) spectra, and considering that a material linear elastic analysis is performed for the verification of the structure, a factor can be used to determine the displacements for the "damage limitation" criteria (EN-1998-1). This approach is followed in defining the damage limitation criteria. The damage limitation criteria adopted for the Code have been obtained from the UBC in view of the fact that behaviour factors have mostly been based on the same reference. The damage verification is therefore performed using the results from the linear elastic analysis in the ultimate limit state, but verified against damage criteria which have been adjusted to account for the reduced return period.

7.3 Redundancy factor

A redundancy factor is included in the Code with the purpose of awarding designers when longer or more shear walls are used, enabling them to use a lower lateral design force for longer walls. The formula is obtained from the UBC with lower and upper limits in the revised Code of [1.2 < redundancy factor < 1.5] (as opposed to the lower limit of 1.0 in the UBC). The reason for the lower limit value being taken as 1.2 is that this factor provides an additional factor on the value of the peak ground acceleration chosen as 0.1g (whilst the values shown in Figure 2 have a maximum of 0.15g for the Western Cape). Furthermore, although the redundancy factor is generally smaller for multiple columns (moment resisting frames) it is considered that this type of construction is not effective for seismic design and is very seldom used as a structural system.

7.4 Design method

The design method for buildings subject to seismic loads in the existing Code is based on the equivalent static lateral force procedure. A procedure is given to calculate the total horizontal nominal seismic base shear force together with a method which is also used in several other codes (EN 1998-1: UBC: NZS:), to determine the vertical distribution of the load over the height of a building. In the existing code no other design method is mentioned and no limitations are set for the use of this method. The result is that structures can potentially be designed on the basis of this method, regardless of structural concept, possible irregularities in elevation, plan or other characteristics.

On the other hand, EN 1998-1, UBC and NZS, provide clear limitations on the use of the equivalent static lateral force procedure. These codes all require dynamic analyses either in the form of the modal response spectra method or a time history analysis for those structures which do not meet the limiting criteria of the equivalent static lateral load method. The requirements relate to regularity in plan and regularity in elevation, to building height, and to the magnitude of the fundamental period of vibration of the building. The UBC allows the use of the equivalent static lateral force procedure for some categories of buildings in seismic zones with low nominal peak ground accelerations, regardless of regularity, or fundamental period of vibration.

In the revised Code the limitations of the equivalent lateral static load procedure are now clearly defined. For buildings outside the scope of these limitations, designers are referred to specialist literature for other more advanced design methods. It is expected that designers would refer to EN 1998-1 for more information. The purpose of not providing details for alternative design methods is an attempt to force designers to design buildings based on acceptable conceptual principles for buildings in seismic zones. It is also an attempt to prevent designers without the necessary knowledge and training, from blindly using available software packages where such possibilities are offered.

It is important to remember that the equivalent static lateral load method is based on the assumption that the mode shape of the first natural frequency dominates the dynamic behaviour. It also assumes that this mode shape follows the displaced shape of a cantilever subjected to a uniform distributed load (Clough and Penzien 1993). Buildings for which the mode shape of the first natural frequency does not comply with these assumptions can not be designed using this method.

The following limitations are set in the revised Code for the use of the equivalent static force method:

- The fundamental period of vibration $T_a \leq 4.T_0$, or $T_a < 2.0\,\mathrm{s}$
- All lateral load-resisting systems (cores, walls, frames) run without interruption from their base to the top of the building, or if setbacks are present at different heights, to the top of the relevant zone of the building.
- Both lateral stiffness and the mass of the individual storeys remain constant or reduce gradually, without abrupt changes, from the base to the top.
- The sum of setbacks at any storey is less than 30% of the plan dimension at the first storey and less than 10% of the previous plan dimension.
- The plan layout of the building and the distribution of mass are approximately symmetrical about two orthogonal directions and without significant discontinuities throughout the height of the building.

The method provided in the revised Code for distribution of the total base shear over the height of the building is similar to those in international codes EN 1998-1, UBC and NZS. It differs only marginally from the formula in the existing Code, with the notable omission of some power factors in the calculation of the distribution of the lateral loads over the height of the building.

8 CODE PARAMETERS

8.1 Importance of buildings

It is internationally recognized that reliability differentiation needs to be made between structures with different occupations. This differentiation is implemented by classifying structures in different importance classes. Each importance class corresponds to a higher or lower value of the return period of the seismic event. Importance factors in SABS 0160 [1989] varied between 0.9 and 1.2. In EN 1999-1 and in NZS, structures are classified into different risk categories. To each category an importance factor is assigned which varies between 0.8 and 1.4 (EN 1998-1) and between 0.6 and 1.3 (NZS). The categories as defined in EN 1998-1 are shown in Table 1 and have been adopted for the revised Code.

Table 1 Importance classes for buildings (from EN 1998-1:2004)

1	2	3
Importance class	**Buildings**	**Importance factor** γ_I
I	Buildings of minor importance for public safety, e.g. agricultural buildings, etc.	0,8
II	Ordinary buildings, not belonging to the other categories.	1,0
III	Buildings for which seismic resistance is of importance in view of the consequences associated with a collapse, e.g. schools, assembly halls, cultural institutions, etc.	1,2
IV	Buildings for which integrity during earthquakes is of vital importance for protection, e.g. hospitals, fire stations, power plants, etc.	1,4
Note — The numbering of importance classes differ from those in the Eurocode where from these definitions were taken.		

8.2 Response spectra and soil types

The response spectra in the existing SABS 0160 were obtained from the ATC 3-06 (1978) code together with three soil types described in the same document.

Response spectra in international codes have since been revised (EN 1998-1, UBC and NZS) and are presented for more soil types. A comparison of the response spectra for different soil types are shown in Figure 6 where the spectra from EN 1998-1 (Type 1), UBC and the existing Code SABS 0160 are compared.

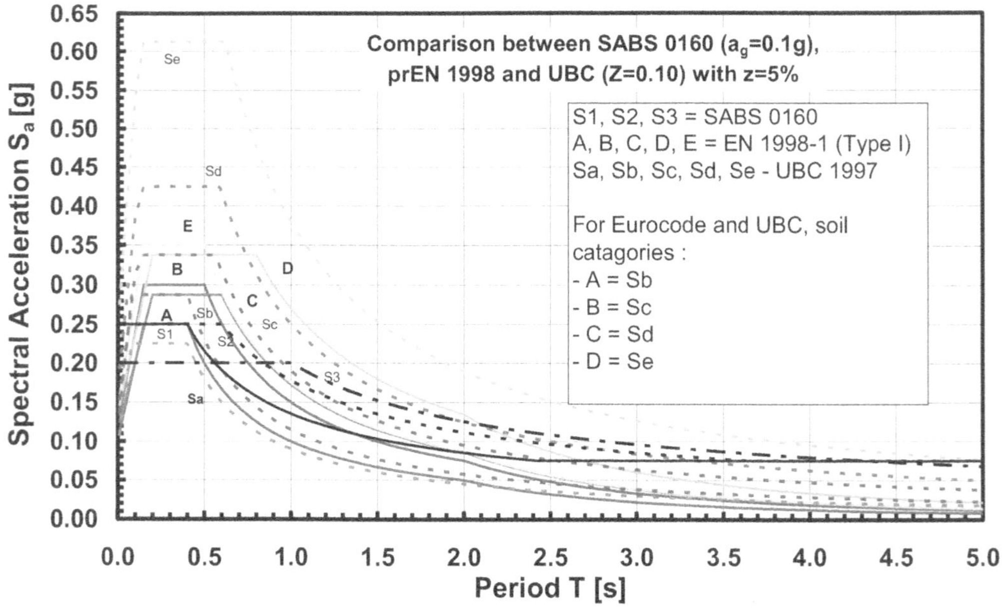

Figure 6 : Comparison of the response spectra from different codes and soil types.

From Figure 6 it can be seen that the spectra from the existing Code are vastly different from those in the other codes, especially for softer soils. Until better information becomes available about spectra for South Africa, it was decided to adopt the spectra from EN 1998-1 for the simple reason that the maximum values are less than those of the UBC.

EN 1998-1 provides Type I and Type II response spectra. A comparison between the response spectra for Type 1 and Type 2 soils is presented in Figure 7. EN 1998-1 distinguishes between the two response types as follows :

> *"If deep geology is not accounted for, the recommended choice is the use of two types of spectra: Type 1 and Type 2. If the earthquakes that contribute most to the seismic hazard defined for the site for the purpose of probabilistic hazard assessment have a surface-wave magnitude, Ms, not greater than 5,5, it is recommended that the Type 2 spectrum is adopted."*

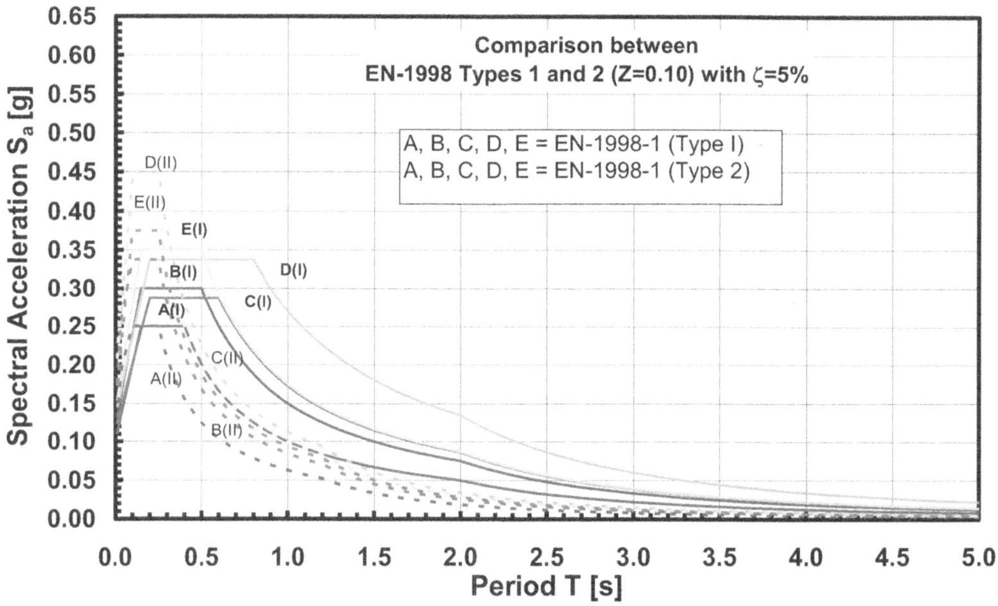

Figure 7: Comparison between response spectra for Type 1 and Type 2 soils from EN-1998-1.

It could not be ascertained which of the two response spectra types would be the most appropriate for South African soil and seismicity conditions. However, considering the high peaks at low periods in the Type 2 response spectra and the rapid reduction of values at higher periods of the same spectra, it was considered more appropriate to choose the response spectra Type 1 for the South African Code. It would appear as if Type 2 is closer to the mining type event. This may however not be specifically for mining.

Furthermore, after comparing Type 1 response spectra of EN 1998-1 with the Swiss Code (SIA 261:2003, UBC, and the existing SABS 0160, it was decided to base the revised response spectra for the Code on the Type 1 curves of EN 1998-1.

In order not to expose the designer to more uncertainty, it was decided to adopt the response spectra for a damping ratio of 5%. These are the curves now presented in the revised Code. Designers could however refer to EN 1998-1 for the equations which would allow other damping ratios to be used.

It must be noted that whilst the design spectra of EN 1998-1 were adopted for the revised Code, in these spectra allowance is made for the ductility of the structure to enable energy dissipation through plastic behaviour. EN 1998-1 also provides the elastic spectra which can be used for more advanced methods such as push-over analyses which are useful for evaluating existing buildings. These procedures are however not included in the revised Code.

8.3 Behaviour factors

In order to avoid explicit inelastic structural analysis in design, design codes allow the capacity of the structure to dissipate energy mainly through ductile behaviour of its elements. This is achieved by performing elastic analyses based on forces from seismic response spectra, but which have been reduced using *behaviour factors. Behaviour factors* are representative of the ductility of members that make up the structural system. The *behaviour factor* is basically the ratio between the ductile non-elastic deformation capacity and the linear elastic deformation capacity of a member (Chopra 2002).

The appropriate *behaviour factors* for a structural system are determined for reinforced concrete structures by careful detailing of the members. The correct reinforcement detailing enables a structural element to achieve a required degree of plastic deformation. Although reference is made in the existing SABS 0160 to a need for "sufficient stirrups" in beam and column elements to achieve ductility, no guidance is provided to the designer on what is considered to be "sufficient".

A comparison between behaviour factors in the existing code SABS 0160 and those in some international codes (EN 1998-1, UBC and SIA 262, shows that some of the values in SABS 0160 are higher and some lower than those of other codes (see Table 2). In general, the values in international codes are dependent on the reinforcement detailing rules, minimum sectional dimensions, and on the amount of ductility which is achieved by following the rules. More stringent detailing rules allow the use of higher behaviour factors. Behaviour factors also depend on the choice of response spectra, and load and material factors of the different codes. A direct comparison as shown in Table 2 is therefore not always possible and the selection of appropriate behaviour factors needs to be considered with care. In the existing code SABS 0160, behaviour factors for reinforced and prestressed concrete construction of moment frame systems correspond generally to behaviour factors in EN 1998-1, for the case when no special detailing rules are applied. The opposite is however true for buildings with shear walls. For such structural systems, the existing code SABS 0160 uses a significantly higher behaviour factor, although no special detailing rules are specified. From comparison with the codes EN 1998-1 and UBC, it is evident that special detailing rules are required to justify the behaviour factors as used in the existing code SABS 0160. This omission in the existing code needs to be addressed in the revised Code.

Ideally the reinforcement detailing rules should form part of materials codes (SABS 0100:2003). Unfortunately, a recent revision of SABS 0100-1 was completed before these rules could be proposed. For this reason, it is considered appropriate to present the reinforcement detailing guidelines as a normative Annex in the revised Code.

Ductility of members is achieved by using reinforcement with sufficient ability to deform beyond the elastic limit. For this reason, design codes require a specific ratio between the tensile strength and the yield strength. South African reinforcement is required to have a ratio of at least 1.15 between the tensile and yield strengths (SABS 920:1985). It is of interest to note here that the requirement from the ACI 318-02 Code (referred to subsequently as the ACI Code for brevity) is that this ratio should not be less than 1.25. The Swiss Code (SIA 262:2003) distinguishes between different steels with the ratio between yield strength and tensile strength varying between 1.05 and 1.15. More background on the need for the ratio between steel yield stress and steel tensile stress can be found under the materials discussion in the book by Paulay and Priestley (1992).

Table 2 Behaviour factor from different codes

System	SANS 10160	SABS 0160	UBC	EN 1998-1	SIA 262
Reinforced concrete shear walls	5	5	5,5	3 (medium ductility) 4 (high ductility)	4
Concrete Moment resisting frame	3	2	3,5	3 (medium ductility) 4,5 (high ductility)	4
Steel moment resisting frame	4,5	5	4,5	4 (medium ductility) 5 (high ductility)	5 (Class 1) 4 (Class 2) 2 (Class 3)
Non ductile structures	1	1	-	1,5	

Notes :

1. Provision is made for the multiplication of these values with a factor ranging between 1.0 and 1.5, depending on the load level at which first member flexural resistance is reached, and the load level at which plastic hinges form which can create instability in the structure.

2. Values presented are those for reinforcement steel with a ratio of > 1.15 between the tensile and yield strengths.

The ductility factor for reinforced concrete moment resisting frames used in the revised Code is similar to that in EN 1998-1, justified by the improved detailing rules in the revised Code. The value for reinforced concrete shear walls is higher to be consistent with the value in the existing code SABS 0160, and also because the detailing rules from ACI Code were adopted (see Paragraph 10 below).

It is clear that the choice of values for behaviour factors for South Africa needs to be justified by in-depth studies supported by experimental evidence using South African reinforcement and concrete characteristics; this will need to be undertaken before an amendment of the Code is considered.

8.4 Calculation of the first natural period

The formulae for calculating the first natural period of structures is similar in both EN 1998-1 and UBC. These simplified formulae for steel framed structures and for moment resisting frames are also similar to those in the existing code and these have thus been retained for the revised code.

The formula for shear wall structures has however been revised in the revised Code. The formula now takes into consideration the length of wall and the surface area of the building footprint. This approach seems to be more logical than the formula in the existing code where the first natural period was a function of the building length and not the length of shear wall.

It must be noted however that these formulae are approximations, based on research of a large number of existing structures (Goel and Chopra 1997; 1998; see also Wand and Wang 2005).

Existing structures are not damaged, as would be the case for structures subject to seismic loading, and the formulae are therefore more likely to have been based on behaviour of structures under wind loading in the elastic range.

A more accurate method to determine the first natural period would be by use of the Rayleigh method (EN 1998-1; UBC; SIA 261:2003), which is not presented in the revised Code. Refer also to the section on the Young's modulus of elasticity below.

8.5 Displacements

The revised Code provides a means for verification of the lateral displacements of structures. It requires a check on the P-delta effect and on the magnitude of lateral displacements. Both these criteria were obtained from the UBC. The concept of a reduced return period seismic event is discussed in Paragraph 4 above. The damage limitation criteria presented in the revised Code makes allowance for the reduced return period earthquake by providing damage limitation criteria to be used with the inelastic displacement results obtained from the design earthquake event (probability of 10% in 50 years event). The damage limitation criteria are already modified to allow for a factor of 0.4 to 0.5 on the magnitude of the design earthquake EN 1998-1.

The requirement of the existing Code that expansion joints in buildings should not be less than 40mm is not repeated in the revised Code. The designer now has the guidance to determine this requirement for the structure being considered. It is important that the magnitude of displacements be taken as the inelastic displacements. This can be obtained by multiplication of the elastic displacements by the behaviour factor used in an elastic analysis as shown in the revised Code.

8.6 Young's Modulus of Elasticity

Guidance is given for defining the appropriate Young's modulus of elasticity to be used in the case of concrete structures to make allowance for concrete cracking. This guideline is adopted from EN 1998-1, which gives a value of 0.5 times the short term concrete modulus of elasticity. More appropriate values can be found in the New Zealand Code where the equivalent Young's modulus of elasticity is expressed as a function of the axial force in a member and as a function of the type of structural element. It is also important for designers to remember that a seismic event is of short duration and that the reference to Young's modulus of elasticity for concrete relates to the short term value, and not the long term value as is sometimes mistakenly used.

9 STRUCTURES WITH MASONRY INFILL PANELS

In South Africa it is common to construct buildings with a structural frame and infill masonry panels. Designers often argue that additional strength is provided to the structure by these panel elements which in general are not considered as part of the lateral force resisting system.

Masonry infill panels can however significantly modify the dynamic behaviour of the structure if they interact with the structural frame. The interaction can modify the magnitude of the first natural period, resulting in a stiffer structure and an increased response spectrum value. It can also increase shear forces on columns, create torsional behaviour of the structure, and can also cause significant damage if there is local failure (often laterally) under seismic loads.

For the above reasons, a section is introduced into the revised Code to draw the attention of designers to aspects to be considered for masonry infill panels. Ideally, infill masonry panels should be isolated from the structural frame.

An investigative evaluation was made of building lateral displacements in a research study at the University of Stellenbosch (Loots 2005). It was shown that lateral displacements of the structural frame can be small for systems with reinforced concrete shear walls. It showed that the normal practice of allowing a 10mm gap may be sufficient to isolate infill masonry from the structural frame for the magnitude of design events in South Africa. This is however not the case for moment resisting frames. It would also be prudent to perform such verifications for individual buildings regardless of the structural system.

10 REINFORCEMENT DETAILING

The principles for the design of structures subjected to seismic loads rely heavily on the ability of a structure to undergo inelastic displacements. In general, for structural steel sections this behaviour is achieved by the ductile nature of steel. The structural steel design code (SANS 10162:2005) provides rules to verify that steel sections will behave in a ductile manner before local failure or local buckling occur. On the other hand, reinforced concrete can only achieve ductile behaviour if the concrete is sufficiently contained using containment reinforcement in elements undergoing compression stresses.

The existing Code SABS 0160 refers to structural elements that are appropriately detailed but no guidance is given on how to achieve such detailing. The revised Code now provides information for detailing of reinforced concrete elements to achieve the levels of ductility which is required for adopting the values of the listed behaviour factors. Ideally this information on reinforcement detailing should form part of the materials codes and these rules should eventually be incorporated into the concrete design code (SABS 0100-1:2003).

The following directives are presented in the Swiss reinforced concrete design code (SIA 262:2003) for detailing of reinforced concrete elements:

- During the design of a structure, a distinction shall be made between plastic and elastic regions of the structure.
- Elastic regions shall remain elastic under the action of earthquakes and it must be possible that plastic hinges form in the plastic regions.
- The position of the plastic regions shall be chosen such that energy dissipation is favoured.
- The detailing of the plastic regions shall ensure a high deformation capacity.

The provisions in the revised Code address the detailing of reinforced concrete elements to comply with the requirements in plastic regions. The following items are covered in the revised Code with the provision of detailing rules:

- **Reinforced concrete moment resisting frames: columns and beams:**

The detailing requirements are taken from intermediate moment frame requirements in the ACI Code. Rules are given for the spacing, size and shape of stirrups and links in columns and beams.

An important revision in the revised Code is the introduction of stirrups and hoops with "seismic hooks". The purpose of seismic hooks is to prevent the end hooks of containment reinforcement from opening up upon the loss of concrete cover during seismic actions. When the hooks open up, the containment effect is lost with a resultant loss in the ductility of the member in the plastic region.

An item not specifically addressed in the code is the practice of designing frame systems based on the weak beam strong column principle. This is considered to be part of the good design practice that engineers need to apply.

- **Reinforced concrete shear wall structures:**

The detailing rules are taken from the ACI Code where no distinction is made between walls in intermediate and high ductility classes. The containment of reinforcement in end or boundary walls in shear walls is similar to those of high ductility class columns.

The definition of the height of the plastic region in reinforced concrete shear walls is taken from the Swiss Code (SIA 262:2003). An important item in the code is the requirement that reinforcement bars should not be lapped in the plastic region. If this occurs, the height over which the plastic deformation develops is seriously compromised and the concept of an extended plastic region is defeated. This would particularly relate to starter bars in shear wall foundations. It is generally recommended that the first vertical splice not be located in the plastic zone.

- **Two-way slabs without beams:**

The requirements for flat slabs are taken from the ACI Code.

Flat slab systems form a large portion of structural systems in the South African construction industry. The existing code (SABS 0160 requires a reduction of 20% on the behaviour factor for structures with flat slabs. It is of interest to note that EN 1998-1 states that the design of flat slabs as primary seismic elements are "...not fully covered.." by the document. Apart from this phrase, no other mention is made of flat slab construction.

The revised Code now provides an improved description of the application of the 20% factor for flat slabs. The Code requires that when floor slabs are designed for bending moments and shear forces which results from lateral drift under seismic loads, the behaviour factor shall be reduced by a factor of 1.2 for use with structures comprising reinforced concrete flat or waffle slabs, and by a factor of 1.4 for the use with structures comprising pre-stressed concrete flat or waffle slabs. In general the capacity of structures with flat slab systems is limited in resisting lateral loads. Shear walls are required for this purpose in the majority of cases. For structures with shear walls the magnitude of lateral drift is limited with relatively small bending moments and shear forces developing in the slab systems as a result of the lateral loads. The dominant loading condition for flat slab systems is usually the gravity loads.

- **Columns supporting discontinuous systems:**

A discontinuous system occurs where a structural wall is not vertically continuous to the foundation, but transfers its vertical load to a column or columns. This type of construction is considered to be highly irregular and if considered part of the main lateral force resisting system, can not be designed using the equivalent static lateral load procedure. Special precautions are required. The revised Code provides guidelines for the detailing of such discontinuous systems, taken from the ACI Code.

- **Coupling beams:**

Coupling beams are often used between shear walls or lift shafts which act as shear walls. Special detailing rules are provided based on the ACI Code.

- **Foundations:**

Rules for foundations, piles and ground beams have been taken from the ACI Code.

11 UN-REINFORCED LOAD-BEARING MASONRY

Load-bearing masonry is widely used in low rise buildings in South Africa. This type of construction is common in housing and commercial developments and is often used in conjunction with pre-cast floor systems.

Un-reinforced masonry systems are known to be highly brittle by nature. For this reason, the use of load bearing masonry in regions of high seismicity is only allowed in the form of reinforced masonry construction.

The revised Code includes guidelines for the design of un-reinforced masonry structures as taken from EN 1998-1. Some *deem to satisfy* rules are provided, but it may be found that these values are rather conservative and that a designer would be better off carrying out a rational design.

12 COMPARISONS WITH OTHER CODES

A comparison was made between design base shear values calculated using the revised Code and other codes (Wium 2008). For the format of presentation the design base shear is expressed as a ratio of the total seismic weight of the building (design base shear ratio).

Table 3 (shear walls) and Table 4 (moment frames) present a comparison between the revised Code, the UBC and EN 1998-1. Comparisons are made using both response spectra types (1 and 2) of the EN 1998-1. Values are presented for ground type 2 of the revised Code which is the same as type B in the EN 1998-1 and comparable to type S_c in the UBC. For each of these codes the load factor for seismic loads is unity, a direct comparison is therefore possible. Values are all presented for a nominal peak ground acceleration of 0.1g.

Different structural heights and wall lengths were considered. Other variables are identified in the tables. From Table 3 it can be seen that values calculated for *shear wall structures* using the revised Code compare well with those of the other codes, except for the Type 2 response spectra in EN 1998-1. Table 4 shows that the revised Code values for *moment resisting frames* are higher than the other codes for low structures, but similar for medium to high structures. The Type 2 response spectra values from EN 1998-1 show the biggest difference with the revised Code.

A comparison is also presented between the current code SABS 0160 and the revised Code. The comparison for a variety of parameters, including the structural system, is presented in Table 5 (4 storey), Table 6 (10 storey), and Table 7 (20 storey). It can be seen that the values from the revised Code are significantly lower than those using the current code SABS 0160, mostly as a result of the load factor of 1.6 used by the current code SABS 0160 and the revised response spectra which have a more notable influence on taller structures. The values are presented for a nominal peak ground acceleration of 0.1g.

Table 3 Comparison of design base shear ratios between revised Code and international codes for shear wall structures (ground type 2)

Building Height (m) [storeys]	Structural Type	Nr of Walls	SANS 10160 (2008)	UBC (1997)	EN 1998 (Type 1)	EN 1998 (Type 2)
			Design base shear ratio			
13.2m [4]	Shear wall (5m)	2	0,088	0,092	0,076	0,050
13.2m	Shear wall (5m)	4	0,062	0,062	0,077	0,060
[4]	Shear wall (7m)	2	0,078	0,080	0,077	0,060
33m [10]	Shear wall (6m)	2	0,038	0,043	0,041	0,023
33m [10]	Shear wall (6m)	4	0,046	0,044	0,049	0,032
66m [20]	Shear wall (7m)	4	0,028	0,027	0,035	0,018

Table 4 Comparison of design base shear ratios between revised Code and international codes for moment frame structures (ground type 2)

Building Height (m) [storeys]	Structural Type	Nr of Cols	SANS 10160 (2008)	UBC (1997)	EN 1998 (Type 1)	EN 1998 (Type 2)
			Design base shear ratio			
13,2m [4]	Moment frame	40	0,118	0,096	0,076	0,043
33m [10]	Moment frame	40	0,058	0,054	0,044	0,025
66m [20]	Moment frame	40	0,035	0,032	0,026	0,010

Table 5 Comparison of design base shear ratios between the revised Code and current code for 4 storey buildings

Building Height (m) [storeys]	Structural Type	Building length (m)	Soil Type (2008/1989)	SANS 10160 (2008)	SABS 0160 (1989)
				Design base shear ratio	
13,3m [4]	Shear wall 4 x 5m	35	1/S1	0,060	0,080
13,2m [4]	Shear wall 4 x 5m	35	2/S2	0,072	0,080
13,2m [4]	Shear wall 4 x 5m	35	3/S2	0,069	0,080
13,2m [4]	Moment frame	35	1/S1	0,078	0,195
13,2m [4]	Moment frame	35	2/S2	0,118	0,200
13,2m [4]	Moment frame	35	3/S2	0,115	0,200

Table 6 Comparison of design base shear ratios between the revised Code and current code for 10 storey buildings

Building Height (m) [storeys]	Structural Type	Building length (m)	Soil Type (2008/1989)	SANS 10160 (2008)	SABS 0160 (1989)
				Design base shear ratio	
33m [10]	Shear wall 4 x 6m	35	1/S1	0,030	0,069
33m [10]]	Shear wall 4 x 6m	35	2/S2	0,046	0,080
33m [10]	Shear wall 4 x 6m	35	3/S2	0,052	0,080
33m [10]	Moment frame	35	1/S1	0,039	0,123
33m [10]	Moment frame	35	2/S2	0,058	0,161
33m [10]	Moment frame	35	3/S2	0,067	0,161

Table 7 Comparison of design base shear ratios between the revised Code and current code for 20 storey buildings

Building Height (m) [storeys]	Structural Type	Building length (m)	Soil Type (2008/1989)	SANS 10160 (2008)	SABS 0160 (1989)
				Design base shear ratio	
66m [20]	Shear wall 4 x 7m	35	1/S1	0,018	0,043
66m [20]	Shear wall 4 x 7m	35	2/S2	0,028	0,057
66m [20]	Shear wall 4 x 7m	35	3/S2	0,032	0,057
66m [20]	Moment frame	35	1/S1	0,023	0,087
66m [20]	Moment frame	35	2/S2	0,035	0,.114
66m [20]	Moment frame	35	3/S2	0,040	,.114

13 RESEARCH NEEDS

The following items have been identified as items for further research in order to improve the provisions of the revised Code. In many instances, this refers to an evaluation of the proposed guidelines in the South African context of local building practice and materials:

- quantification and confirmation of the behaviour factors for reinforced concrete elements, structural steel, un-reinforced and reinforced masonry
- development of details and guidelines to fix pre-cast floor slabs to load bearing masonry
- study of the effect of infill masonry panels on the behaviour of structural frames
- a possible alternative design approach to limit foundation size of reinforced concrete shear walls
- the increase in punching shear in floor slabs at columns located close to shear walls, including the additional force in columns in these locations
- development of rules to determine when punching shear in floor slabs becomes dominated by seismic load conditions
- the definition of the design magnitude of peak ground acceleration for Zone I areas (10% in 50 years)
- confirmation of the research of 1990 which shows typical response spectra for seismic events of mining origin.
- verification and improvement of the proposed guidelines in the code for un-reinforced masonry
- development of retrofitting procedures for existing structures with soft storey construction
- development of retrofitting procedures for structures with infill masonry wall panels of full and medium height
- development of a design approach for structures on piled foundations which penetrate a soft layer to be founded on bed rock at levels between 5 and 20 m depth.
- the requirements for lateral out-of-plane stability of masonry wall panels.
- the effect of staircases on the lateral stability of a building, the effect on torsional behaviour and the first natural period, including the requirements for structural details at stair-slab connections.

14 CONCLUSION

This chapter presents an overview of the background to the stipulations on seismic loads in the revised South African loading code for building structures.

In the revision of the code, reference was made to other existing codes. The South African construction practice and materials however require that certain items be investigated for local conditions. A list of items to be investigated is included.

The South African industry has had limited exposure to the design of building structures for seismic loading. It is important that the revised code and the correct design and construction procedures be adopted for all designs in regions of defined seismicity. To this effect, a program of awareness and training is required for the local profession.

15 ACKNOWLEDGEMENT

The time, effort and contributions made in the revision of the part on seismic loading in the loading code by the following individuals are highly appreciated and acknowledged:

Mr Keith Bokelman (Africon)
Mr Frik Riedeman (Bergstan)
Mr Colin Koen (Sutherland Consulting Engineers)
Mr Ferdie Heymann (De Villiers Partnership)
Mr Alten Hulme (Hulme Consulting Engineers)

The comments by Prof Alessandro Dazio of the Swiss Federal Institute of Technology in Zurich are also highly appreciated.

16 REFERENCES

ACI 318-02 (2002), *Building code requirements for structural concrete*. American Concrete Institute.

Chopra, A (2002). *Dynamics of structures*. Second ed. Prentice Hall.

Clough RW and Penzien J (1993). *Dynamics of structures*, 2nd edition, McGraw-Hill, New York.

ATC 3-06 (1978). Applied Technology Council, *Tentative Provisions for the Development of Seismic Regulations for Buildings*, NBS Special Publication 510, US National Science Foundation.

Earthquake Hazard Centre *Newsletter*(1999) Vol. 3 No. 1 page 4, July 1999 ISSN 1174-3646.

Eurocode EN 1990:2002. *Basis of Structural Design*, European Committee for Standardization, April 2002.

Eurocode 8 (2003): prEN 1998-1 : *Design of structures for earthquake resistance* Part 1: *General rules, seismic actions, and rules for buildings*. Draft No 6 January 2003, European Committee for Standardization.

Eurocode 1 EN 1991 (2002+).*Actions on structures*, European Committee for Standardization.

Eurocode EN 1998-1:2004. *Design of structures for earthquake resistance* Part 1: *General rules, seismic actions, and rules for buildings*. European Committee for Standardization, December 2004.

Goel RK, and Chopra AK (1997). *Period Formulas for Moment-Resisting Frame Buildings*. Journal of Structural Engineering, Vol. 123, No. 11, November 1997.

Goel RK, and Chopra AK (1998). *Period Formulas for Concrete Shear Wall Buildings*. Journal of Structural Engineering, Vol. 124, No. 4, April 1998.

Kijko A, Graham G, Bejaichund M, Roblin DL, and Brandt MBC (2003). *Probabilistic Peak Ground Acceleration and Spectral Seismic Hazard Maps for South Africa*, Report Number 2003-0053, Council for Geoscience, Pretoria, 2003.

Kijko A (2008). Personal discussions, Council for Geoscience, Pretoria.

Loots JJ (2005). *Computational assessment of seismic resistance of RC framed buildings with masonry infill*. MScEng thesis, University of Stellenbosch.

Milford, RV and Wium DJW (1991). *The impact of seismic events on buildings in mining areas*. The Civil Engineer in South Africa, December 1991.

NZS 4203:1992. New Zealand Standard *Code of Practice for general structural design and design loading for buildings*. Volume 1 Code of practice. Standards New Zealand.

Paulay T, Priestley MJN (1992). *Seismic Design of Reinforced Concrete and Masonry Buildings*, 1st ed. New York: Wiley.

SABS 0100-1:2003: Code of Practice: *The structural use of concrete*, Part 1: Design, Pretoria: South African National Standards, 1992 & 2003.

SANS 10162-1:2005. *The structural use of steel* Part 1: *Limit-state design of hot-rolled steelwork*. Standards South Africa.

SABS 0160:1989 (Rev 1993). South African National Standard. *Code of practice for: The general procedures and loadings to be adopted in the design of buildings*. Pretoria: South African National Standards, 1989 and 1993.

SANS 10160 (2008). *South African National Standard. Basis of structural design and actions for buildings and industrial structures*. Pretoria: South African National Standards, Committee Draft Standard 2008.

SANS 920-2005. *Steel bars for concrete reinforcement*. Standards South Africa.

SIA 261:2003. *Actions on Structures*, Swiss Society of Engineers and Architects, Swiss Standard 505 261, Zurich.

SIA 262:2003. *Concrete Structures*, Swiss Society of Engineers and Architects, Swiss Standard 505 262, Zurich.

UBC:1997. *Uniform Building Code* 1997. International Conference of Building Officials, Witthier, California.

Wang Q and Wang LY (2005). *Estimating Periods of Vibration of Buildings with Coupled Shear Walls*, Journal of Structural Engineering, Vol. 131 No. 12, December 2005.

Wium JA (2008). *Background to Draft SANS 10160 (2008), Part 4 Seismic Loading*. Paper submitted in April 2008 for publication in Journal of the South African Institution of Civil Engineering.

Part 5 – Basis of Geotechnical Design and Actions

5-1 Provision for Geotechnical Design in SANS 10160

Day PW & Retief JV

1. BACKGROUND

Over the years, the Geotechnical Division of the South African Institution of Civil Engineering (SAICE) has made numerous efforts to produce a South African geotechnical design code. In the mid-1980's, the Division appointed a sub-committee to draft such a code. The committee concluded that the resources of the Division were too limited for the task and recommended the adoption of a recognized international code. At this time, drafting of the Eurocodes, including Eurocode 7 (then ENV 1997 *Geotechnical Design*), was in full swing.

In 1995, the Division arranged a seminar in Johannesburg on limit states design in geotechnical engineering. They invited the Dr Niels Krebs Ovesen of Denmark, then chairman of the Eurocode 7 drafting committee, and his vice chairman, Dr Brian Simpson of the UK, to lead the discussions. One of the outcomes of this seminar was the recognition that differences between the draft European and South African "basis of design" codes (ENV 1991-1 and SABS 0160-1989 respectively) would have to be resolved before Eurocode 7 (ENV 1997) could be adopted as a South African design standard.

This recognition, and the need for a common basis for structural and geotechnical design, lead to the South African National Conference on Loading in 1998, organized jointly by the Geotechnical and Structural Divisions of the SAICE. International experts from Australasia (Dr Lam Pham), North America (Dr Laurie Kennedy) and Europe (Dr John Menzies and Dr Brian Simpson) were invited to participate. At the conclusion of this event, a decision was made to revise SABS 0160-1989 ensuring that adequate provision was made for geotechnical design and that the code was harmonised with international standards. These objectives have been largely realized in the new code, SANS 10160.

The purpose of this Chapter is to describe the provisions made in SANS 10160 for geotechnical design and the opportunities that this creates for the use of an international geotechnical design code, specifically EN 1997-1: 2004, in South Africa. The Chapter describes the changes made to Part 1 *Basis of Design* to accommodate geotechnical limit states and the introduction of new provisions for geotechnical design as Part 5 *Basis of Geotechnical Design and Actions*. It also provides the background to decisions taken during the revision process.

2. REVISION OF SABS 0160-1989

At the South African Conference on Loading in September 1998, the following resolutions were adopted (Day and Kemp, 1999):

- to establish of a working group to review SABS 0160,
- to increase co-operation with ISO through the South African Bureau of Standards (SABS),
- that the SAICE technical divisions should establish technical committees to review the materials codes, and
- that co-operation with SADC countries should be increased.

The working group held its first meeting in March 1999 and has met two or three times a year since. The group consists of members of the SAICE Geotechnical and Joint Structural Divisions, academics, representatives of the various materials codes (steel, timber, masonry, concrete, etc.) and practising engineers.

The revision of the various sections of the code was assigned to sub-committees each with its own champion. Each part of the code has been through a proposal stage, a first draft and a final committee draft. The final committee drafts were forwarded to readers in the profession for comment and most parts are now ready for inclusion into the Committee Draft code (CD). Once compiled into a correctly formatted document by the SABS, the draft code will be issued to the profession for comment as a Draft South African Standard (DSS).

Throughout the revision process, cognisance has been taken of international codes particularly ISO 2394:1988 *General principles on reliability for structures*, EN 1990:2002 *Basis of structural design* and EN 1997-1:2004 *Geotechnical design – General rules*. Unless there were compelling reasons to the contrary, the committee has tended towards compatibility with EN 1990 and other Eurocodes.

Throughout this Chapter, the old (existing) code is referred to as SABS 0160 or SABS 0160-1989 while the new code is referred to as SANS 10160. Part 5 of SANS 10160 deals with *Basis of Geotechnical Design and Actions* and is referred to as SANS 10160-5.

3. SCOPE OF SANS 10160

Part 1 of SANS 10160 (SANS 10160-1) states that the standard covers *building structures and industrial structures utilising structural systems similar to those of buildings*. It specifically excludes structures subject to internal pressure from the contents (e.g. silos and reservoirs), chimneys, towers and masts and bridges. Part 5 amplifies the scope of the code as providing design guidance on the determination of geotechnical actions on buildings and industrial structures including:

- vertical earth loading,
- earth pressures,
- ground water and free water pressure,
- downdrag or uplift caused by ground movements, and
- deformations caused by ground movements.

Geotechnical structures such as slopes, embankments and retaining structures are not covered by the code. The term "not covered" has been chosen in preference to the word "excluded". In fact, an attempt has been made to achieve as much compatibility between SANS 10160-5 and EN 1997-1 as possible thus opening the way to using SANS 10160 in conjunction with EN 1997, or a future South African geotechnical design code based on EN 1997, for any geotechnical structures.

It may seem strange to include earth pressures in a code that does not cover retaining structures. The rationale is to provide sufficient information for the design of building components such as basement walls which are subject to earth loading but not to cover free-standing retaining walls that rely mainly on the resistance of the ground for their stability.

4. CLASSIFICATION OF GEOTECHNICAL ACTIONS

Geotechnical actions are classified as permanent or variable, fixed or free. The classification of actions as permanent or variable is based not only on the length of time for which these effects act but also the uncertainty attached to their prediction. In certain cases, the designer may elect to classify certain long term actions such as uplift on piles due to heave as a variable action due to the uncertainty associated with the prediction of the uplift force.

Vertical earth loading and earth pressure are regarded as permanent fixed actions. Temporary stockpiles of earth are, however, regarded as variable free actions. Ground water pressure is not regarded as a separate action but as a component of vertical earth loading or lateral earth pressure. Free water above the ground surface including the additional water pressure within the ground caused by such water is classified as a variable action. Water pressure arising from temporary flooding is classified either as a variable action or an accidental action depending on circumstances.

Actions caused by ground movement include those that give rise to additional loading on the structure (e.g. downdrag on piles) or those that impose deformations on the structure (e.g. differential settlement). The proposal is that both should be classified as permanent actions in view of the length of time over which they act.

5. GEOTECHNICAL AND GEOMETRICAL DATA

5.1 Geotechnical parameters

5.1.1 *General*

Ground properties are different to the properties of other structural materials in that, rather than having a single measurable value with an assessable statistical variation, the value of a ground property is often dependent on a variety of factors such as stress level, mode of deformation, rate of loading, stress history and moisture content. Ground properties may also be affected by the method of sampling and measurement. Spatially, variations in ground properties may be localised or general, random or systematic. Although the code lists the factors that should be taken into account in assessing the properties of the ground and emphasises the need for an appropriate level of site investigation, it is expected that the designer has sufficient knowledge of geotechnical engineering to take these factors into account in selecting parameter values. Further elaboration would be more appropriate to a geotechnical design code than a basis of design code.

5.1.2 *Characteristic Values*

The characteristic value of a geotechnical parameter is defined in 5.3.1.6 as *the value so determined that the probability of a worse value governing the occurrence of the limit state under consideration is not greater than 5%.* In most practical design situations, the designer has insufficient test results to justify a statistical analysis of the data. In such instances, the code permits the use of a more intuitive definition, namely *a cautions estimate of the value affecting the occurrence of the limit state under consideration.*

The reference in both the above definitions to "*the occurrence of the limit state under consideration*" requires explanation as it could lead to the selection of two different characteristic values in the same geotechnical conditions. For example, the bearing capacity of an end bearing pile is governed by the properties of the ground in close proximity to the base of the pile. Where the soil is variable, the possibility of a low value being encountered in the localised area around the pile base is significant and the chosen value should thus be on the conservative side of the mean. On the other hand, the behaviour of a friction pile is governed by the average properties of the ground along the full length of the pile shaft. As a result, the effect of high and low values is averaged out along the pile shaft and a value closer to the mean would be appropriate.

It is also noteworthy that reference is made to a "*worse value*" rather than a "lower" value. This is because, in certain design situations, it is the upper characteristic value that governs the performance of the structure. An example of this is the case of downdrag forces on piles where the maximum average strength of the soil around the pile shaft determines the magnitude of the downdrag force for which the pile should be designed.

5.1.3 *Design Values*

The design value of a geotechnical parameter is determined by dividing the characteristic value by a partial material factor. Values of these partial factors are given in Annex B of the code. In instances where the upper characteristic value applies, provision is made in Annex B of the code (B.2.3) for the characteristic value so obtained to be divided by the reciprocal of the partial factor to obtain the design value where this produces a more onerous effect than dividing by the partial factor itself.

The code also allows for the direct determination of design parameters where this is considered appropriate.

5.2 Geometrical data

Ground level and slope, water levels, depth of interfaces between strata and excavation levels are all examples of geometrical data used in geotechnical design. The code defines the characteristic value of geometrical data to be a measured, nominal or estimated upper or lower value. Design values may be derived from characteristic values by adjusting them up or down by a specified tolerance (e.g. ±75 mm on the plan position of a vertical pile) or may be determined directly.

In cases where the level of groundwater has a significant effect on the reliability of the structure, the design value of the ground water level is often determined directly taking account of variations in ground permeability and physical controls on the level of the water surface.

6. VERIFICATION OF ULTIMATE LIMIT STATES

6.1 Design Approaches

Part 1 of the code requires that, when considering the ultimate limit state, it shall be verified that:

$$E_d \leq R_d$$

where E_d is the design value of the effect of actions, and
R_d is the design value of the corresponding resistance.

In the case of most structural materials, it is a relatively simple matter to verify whether this condition is met. One simply "factors up" the loads and "factors down" the strength and then checks that the design resistance exceeds the design action effect. This is because the resistance and the actions are sensibly independent of one another. However, this is not the case with geotechnical materials where strength is stress dependent, i.e. the resistance is influenced by the load, or the value of a geotechnical action (e.g. earth pressure) varies according to the strength of the material, i.e. the load is influenced by the resistance.

To complicate matters further, the self weight of the soil may be a destabilising action in one part of the soil mass and a stabilising action in another. For example, the weight of soil near the top of a slope has the effect of driving the failure whereas that near the toe resists failure. Frank et al (2004) describe the problem along the following lines: *In geotechnical design, the self weight of the ground is usually the dominant action; however, it is very often difficult to determine which part of the ground contributes to favourable action and which to unfavourable action.*

In the early versions of Eurocode 7 (ENV 1997) these difficulties were resolved by using two design calculations, each with a different set of partial factors:

Calculation 1: in which the partial factors were applied to permanent and variable actions while the ground strength was not factored.

Calculation 2: in which partial factors were applied to ground strength while permanent actions (including self weight) were not factored.

This led to considerable debate in Europe and pressure to reduce the perceived number of calculations from two sets to one and to simplify the calculations by applying partial factors to resistances and the effects of actions rather than to the material properties and the actions themselves. In the end, three separate design approaches for ultimate limit state design were included in EN 1997-1:2004 (Frank et al, 2004):

Design Approach 1: which is similar to the approach described above and involves two separate calculations. This is an "action and material factor" approach.

Design Approach 2: which requires a single calculation with partial factors applied to actions (or the effect of actions) and to resistances. This is an "action (or action effect) and resistance factor" approach. This approach, with the application of partial factors to the effect of actions and to resistances, is similar to the traditional factor of safety approach.

Design Approach 3: also requires a single calculation in which partial factors are applied to actions or the effect of actions and to ground strength parameters. Design Approach 3 is an "action (or action effect) and material factor" approach.

SANS 10160, has opted to follow Design Approach 1 for two main reasons. Firstly, the partial factors are applied at the source of the uncertainty, i.e. to the individual actions rather than the effect of the actions and to material properties rather than to resistances. Secondly, the results obtained using this method agree reasonably with current South African practice based on an overall factor of safety approach.

When compared with Design Approach 1, Design Approaches 2 and 3 tend to produce less conservative and more conservative results respectively. These observations by the first author are supported by the design examples given in the proceedings of the international workshop held in Dublin in 2005 (Orr, 2005).

6.2 Limit States and Partial Factors

6.2.1 *STR and STR-P Limit States*

Calculation 1, in which partial factors are applied to permanent and variable actions while the ground properties are not factored, is that required by Design Approach 1, Combination 1 as described in 2.4.7.3.4.2 of EN 1997-1

The action combination schemes used in SANS 10160 are based on Equations 6.10a and 6.10b of EN 1990:2002. These equations produce a similar result to the action combinations schemes used in SABS 0160-1989 as given by the familiar combinations of permanent load G_k, and imposed load Q_k:

$1{,}2\ G_k$ "+" $1{,}6\ Q_k$ for structures with significant imposed loads

and $1{,}5\ G_k$ for self-weight dominated structures

where "+" indicates "combined with".

These two combinations were found to produce a more uniform level of reliability than the use of a single load combination (Kemp *et al* 1987).

In SANS 10160, these two action combinations are used in the verification of the STR and STR-P limit states and are analogous with Equations 6.10b and 6.10a respectively. As the mnemonic suggests, STR implies a limit state to be used in verifying structural resistance and the suffix P indicates the dominant effect of permanent actions. The combination scheme for structures with significant imposed loads (STR - first of above equations) has been carried forward unaltered into SANS 10160. However, as explained in Chapter 1-2, the self-weight dominated combination (STR-P - second of above equations) has been modified to improve the uniformity and consistency of reliability to the following:

$$1{,}35\ G_k \text{ "+" } 1{,}0\ Q_k$$

One of the reasons for the modified STR-P combination was to provide specifically for geotechnical design situations where permanent actions frequently dominate.

In both the STR and STR-P limit states, partial factors are applied to the actions and the ground parameters are not factored. For example, the earth pressure acting on a basement wall is calculated using the characteristic (i.e. unfactored) values of soil parameters and the resulting pressure is then treated as a permanent action in the analysis of the structure attracting the appropriate partial action factor.

6.2.2 *GEO Limit State*

The second of the two calculations is based on Design Approach 1, Combination 2 as described in 2.4.7.3.4.2 of EN1997-1. It requires the introduction of a new limit state in SANS 10160 known as the GEO limit state. This limit state generally governs the design where failure occurs in the ground, e.g. bearing capacity or slope stability.

In this limit state, partial factors are applied to the soil strength parameters (or resistances in the case of piles or anchors) but permanent actions are not factored.

The exception to the above rule is in the design of piles and anchors where a resistance factor is used in preference to factoring the material properties. In the Eurocodes, different resistance factors are applied to different types of piles and to the contributions of shaft friction and end bearing. Analyses of model uncertainty carried out on South African data (Dithinde, 2007) do not support the adoption of different partial resistance factors for the various pile types. In addition, the data are insufficient to permit differentiation between the contributions of shaft and base resistance in assigning resistance factors.

6.2.3 *Partial Action, Material and Resistance Factors for STR, STR-P and GEO Limit States*

The partial action factors used in SANS 10160 are specified in SANS 10160-1 in the normative section of the code. The partial material and resistance factors for geotechnical design should ideally be specified in a geotechnical design code. In the absence of such a code, a set of partial factors compatible with the action factors in SANS 10160-1 has been given in Annex B of SANS 10160-5. As these factors may be amended when a geotechnical South African geotechnical design code Africa has been compiled, this Annex is regarded as informative rather than normative.

Table 1 summarises the partial action and material factors for persistent and transient design situations in the ultimate limit state that are applicable to the STR, STR-P and GEO limit states. For completeness, the partial factors for the equilibrium (EQU) and accidental (ACC) limit states are also given. In each case, the values given in SANS 10160 are compared with the corresponding values from EN 1990 and EN 1997. The symbol ψ is used for the action combination factor.

Table 1 Summary of partial factors for persistent and transient design situations in the ultimate limit state

	Limit State	EN 1990 reference:	Partial factor γ				
			EQU	**STR** Eq 6.10b	**STR-P** Eq 6.10a	**GEO**	**ACC** Eq 6.11b
Partial action factors	**Permanent Actions**			Set A1[8]	Set A1[8]	Set A2[8]	
	Unfavourable						
		SANS 10160	1,2	1,2	1,35	1,0	1,0
		EN 1990/1997	1,1/1,0[3]	1,15	1,35	1,0	1,0
	Favourable						
		SANS 10160	0,90	0,90	-	1,0	1,0
		EN 1990/1997	0,90	1,0	-	1,0	1,0
	Variable Actions						
	Leading action – Unfavourable						
		SANS 10160	1,3/1,6[1]	1,3/1,6[1]	1,0	1,3	1,0 [4]
		EN 1990/1997	1,5	1,5	$1,5\psi_{0,1}$	1,3	$\psi_{1,1}$ or $\psi_{2,1}$
	Accompanying action – Unfavourable						
		SANS 10160	$1,3\psi_i$	$0/1,6\psi_i$ [1]	0[2]	$0/1,3\psi_i$ [1]	$0/1,0\psi_i$ [1]
		EN 1990/1997	$1,5\psi_{0,i}$	$1,5\psi_{0,i}$	$1,5\psi_{0,i}$	1,3	$\psi_{2,i}$
	All variable actions – Favourable						
		SANS 10160	0	0	0	0	0
		EN 1990/1997	0	0	0	0	0
Partial material & resistance factors	**Soil Parameters**	SANS 10160 and EN 1990/1997	Table A2	Set M1[8]	Set M1[8]	Set M2[8]	
	Angle of shearing resistance [5]	ϕ'	1,25	1,0	1,0	1,25	1,0[7]
	Effective cohesion	c'	1,25	1,0	1,0	1,25	1,0[7]
	Undrained shear strength	c_u	1,4	1,0	1,0	1,4	1,0[7]
	Unconfined strength	q_u	1,4	1,0	1,0	1,4	1,0[7]
	Weight density	γ	1,0	1,0	1,0	1,0	1,0[7]
	Resistances			Set R1[8]	Set R1[8]	Set R4[8]	
	Pile – compression	SANS 10160	-	1,0	1,0	1,6[6]	1,0[7]
		EN 1990/1997	-	1,0-1,15	1,0-1,15	1,3-1,5[6]	1,0
	Pile – tension	SANS 10160	1,4	1,25	1,25	1,7[6]	1,0[7]
		EN 1990/1997	1,4	1,25	1,25	1,6[6]	1,0
	Anchors	SANS10160	1,4	1,1	1,1	1,1[6]	1,1[7]
		EN1997	1,4	1,1	1,1	1,1[6]	1,0

Notes:

(1) Values apply to wind actions and to variable actions other than wind respectively.

(2) For the STR-P combination, only permanent actions and the leading variable action are combined. Accompanying variable actions not considered.

(3) Values apply to EQU and to UPL limit states respectively.

(4) Design value of accidental action regarded as a leading variable action.

(5) Factor applies to tan ϕ'.

(6) Resistance factor applied to pile and anchor capacities calculated using unfactored ground parameters. In the case of unfavourable actions on piles (e.g. due to downdrag or transverse loading on the piles) the resistance factor is applied to the actions calculated using factored ground parameters.

(7) See B.2.5 of SANS 10160-5.

(8) Set A1, Set M1, etc. refer to the sets of partial factors given in Annex A of EN1997-1.

6.2.4 *Partial Factor for Uncertainty in the Resistance Model*

In Clauses 2.4.1(6) - (9), EN 1997-1 introduces a model factor to account for:

- the range of uncertainty in the results of the method of analysis
- any systematic errors known to be associated with the method.

The use of a model factor is specifically mentioned in Clause 7.6.3.2 (2) of EN 1997-1 which deals with the determination of the compressive resistance of piles from ground tests. EN 1997-1 is, however, silent on the value to be used for this important parameter, preferring instead to leave the assignment of suitable values to the National Annexes. This is probably due to the wide range of design methods used by European member states and the differences in the results obtained (see De Cock et al, 1999).

The UK National Annex recommends a value 1,4 which may be reduced to 1,2 if the resistance is verified by a maintained load test taken to the calculated, unfactored ultimate resistance of the pile. These values must, however, be read in conjunction with the values recommended in the UN National Annex for the partial resistance factors to be used in pile design which vary from 1,7 to 2,0 for piles in compression compared to a value 1,6 recommended in SANS 10160-5.

Dithinde (2007) gives the results of an analysis of 174 documented South African pile load tests for which sufficient data was available to calculate the resistance of the pile based on ground test results. Statistical analysis of these data has indicated that a model factor of 1,5 used in conjunction with the partial resistance factor of 1,6 for piles in compression will be sufficient to achieve the target reliability index of β = 3,0 as aimed for in SANS 10160. Accordingly, a value of 1,5 has been recommended in B.3.3 of SANS 10160-5 for the design of piled foundations by calculation using ground test results.

7. VERIFICATION OF SERVICEABILITY LIMIT STATES

The serviceability limit state is verified in accordance with the criteria set out in SANS 10160-1. Partial action factors and action combination factors are given for three serviceability limit states, namely irreversible, reversible and long term.

In selecting which of the combinations of actions to use in the analysis of the various serviceability limit states, SANS 10160-5 recommends that account should be taken not only of the reversibility of the effect but also the period of time required for the effect to take place. The irreversible serviceability combination is one which will occur infrequently and probably for limited periods of time but its occurrence cannot be excluded. The reversible and long term combinations are more representative of the average conditions likely to occur over a period of time. As such, SANS 10160-5 requires that the irreversible action combination be used in the assessment serviceability conditions that develop rapidly, such as settlement of granular soils. The long term action combination is to be used in the assessment of conditions that develop over a longer period of time such as consolidation, settlement or creep.

An alternative approach to the verification of the serviceability limit state included in SANS 10160-5 is to ensure that a sufficiently low fraction of the strength of the ground is mobilised to keep deformations within the required limits. This approach, which has been adopted from EN 1997-1, applies only when a quantification of the deformation is not required and where comparable experience exists under similar circumstances.

Movement caused by heave or collapse of soils is not dealt with specifically. However, the code does require that the effect of changes in ground properties that may occur during the life of the structure should be taken into account, including desiccation of the soil, saturation, ground water lowering, etc.

8. DETERMINATION OF GEOTECHNICAL ACTIONS

8.1 Vertical Earth Loading

The requirements of the code for the determination of vertical earth loading amount to little more than common sense and sound engineering practice. Attention is drawn to the following:

- The bulk unit weight of the ground is used in the calculation thereby including the self weight of the groundwater with the self weight of the ground and not regarding this as a separate action. Allowance should be made in the assessment of the bulk unit weight for variations in moisture content.
- In the case of non-uniform loading, e.g. conical stockpiles, allowance may be made for redistribution of the loads due to arching.
- On non-uniform subgrades, cognisance should be taken of the tendency for stiffer areas to "attract" load, e.g. as is the case with positive projection culverts below fill embankments.
- Any surcharge on the ground surface or free water above the surface should be regarded as a variable action and should be considered separately from the vertical earth loading.

8.2 Earth Pressures

The code draws attention to the effect of the magnitude, mode (rotation, translation, etc) and direction of wall movement (into or away from the retained material) on the magnitude and distribution of the earth pressure. It also lists factors to be taken into account in the determination of earth pressure including wall roughness, inclination of the wall and the ground surface, compaction forces, swelling pressures, etc. The normative sections of the code do not provide methods for the calculation of earth pressure as this information is better suited to a geotechnical design code than a basis of design code. Guidance is, however, given in informative Annex C for the determination of earth pressures exerted by granular backfills on vertical basement walls.

One somewhat contentious requirement included in SANS 10160 from EN 1997 is that, unless a reliable drainage system is provided or infiltration into the soil is effectively prevented, the water table in retained earth of low or medium permeability should be assumed to be at the surface unless indicated otherwise by comparable local experience.

8.3 Actions due to ground movement

Actions due to ground movement are considered in two categories, namely those that exert a force on the structure and those that cause the structure to deform.

The first category includes uplift and downdrag forces on piles caused by heave or settlement of the surrounding ground respectively. In such cases, the upper characteristic value of the shear strength (above the mean value) should be used in the design. In addition, the code draws attention to the fact that the shear stresses mobilised in the ground strata overlying the expansive or compressible horizon must also be considered in the calculation of uplift or downdrag forces.

The second category includes the effects of ground movements such as differential heave or settlement on structures and services. It is pointed out that predictions of differential movement which ignore the stiffness of the structure tend to overestimate the distortion of the structure. The code cautions that the accuracy of settlement predictions tends to be poor.

9. GEOTECHNICAL CATEGORIES

Annex A of SANS 10160-5 introduces the concept of Geotechnical Categories based on the nature of the ground, the complexity of the structure, the intensity of loading and the associated risks. It goes on to stipulate the minimum requirements for investigation, design, construction control and monitoring for each category. The content of this annex is based on a similar section in EN 1997, modified for South African conditions, legislation and practices.

This information is included as an informative annex rather than being placed in the main body of the report as many sections of the Annex pertain more to geotechnical design than to basis of design. In all probability it will be moved to the geotechnical design code once such a code is written. Its inclusion in the code was considered warranted by the committee in view of the widespread tendency in South Africa to limit the scope of geotechnical investigations and monitoring during construction to levels that are not compatible with the scope of the project and the risks involved. This Annex supplements the requirements of the Construction Regulations of the Occupational Health and Safety Act and provides a standard of good practice against which the requirements of the Act can be adjudicated.

Four Geotechnical Categories are applied, similar to the four level classification applied in Part 1 for buildings and the consequences of accidental actions. This scheme is somewhat different from the three level classification generally applied in the Eurocodes which often requires a sub-division of the second class. The following four Geotechnical Categories are presented in Annex A of SANS 10160-5:

- **Category 1** includes small and relatively simple structures constructed on non-problematic ground where there is negligible risk of instability or of significant ground movements (e.g. houses or other simple structures on stable soil profiles). In such cases, the investigation may be limited to a qualitative assessment based on a systematic description of the soil profile. Deemed-to-satisfy design procedures will suffice and only routine inspections at critical stages of construction are required.

- **Category 2** includes conventional structures and foundations for which design methods are well established, where there are no exceptional risks in terms of overall stability or difficult ground conditions (e.g. conventional buildings on spread footings, rafts or piled foundations). Here, the geotechnical investigation should include routine field and laboratory tests producing quantitative geotechnical data for design purposes. In these cases, design calculations are required including the assessment of bearing capacity, settlement, earth pressure, etc. Systematic checking by the designer is required during construction to confirm the validity of the design assumptions coupled with periodic inspections by the geotechnical engineer. Additional field and laboratory testing may be required during construction. Monitoring will generally be limited to ensuring that critical performance criteria are met (e.g. total settlement of foundations or movement of retaining structures).

- **Category 3** structures include conventional structures and foundations with no exceptional risks or loading conditions, but for which the nature of the ground or complexity of the design requires specialist geotechnical input (e.g. anchored retaining systems, deep excavations below the water table, problems requiring soil-structure interaction analysis, etc). The geotechnical investigation requirements are similar to those for Category 2 supplemented by specialised field and laboratory tests as specified by the geotechnical engineer. Specialised geotechnical design and cooperation between the geotechnical and structural engineers is required as are regular and detailed monitoring by the geotechnical engineer with additional field and laboratory tests as appropriate. A rigorous construction quality control programme is essential. Detailed monitoring will be required including, as appropriate, piezometer levels, ground movements and anchor loads often coupled with the use of the observational method. Ongoing monitoring of the structure may be required after completion.

- **Category 4** includes structures or parts of structures that lie outside Categories 1 to 3, e.g. very large or complex structures, structures involving abnormal risks or in unusual, unstable or exceptionally difficult ground conditions. Such projects require the application of the requirements of Categories 1 – 3 supplemented by requirements in addition or alternative to those in the Code.

Annex A includes a brief description of the observational method as a means of adapting the design to suit conditions encountered on site. Its inclusion is intended to establish the "legitimacy" of a procedure that is regarded by some as an excuse by geotechnical engineers to change their minds during construction.

10. GUIDANCE FOR STRUCTURAL DESIGNERS

Annex C of SANS 10160-5 provides guidance for structural engineers on typical geotechnical aspects of the design of buildings and industrial structures. This includes charts for assessing the bearing capacity of soils, values of earth pressures (including compaction pressures) exerted by granular soils on vertical walls and guidance on the design of piles. This informative Annex is provided in the absence of a South African geotechnical design code.

10.1 Design of Spread Footings

Annex C.2.2 gives charts for determining the design (bearing) resistance of shallow foundations founded on two classical soil types, namely an undrained normally consolidated clay (ϕ = 0) and a non-cohesive granular soil (c' = 0).

In the case of the granular soils, separate charts are given for various positions of the water table, namely at the surface, at founding level and below the depth of influence of the foundation. The charts are in non-dimensional form to allow for the evaluation of the design resistance of any size of footing and at depths of founding ranging from surface (Z/B = 0) to Z/B = 1,0. The assessment of the bearing pressure is based on the method given in EN 1997-1, Annex D.4.

For undrained conditions, only one chart is required as the value of the bearing capacity factor N_c is independent of foundation size and depth of founding. This chart is based on the classical Skempton (1951) bearing capacity equation.

10.2 Design of Axially Loaded Piles and Pile Groups

Annex C.3 summarises the various approaches given in EN1997-1 to the design of piles namely load testing (static or dynamic), analysis of pile driving records and calculations based on ground test results. Details of the first two methods are, however, too lengthy to be dealt with in a basis of design code and correctly belong in a geotechnical design code. Although some pointers are given on the latter method, reference to EN1997-1 will still be required to apply the method.

10.3 Earth Pressures

Unlike the design of piles or spread footings which are often left to the geotechnical engineer, structural engineers are frequently required to estimate the earth pressure exerted on a buried structure as an input to the structural design. In order to facilitate such calculations, an approach similar to that contained in TMH 7 is adopted where two types of backfill are defined and typical parameters are assigned to these materials. These parameters are then used to determine the earth pressure acting on yielding and rigid structures (active and at-rest conditions respectively).

The approximate earth pressure distribution given ignores cohesion in the backfill and wall friction and its application should be limited to walls lower than 7,5m. The distribution given in Annex C.4 includes the effect of compaction and a water table within the retained material. Compaction pressures are calculated using the method recommended by Clayton et al (1993).

11. COMPATIBILITY WITH THE EUROCODES

One of the key objectives in re-writing the code was to achieve compatibility of SANS 10160 with international standards. The extent to which the provisions of SANS 10160-5 for geotechnical design are compatible with those given in EN 1990 and EN 1997-1 is explored below.

Table 1 above summarises the action combinations, partial material factors and resistance factors for persistent and transient design situations in the ultimate limit state as included in SANS 10160 and compares them with those given in the informative Annexes to Eurocodes EN 1990 and EN 1997. Table 2 below gives typical load combinations for a residential building in which the appropriate numerical values of the action combination factors ψ have been included.

Table 2 Typical load combinations of G_k, Q_k and W_k (illustrative only)

Limit State		Action Combination		Comments
EQU	SANS 10160	$0{,}9G_k$ "+" $1{,}3W_k$		G_k and Q_k both assumed to be favourable actions
	EN 1990/1997	$0{,}9G_k$ "+" $1{,}5W_k$		
STR	SANS 10160	$1{,}2G_k$ "+" $1{,}6Q_k$ $1{,}2G_k$ "+" $0{,}48Q_k$ "+" $1{,}3W_k$	and	Q_k leading ($\psi_{wind} = 0{,}0$) W_k leading
	EN 1990/1997	$1{,}15G_k$ "+" $1{,}5Q_k$ "+" $1{,}05W_k$ $1{,}15G_k$ "+" $1{,}05Q_k$ "+" $1{,}5W_k$	and	Equation 6.10b – Q_k leading Equation 6.10b – W_k leading
STR-P	SANS 10160	$1{,}35G_k$ "+" $1{,}0Q_k$ $1{,}35G_k$ "+" $1{,}0W_k$	and	Q_k leading W_k leading (no accompanying variable action)
	EN 1990/1997	$1{,}35G_k$ "+" $1{,}05Q_k$ "+" $1{,}05W_k$		Equation 6.10a
GEO	SANS 10160	$1{,}0G_k$ "+" $1{,}3Q_k$ $1{,}0G_k$ "+" $0{,}39Q_k$ "+" $1{,}3W_k$	and	Q_k leading ($\psi_{wind} = 0{,}0$) W_k leading
	EN 1990/1997	$1{,}0G_k$ "+" $1{,}3Q_k$ "+" $0{,}78W_k$ $1{,}0G_k$ "+" $0{,}91Q_k$ "+" $1{,}3W_k$	and	Equation 6.10 – Q_k leading Equation 6.10 – W_k leading
ACC	SANS 10160	$1{,}0G_k$ "+" A_d "+" $0{,}3Q_k$		Q_k or W_k both accompanying variable actions ($\psi_{wind} = 0{,}0$)
	EN 1990/1997*	$1{,}0G_k$ "+" A_d "+" $0{,}5/0{,}3Q_k$ $1{,}0G_k$ "+" A_d "+" $0{,}3Q_k$ "+" $0{,}2/0W_k$	and	Equation 6.11b – Q_k leading Equation 6.11b – W_k leading

As far as the new GEO limit state is concerned (generally associated with failure in the ground), the differences between SANS 10160 and EN 1990 / EN 1997-1 pertain mainly to the effect of wind. For most geotechnical structures, apart from foundations of tall buildings, wind loading plays a relatively minor role. If wind loading is omitted, the combinations of the permanent actions and variable actions are identical. All the partial material factors used in SANS 10160 have been taken directly from the values recommended in Annex A of EN 1997-1 and are therefore compatible. The resistance factors used are also similar. Thus, as far as the GEO limit state is concerned, SANS 10160 is largely compatible with the Eurocodes.

With regard to the STR and STR-P limit states, the differences in the partial action factors are either minor or arise from the treatment of the accompanying variable action. The factoring of accompanying variable actions is in accordance with SABS 0160:1989 which includes accompanying variable actions at their "arbitrary point in time" values in accordance with the principles of the Turkstra rule (Kemp 1987). This was a well researched decision based on reliability analyses and is maintained in the SANS 10160, as is discussed in Chapter 1. The partial action factors and the combination factors used in SANS 10160 are thus rational values regarded by the committee as preferred alternatives to the values given in the Eurocodes yet fully compatible with the Eurocode approach as embodied in Equations 6.10(a) and 6.10(b) of EN1990.

It should be noted that the partial factors given in EN 1990 and EN 1997 are all classified as National Determined Parameters (NDP) and Eurocode member states are required to select appropriate values by means of a National Annex. In addition, member states are free to select among any one (or a combination) of the Design Approaches 1, 2 or 3. These selections can lead to wide variation in the results obtained. Against this background, the differences reflected in Table 2 are well within the range of values likely to be chosen by European member states.

12. APPLICATION OF SANS 10160-5

12.1 Status

SANS 10160-5, and SANS 10160 in general, is a standard written by engineers for engineers. It is intended to assist designers with the application of reliability based design methods that are in line with international practice. The methods given in the code represent a logical and harmonised approach to limit states design but are not the only methods that may be used nor are they necessarily applicable to every design situation. The standard is an aid to the application of engineering judgement and not a substitute for it.

Along with the majority of South African National Standards, SANS 10160 is a statement of good practice. Despite the prescriptive language used in the normative sections, the standard itself is not mandatory and has no legal status. Specifically, SANS 10160 is not a document written by engineers to be used against them in a court of law. Compliance with the standard will, in general, be sufficient to demonstrate the acceptability of the design approach used. However, the converse does not necessarily apply.

As indicated earlier in this Background Report, SANS 10160 has used the Eurocodes as reference documents. There are, however, significant differences in the application of the Eurocodes among the European member states and the application of SANS 10160 in South Africa.

The Eurocodes are sponsored by the European Commission with a view to eliminating technical obstacles to trade and harmonisation of technical standards among the member states. As safety is regarded as a national issue under the control of each individual member state, the selection of parameters which relate to safety (or more correctly, to reliability) of structures are not prescribed in the normative sections of the Eurocodes but fall to be defined in National Annexes prepared by the member states. The parameter values contained in the informative annexes to the various Eurocodes are for guidance only. By treaty, all European member states are obliged to afford the Eurocodes the status of national standards, compile National Application Documents and to withdraw all conflicting codes and standards within a prescribed timeframe.

Although South Africa has elected to model its basis of design code on the Eurocodes, it is not obliged to adhere to any of the requirements imposed on European member states with regard to the implementation of the code. It is free to change, omit or add whatever it deems fit in the compilation of its own national standards. There are, however, many benefits for the country in remaining reasonably aligned with the basis of design embodied in the Eurocodes. In particular, this will facilitate technical exchange and trade with one of our biggest trading partners. This will also permit the use of Eurocodes where no equivalent South African Standard exists.

12.2 Application

SANS 10160-5 provides design guidance sufficient for projects that lie in Geotechnical Categories 1 and 2. Simplified design rules, including deemed-to-satisfy requirements may be used for Category 1 structures as stated in Annex A. Category 3 and 4 projects will generally require alternative or additional design rules to those contained in SANS 10160-5.

As indicated above, the use of the code is not mandatory for any Geotechnical Category. Designers may choose to use other established design approaches as they see fit. However, should they do so, they may be called on to defend the approach adopted and would no longer enjoy the protection that compliance with the code affords.

12.3 Use of SANS 10160 in conjunction with Eurocodes

Two of the objectives of rewriting of SABS 0160 were to ensure that there is adequate provision for geotechnical design and that the code is harmonised with international standards. The intention was to facilitate the adoption of an internationally accepted geotechnical design code or the use of such a code as a model for compiling a South African geotechnical design code.

At a meeting of the Geotechnical Division of the South African Institution for Civil Engineering held in May 2008, the geotechnical fraternity was requested to choose between three alternatives:

- Adopting EN 1997-1 as a South African design code. This will entail writing what amounts to a South African National Annex to the code.
- Writing a South African design code based on SANS 10160 and EN 1997. This code would contain only those aspects of the Eurocode relevant to South African conditions.
- The laissez-faire approach. This is effectively the current situation where, in the absence of a geotechnical design code, designers use whatever design method is best suited to the problem at hand.

Option 3 above was seen as the easy way forward but was rejected as it holds few long term benefits for the profession.

The meeting acknowledged that drafting a South African design code will be beneficial. The new code would strive to be a practical design code, written by engineers for engineers. It would be concise in that it would contain only those aspects of the EN 1997 relevant to South African conditions and the selected design approach. The main drawbacks are the amount of time required to write such a code and that it would be difficult to write a code of this nature before the profession has more experience in the use of limit states design in geotechnical engineering.

The meeting agreed that geotechnical designers should be encouraged to use EN1997-1 in conjunction with SANS 10160 over the next few years. Thereafter, a more informed decision can be taken whether to adopt or adapt EN 1997-1 (or another international design code). This agreement has been ratified by the Geotechnical Division Committee.

In summary, as demonstrated in section 11 of this chapter, SANS 10160-1 and SANS 10160-5 are sufficiently compatible with EN 1990 and EN 1997-1 to permit the use of EN 1997-1 as a geotechnical design code in conjunction with SANS 10160 in South Africa. The use of EN 1997-1 is seen as an interim step to either writing a South African geotechnical design code or formally adopting EN 1997-1 as a South African standard.

13. CONCLUSION

The process of amending SABS 0160-1989 was initiated partly by the conclusions reached at the SAICE seminar on Limit States Design in Geotechnical Engineering in 1995 where the incompatibility between SABS 0160 and the draft Eurocodes was seen as an obstacle to the use of Eurocode 7 as a geotechnical design code in South Africa. The need for revision of SABS 0160 was endorsed at the ensuing South African National Loading Conference in 1998. Among the conclusions reached at this conference was that the code should be harmonised with international standards and that provision should be made for geotechnical loading. In the opinion of the authors, these two objectives have been met with the production of the SANS 10160. The new code is compatible with the Eurocodes and adheres to the principles contained in ISO 2394:1998. This has opened the way for revision of existing South African materials codes in line with international standards and for the use of the various structural Eurocodes in instances where no equivalent South African code exists. This applies particularly to geotechnical design where EN 1997-1 can be used as a geotechnical design code in conjunction with SANS 10160 until such time as a local geotechnical design code is produced.

14. REFERENCES

Clayton C R I, Milititsky J and Woods R I (1993). *Earth Pressure and Earth-retaining Structures*. Chapman & Hall, London.

Day, P W and Kemp A R (1999). South African loading code: Past, present and future. *Civil Engineering*, 7(4), 11-12. SAICE, April 1999.

De Cock F, Legrand C and Lehane B (editors) (1999). *Survey report on present-day design methods for axially loaded piles. European Practice.* Report of European Regional Technical Committee 3, International Society of Soil Mechanics and Geotechnical Engineering.

Dithinde M (2007) *Characterisation of model uncertainties for reliability-based analysis of piled foundations*. PhD Thesis, University of Stellenbosch.

EN 1990:2002. Eurocode – *Basis of structural design*, European Standard. European Committee for Standardisation, Brussels.

EN 1997-1:2004. Eurocode 7: *Geotechnical Design* – Part 1; General Rules, European Standard. European Committee for Standardisation, Brussels.

Frank, R, Bauduin, C, Kavvadas, M, Krebs Ovesen, N, Orr, T and Schippener, B (2004). *Designers' guide to EN 1997-1, Eurocode 7: Geotechnical Design – General Rules*. Thomas Telford, London.

ISO 2393:1998. *General principles on reliability of structures*. International Standard. International Standards Organisation. Adopted by Standards South Africa as SANS 2394:2004 (Ed 1)

Kemp, A R, Milford, R V and Laurie, J A P 1987. *Proposals for a comprehensive limit states formulation for South African structural codes*. The Civil Engineer in South Africa, September 351-360.

NA to BS EN 1997-1:2004 *UK National annex to Eurocode 7: Geotechnical design – Part 1: General rules*.

Republic of South Africa. *Occupational Health and Safety Act*. Act 85 of 1993 as amended, including Regulations.

Orr, T L L (editor) 2005. *International workshop on the evaluation of Eurocode 7*, Proceedings. Trinity College, Dublin.

SABS 0160-1989. *The general procedures and loadings to be adopted in the design of buildings, Code of Practice*. South African Bureau of Standards, Pretoria.

SANS 10160. *Basis of structural design and actions for buildings and industrial structures, Code of Practice*. South African Bureau of Standards, Pretoria.

Skempton A.W. (1951) *The bearing capacity of clays*. Building Research Congress, London.

TMH7:1981 *A code of practice for the design of highway bridges and culverts in South Africa*. NITRR, CSIR, Pretoria.

Part 6 – Actions Induced by Cranes & Machinery

6-1 Investigation into Crane Load Models for Codified Design

Dymond JS & Dunaiski PE

1 INTRODUCTION

The provision for crane induced actions is one of the most important provisions for industrial structures in a code of practice for loadings on buildings. Problems have been encountered with crane installations in practice, indicating that the crane loading provisions in SABS 0160:1989 (Amended 1993) are not sufficient. It is therefore imperative that crane-induced loads are accurately represented in the updated South African loading code SANS 10160. In this chapter the crane load models and their application to requirements and procedures implemented by international design codes for crane supporting structures are surveyed. This information serves as basis for the formulation of the procedures incorporated in the revised SANS 10160 Part 6 *Actions induced by cranes and machinery*, as presented in subsequent chapters. For conciseness, the previous SA loading code is referred to as SABS 0160, the revised code as SANS 10160 in general, and the provisions for crane induced actions as SANS 10160-6.

The crane load models in SABS 0160 appear over-simplistic in comparison with other international codes of practice for crane loadings; namely German Standard (DIN 15018-1:1984), International Standard (ISO 8686-1:1989), Australian Standard (AS1418.1-1994), Eurocode (EN 1991-3), British Standard, Netherlands Standard and American Specification (Dunaiski et al 2001). There is also a discrepancy in load models and approach to crane loading between the codes of practice used by crane manufacturers for design of the crane (DIN 15018-1:1984, SABS 1599-4-5:1995) and the crane load models in SABS 0160 used in the design of the support structure, although the loads experienced by the crane are clearly the same as the loads that the crane imposes on the support structure (Krige 1998).

The over-simplicity of the SABS 0160 crane load models and the lack of uniformity in the crane and support structure design procedure prompted a review of the SABS 0160 crane load models and a decision to update these models for the forthcoming revision of the South African loading code, SANS 10160.

It is preferable to base the updated crane load models on a reference code which can be demonstrated to have international recognition and is suitable for solving the problems of over-simplicity and lack of uniformity in the design process of crane and support structure. The revision process of the crane load models includes a thorough investigation of the load models in the reference code and an assessment of their suitability for inclusion in a South African code of practice, with clear criteria as a basis for their inclusion.

2 CRANE LOAD PROVISIONS IN SABS 0160:1989

Cranes are classified into four classes by SABS 0160 based on a description of their use. The classes distinguish among cranes according to their frequency of use and severity of loading.

The vertical and horizontal loads imposed by cranes are calculated by multiplying the static wheel load by load model factors and an impact factor for the vertical load. The magnitudes of the load model and impact factors used for the various loads depend on the class of the crane.

The crane classification and provisions for calculating the loads are shown below in Table 1 which has been modified from that presented by Krige (1998).

Table 1 Load provisions in SABS 0160:1989

Crane class	Description of class	Vertical loads	Horizontal Lateral Loads			Horizontal Longitudinal Loads	
			Acceleration and braking of crab	Misalignment of wheels or rails	Skewing of crane	Acceleration and braking of crane	End stop buffer forces
	Static wheel load:	V_s	$\frac{Q_{cr}+Q_h}{n}$	$\frac{Q_{br}+Q_{cr}+Q_h}{n}$	$\frac{Q_{br}+Q_{cr}+Q_h}{n}$	$V_s \times \frac{n}{2}$	$Q_{br}+Q_{cr}$
1	Hand operated	1,10	0,05	0,05	0,075	0,10	1,0
2	Light duty and maintenance. Design load seldom applied	1,20	0,10	0,12	0,18	0,10	1,0
3	Continuous, low impact operation. Design load often applied	1,25	0,15	0,15	0,225	0,10	1,0
4	Continuous, high impact operation. Design load usually applied	1,30	0,20	0,20	0,30	0,10	1,0
Where:	V_s = Static wheel load Q_{br} = Weight of crane bridge Q_{cr} = Weight of crab Q_h = Weight of hoistload n = number of crane wheels						

The guidance given by SABS 0160 on which combinations of crane loads are to be considered to act together is that the three horizontal transverse actions need not be considered to act simultaneously. This recommendation results in three crane load situations in design for the ultimate limit state and one crane load situation in design for the accidental limit state. The load combinations are given in Table 2.

Table 2 Crane load combinations in SABS 0160:1989

Load		Ultimate limit state			Accidental
		1	2	3	4
Vertical load with impact		X	X	X	X
Horizontal transverse loads	Acceleration and braking of crab	X			
	Misalignment		X		
	Skewing			X	
Horizontal longitudinal loads	Acceleration and braking of crane	X	X	X	
	End stop buffer forces				X

3 INVESTIGATION INTO ALTERNATIVE CRANE LOAD MODELS

An investigation was carried out into international codified crane loads in order to select a reference code as a basis for the updated crane load models in SANS 10160-6. Four crane loading codes were investigated in detail: DIN 15018-1:1984, ISO 8686-1:1989, AS1418.1-1994 and EN 1991-3. These four codes were found to have the same basis for the load calculations with some variations of application.

3.1 Historical development

The German Standard DIN 15018-1:1984 is a code of practice which specifies loads that are to be applied to cranes for the design of the crane itself. The Standard published in 1984 is a corrected edition of the 1974 version of the crane loading code. At the time of its publishing, the International Standard ISO 8686-1:1989 was under development with the purpose of formulating a standard for proof of competence for structural and mechanical components of cranes in order to obtain an internationally approved standard on crane loads and load combinations. That this was a successful venture is implied by the fact that ISO 8686-1:1989 forms the basis for the South African code of practice on loadings on cranes (SABS 1599-4-5:1995). ISO 8686-1:1989 is largely based on the crane load models in DIN 15018-1:1984.

The Australian Standard (AS 1418.1:1994) and European Standard (EN 13001-2:2004) for loading on cranes are based on a combination of the German and International Standards. Though the Australian code specifies the loads to be imposed on the crane, it is also referred to for the calculation of loads by the part of the steel design code for design of crane support structures AS 1418.18.

The Eurocode EN 1991-3 differs from the abovementioned Standards in that it specifies crane actions that are applied to the supporting structure rather than on the crane itself. The crane load models in EN 1991-3 are based on those in EN 13001-2:2004 which in turn have been developed from the German and International standards. EN 1991-3 then has the advantage in that it is based on the codes that are used for design of the crane itself but has been modified and formulated for application to crane imposed loads on support structures.

The Eurocode steel design code contains a section for the design of crane support structures (EN 1993-6). EN 1991-3 was developed in close cooperation with the working groups for the loads on cranes (EN 13001-2) as well as on the design of crane support structures (EN 1993-6) (Sedlacek & Grotmann 1996). This suggests a further advantage of EN 1991-3 in that uniformity and compatibility are ensured throughout the design process of the crane and support structure when designing according to the Eurocode set of Standards.

This uniformity is partly carried through to the South African Standards in that the codes used for the design of the crane have the same basis as EN 13001-2. Should the Eurocode steel design code for crane support structure design also be adopted as a South African standard, the crane and support structure design processes would be completely harmonised.

EN 1991-3 has a further advantage over the German, International and Australian crane loading codes in that it has been simplified to some degree for application to the support structure while still being based on the same principles as the codes for loads acting on cranes.

EN 1991-3 therefore has the advantage that it is based on the codes that specify loads on cranes but has been specially formulated for application to crane loads imposed on the support structure and, combined with the steel design code for crane support structures, ensures a harmonised design process of crane and support structure. For these reasons, the load models from EN 1991-3 were selected as possible load models for the updated loading code SANS 10160-6.

3.2 Basis for crane load provisions in EN 1991-3

Crane loads are inherently dynamic due to the movement of the crane and the loads lifted. The crane wheel loads in EN 1991-3 are treated as quasi-static loads where the effect of the crane dynamics is taken into account by applying dynamic amplification factors to the static wheel loads.

EN 1991-3 divides cranes into four hoist classes based on a description of the use of the crane. The rationale behind the classification of cranes in EN 1991-3 is to distinguish between cranes based on their dynamic characteristics, specifically their dynamic response to lifting a load.

4 VERTICAL WHEEL LOADS IN EN 1991-3:2002

4.1 Load model for vertical wheel loads

The vertical crane wheel loads in EN 1991-3 are calculated using static equilibrium. The assumption is made that the vertical load on one end carriage is evenly distributed between the wheels on that end carriage.

Two loading scenarios are considered for the calculation of the maximum and minimum vertical wheel loads and can be explained with reference to Figure 1(a) & (b). The maximum vertical wheel loads are calculated considering the crane to be carrying the maximum hoistload with the crab at the extreme position of its travel across the bridge, as shown in Figure 1(a). $Q_{r,max}$ is the maximum wheel load on the end carriage closest to the crab and $Q_{r,(max)}$ is the wheel load on the opposite end carriage. The absolute minimum wheel load, $Q_{r,min}$, is attained with an unloaded crane, on the end carriage furthest from the crab, as shown in Figure 1(b). $Q_{r,(min)}$ is then the accompanying minimum wheel load on the end carriage closest to the unloaded crab.

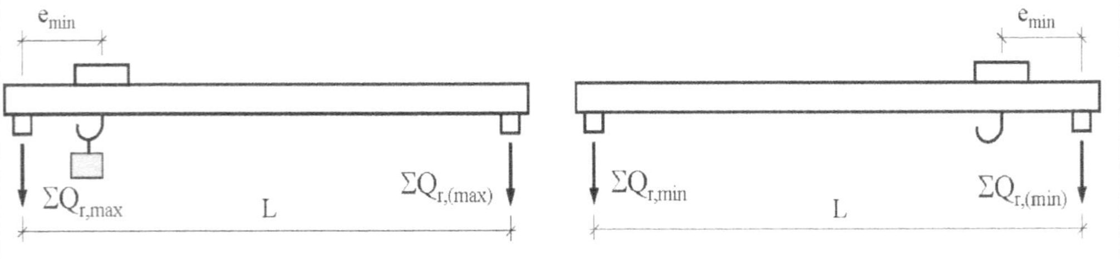

Figure 1 Maximum and minimum vertical wheel loads

The load model for calculating the maximum wheel load on one wheel of the crane is given by:

$$Q_{r,\max} = \frac{1}{2}\left[\frac{Q_{br}}{2} + \frac{L - e_{\min}}{L}\left(Q_{cr} + Q_h\right)\right] \tag{1}$$

Where: Q_{br} is the weight of the crane bridge
Q_{cr} is the weight of the crab
Q_h is the weight of the hoistload

4.2 Dynamic effects

The dynamic effects that are taken into account for the vertical crane wheel loads are those due to hoisting a load off the ground, sudden release of a part of the hoistload or running over uneven rails.

4.2.1 Hoisting a load off the ground

The dynamic effects of hoisting a load off the ground are modelled separately for the hoistload and the crane structure. The dynamic amplification factors are given in Table 3 & 4.

Table 3 Dynamic amplification factors due to hoisting a load off the ground

ϕ_1	Dynamic amplification of the self weight of the crane due to lifting the hoistload off the ground	1,10
ϕ_2	Dynamic amplification of the hoistload due to its being lifted off the ground.	$\phi_2 = \phi_{2,\min} + \beta_2 v_h$ v_h = steady hoisting speed (m/s) $\phi_{2,\min}$ and β_2 depend on the Hoist Class and are given in Table 4

Table 4 Values of $\phi_{2,\min}$ and β_2

Hoist Class	$\phi_{2,\min}$	β_2
HC1	1,05	0,17
HC2	1,10	0,34
HC3	1,15	0,51
HC4	1,20	0,68

The maximum vertical wheel load for the situation when the crane is hoisting a load is given by:

$$Q_{r,\max} = \frac{1}{2}\left[\frac{\phi_1 Q_{br}}{2} + \frac{L - e_{\min}}{L}\left(\phi_1 Q_{cr} + \phi_2 Q_h\right)\right] \tag{2}$$

4.2.2 Sudden release of a part of the hoistload

A dynamic amplification factor ϕ_3 is provided in EN 1991-3 for the dynamic effects caused by the sudden release of a part of the hoistload for cranes equipped with grabs or magnets:

$$\phi_3 = 1 - \frac{\Delta m}{m}\left(1 - \beta_3\right) \tag{3}$$

Where: β_3 = 0,5 for cranes equipped with slow release mechanisms e.g. grab
β_3 = 1,0 for cranes equipped with rapid release mechanisms e.g. magnet
Δm is the part of the hoistload that is released
m is the total hoistload lifted

ϕ_3 will always be less than 1,0 therefore this is unlikely to be a critical load situation for the ultimate limit state but may affect the fatigue analysis.

4.2.3 Travelling on uneven rails

The dynamic amplification factor ϕ_4 provides for the dynamic effects of a crane traversing an uneven rail. ϕ_4 is taken as 1,0 if the rail tolerances in EN 1993-6 are observed. A detailed method based on a single degree-of-freedom system for calculating ϕ_4 in the event that the rail joint tolerances are not observed, is referred to by EN 1991-3, where the dynamic effects of the crane traversing either a step or a gap in the rail are considered.

5 HORIZONTAL WHEEL LOADS IN EN 1991-3:2002

5.1 Acceleration and deceleration of the crane

Horizontal, transverse and longitudinal wheel loads result from the acceleration or deceleration of the crane bridge. The longitudinal loads are caused by the drive force acting at the contact between the wheel and rail. Transverse forces are present when the crane accelerates or decelerates with the crab eccentric to the centre of mass. A moment is induced about the centre of mass of the crane which is resisted by horizontal transverse couple forces at the wheels. The forces are shown in Figure 2.

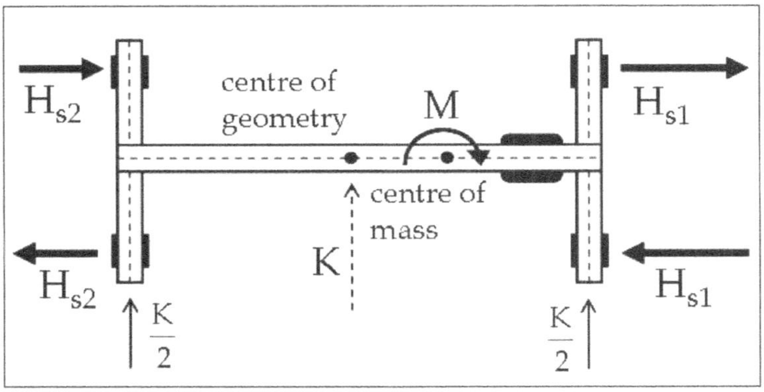

Figure 2 Horizontal loads caused by acceleration or deceleration of crane

The drive forces are calculated on the basis that no wheel spin occurs, i.e. the maximum drive force is equal to the minimum friction force between the rail and wheel. In the case when the crane is equipped with single wheel drives, the drive force on each line of wheels is assumed to be equal and is calculated as follows:

$$K = \mu m_w Q_{r,\min} \tag{4}$$

Where: K is the total drive force
μ is the friction factor between rail and wheel
m_w is the number of single wheel drives
$Q_{r,\min}$ is the minimum vertical wheel load

The resultant of the drive forces acts through the centre of geometry of the crane. In the case where the crab is eccentric to the centre of the crane bridge, the centre of mass will be offset from the centre of geometry. The drive force causes a moment M around the centre of mass. Horizontal transverse forces acting at the wheels form couples that resist the moment M. The magnitudes of the transverse forces are determined by taking into account the distribution of mass of the crane. The forces on the wheels closest to the crab are larger than those on the opposite wheels because greater horizontal friction can be developed due to the larger vertical forces.

A dynamic amplification factor ϕ_5 is included in the calculation of the horizontal, longitudinal and transverse wheel loads due to drive forces, to take into account the dynamic characteristics of the wheel drives. The values of ϕ_5 are shown in Table 5.

Table 5 Dynamic amplification factor for drive forces

$1{,}0 \leq \phi_5 \leq 1{,}5$	Drives where forces change smoothly, e.g. drives equipped with a frequency inverter
$1{,}5 \leq \phi_5 \leq 2{,}0$	Drives where forces change suddenly, e.g. drives equipped with gears
$\phi_5 = 3{,}0$	Drives with considerable backlash

5.2 Skewing of the crane in plan

Horizontal, longitudinal and transverse skewing forces are caused by friction at the crane wheels due to the crane running at an angle to the line of the rails. The mechanism causing the skewing forces and their calculation will be described with reference to Figure 3 which shows a crane equipped with guide rollers.

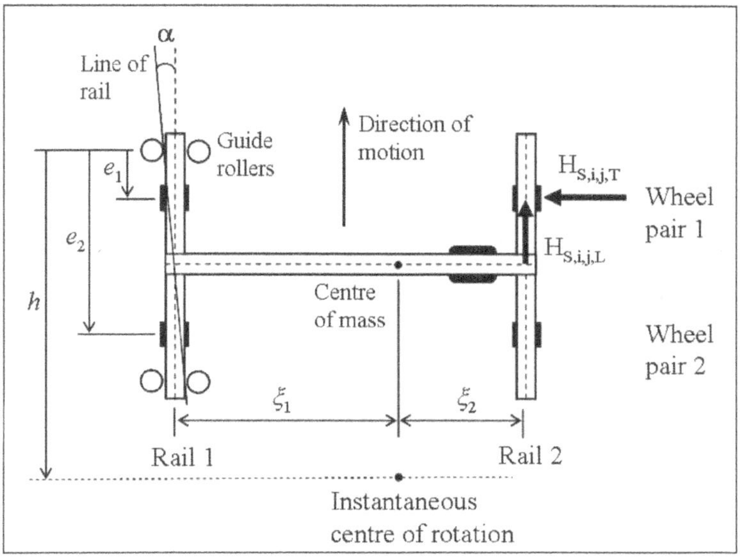

Figure 3 Forces due to skewing of the crane in plan

The direction of motion of the crane is at an angle α to the line of the rails. The deviation of the motion of the crane from the line of the rails is caused by the alignment tolerances of the rails and wheels, the clearance between the rail and guide means (wheel flange or guide rollers) and the wear of the rail and guide means.

The crane is assumed to be travelling at constant speed. Skewing is modelled as the crane rotating about the "instantaneous centre of rotation" the position of which depends on the number and type of wheels, guidance means and the position of the centre of mass. Each wheel will slip transversely and longitudinally to the rail. The amount of slip depends on the position of the wheel relative to the instantaneous centre of rotation. As the wheel slips, it also rolls forwards due to the forward motion of the crane. The forces caused by the slipping and rolling motion of each wheel depend on the sliding distance, the free rolling distance, the wheel load and the surface conditions of the rail and wheel. A generalised, simplified, empirical expression for the effective friction factor f has been developed which depends only on the skewing angle α (ISO 8686-1). This friction factor is referred to in EN 1991-3 as the "non-positive factor" and is given by Equation 5.

$$f = 0.3(1-\exp(-250\alpha)) \leq 0.3 \qquad (5)$$

Where: α is the skewing angle in radians

The transverse and longitudinal wheel loads induced at the wheels due to the wheel slip on the rails, are resisted by a transverse force at the front guidance means. The maximum lateral slip occurs at the front guidance means and a linear distribution of slip is assumed between the guidance means and the instantaneous centre of rotation. The transverse and longitudinal wheel loads are calculated as follows:

$$H_{S,i,j,k} = f\,\lambda_{S,i,j,k} \sum Q_r \tag{6}$$

The guide force is calculated as follows:

$$H_S = f\,\lambda_S \sum Q_r \tag{7}$$

Where: f is the equivalent friction coefficient
$\lambda_{S,i,j,k}$ is the force factor accounting for the wheel position relative to the guidance means and instantaneous centre of rotation
i is the rail i
j is the wheel pair j
k is the direction of the forces, T = transverse, L = longitudinal
ΣQ_r is the combined weight of the crane and hoistload

EN 1991-3 provides formulae for the calculation of the force factors for independent or coupled wheels which are either fixed or moveable in the lateral direction.

Figure 4 shows the position of the instantaneous centre of rotation and the skewing forces for the most common crane configuration: a four wheeled crane guided by wheel flanges, with independently fixed wheels.

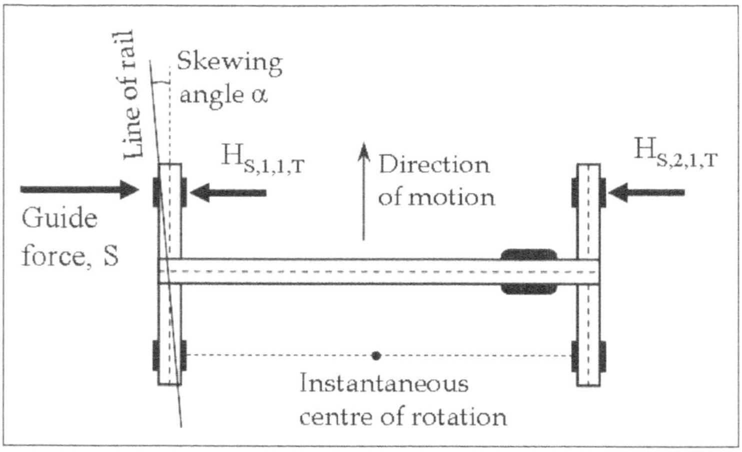

Figure 4 Skewing forces for a four wheeled crane

5.3 Acceleration and deceleration of the crab

EN 1991-3 states that the horizontal transverse forces caused by acceleration and deceleration of the crab may be assumed to be taken into account by the end stop forces caused by the crab running into the end stops at the end of the crane bridge.

The horizontal transverse loads due to acceleration and deceleration of the crab are unlikely to be the critical horizontal loads for the ultimate limit state but do play a role in the fatigue design as they are likely to occur for every load cycle.

6 VERTICAL WHEEL LOADS DUE TO TEST LOADS

Two types of load tests are performed on overhead travelling cranes as part of the safety requirements. A test to 25% overload is carried out on cranes after installation, refurbishment or upgrading by loading the crane to 1,25 times the safe working load (SWL) without moving any of the drives. The second type of test is carried out to 10% overload and is performed annually by loading the crane to 1,10 times the safe working load and moving the crane drives in the way in which the crane is usually operated.

EN 1991-3 makes provision for the two load tests by specifying the value of the hoistload (1,25 SWL or 1,10 SWL) which is to be used in calculating the vertical wheel loads.

The dynamic amplification factor applicable to the test loads is ϕ_6 and is applied to the hoistload, replacing ϕ_2 in Equation 2. A distinction is made between the 25% overload "static" test which has a dynamic factor of $\phi_{6,\text{static}} = 1$, and the 10% overload "dynamic" test which has the dynamic factor given below. The dynamic factor for hoisting the test load is less than that for normal hoisting of working loads because the test loading is carried out under more controlled conditions than normal operating conditions of the crane.

$$\phi_{6,\text{dynamic}} = \frac{1+\phi_2}{2} \tag{8}$$

7 ACCIDENTAL CRANE ACTIONS

Three accidental actions are provided for in EN 1991-3, which are buffer forces related to crane movement, buffer forces related to crab movement and tilting of a crane equipped with a mast. The buffer forces relate to the situation where the crane or crab runs into the end stops at the extreme of their travel on the gantry or crane bridge respectively. This is an accidental situation because most cranes are equipped with limit switches which automatically slow and stop the crane or crab before the end of their travel and an impact would thus only occur in the event of failure of the limit switches.

7.1 Buffer forces related to crane movement

The principle that is given in EN 1991-3 for the determination of the end stop buffer forces related to crane movement is that the forces should be calculated considering "the kinetic energy of all relevant parts of the crane moving at 0,7 – 1,0 times the nominal speed." An application of this principle is given in an equation for the calculation of the buffer force on each end stop as:

$$H_{B,1} = \phi_7 v_1 \sqrt{m_c S_B} \tag{9}$$

Where: ϕ_7 is the dynamic amplification factor for the end stop forces
v_1 is the initial speed of the crane, taken as 70% of the long travel speed [m/s]
m_c is the total mass of the crane and hoistload [kg]
S_B is the spring constant of the buffer [N/m], determined from the energy-deformation and force-deformation buffer curves.

The basis for the equation is simple harmonic motion, where $H_{B,1}$ is the force at the time when the velocity is equal to zero, i.e. the spring has the maximum compression. Simple harmonic motion assumes an elastic spring which would not be the case for most buffers. The dynamic amplification factor ϕ_7 takes into account the non-linearity of the buffer and is calculated by:

$$\phi_7 = 1.25 + 0.7(\xi_b - 0.5) \quad (10)$$

Where: ξ_b is the buffer characteristic, a measure of the non-linearity of the buffer as described in Figure 5.

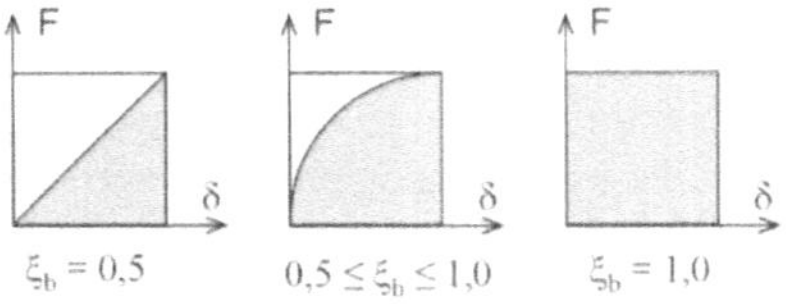

Figure 5 Buffer characteristic

7.2 Buffer forces related to crab movement

EN 1991-3 specifies that if the hoistload is free to swing, the total buffer force resulting from collision of the crab with the end stops may be calculated as 10% of the combined weight of the crab and the hoistload. In the case where the hoistload is not free to swing, the crab buffer forces should be calculated in a similar manner to the crane buffer forces.

The distribution of the buffer forces between the wheels of the crane depends on the wheel configuration and whether the wheels are fixed or free to move laterally.

7.3 Tilting forces

Tilting forces occur in the event that a crane equipped with a rigid lifting attachment, for example a mast, collides with an object causing the crane to tilt and some of the wheels to lift off the rails. EN 1991-3 places the responsibility for determining the resulting forces on the design engineer by simply stating "the resulting static forces shall be considered."

8 TEMPERATURE EFFECTS AND WALKWAY LOADS

EN 1991-3 states that temperature effects on crane support structures should be taken into account where it is necessary, for example in the case of outside gantries. In this case, it is stated that the mean temperature should be taken as 20 °C with a range of 35 °C.

Recommendations are given in EN 1991-3 for vertical and horizontal loads on access walkways, stairs, platforms and guard rails due to people or materials which may be deposited in these areas.

9 CRANE LOAD COMBINATIONS

EN 1991-3 specifies combinations of crane vertical and horizontal loads that are to be considered as one action for combinations with other loads such as wind load or imposed load. The table from EN 1991-3 which defines the crane load combinations together with the relevant dynamic factors is shown in Table 6.

Table 6 Crane load combinations as given in EN 1991-3

	Symbol	Groups of loads									
		ULS							Test loads	Accidental	
		1	2	3	4	5	6	7	8	9	10
Self-weight of crane	Q_c	ϕ_1	ϕ_1	1	ϕ_4	ϕ_4	ϕ_4	1	ϕ_1	1	1
Hoistload	Q_{hl}	ϕ_2	ϕ_3	-	ϕ_4	ϕ_4	ϕ_4	$\eta^{(1)}$	-	1	1
Acceleration of crane bridge	H_L, H_T	ϕ_5	ϕ_5	ϕ_5	ϕ_5	-	-	-	ϕ_5	-	-
Skewing of crane bridge	H_S	-	-	-	-	1	-	-	-	-	-
Acceleration or braking of crab or hoist block	$H_{T,3}$	-	-	-	-	-	1	-	-	-	-
In service wind	F_W	1	1	1	1	1	-	-	1	-	-
Test load	Q_T	-	-	-	-	-	-	-	ϕ_6	-	-
Buffer force	H_B	-	-	-	-	-	-	-	-	ϕ_7	-
Tilting force	H_{TA}	-	-	-	-	-	-	-	-	-	1

(1) η is the part of the hoistload that remains when the payload is removed, but is not included in the self weight of the crane.

In service wind applies only to outside gantries and not those inside a building where the building resists the wind loads.

Each crane load combination models a particular crane activity, for example: ULS combination 1 represents the crane lifting a load (indicated by dynamic factors ϕ_1 and ϕ_2) and accelerating (indicated by dynamic factor ϕ_5), ULS combination 3 represents the crane accelerating (ϕ_5) without a hoistload and ULS combination 5 represents the crane travelling at a constant speed and skewing.

Of the load combinations classified for the ultimate limit state, it can be seen by inspection that only combinations 1, 5, and 6 could be critical for the ultimate limit state design of the support structure. The remaining combinations augment the description of possible crane behaviour and are useful for design of the support structure for fatigue.

10 FATIGUE

Crane loading is cyclic by nature and therefore may induce fatigue failure in the crane girders, columns and other structural elements subject to crane loads. The determination of the range of loads which cause fatigue in the support structure can be complex, requiring information on the distribution of loads lifted, movement of the crane in the building and movement of the crab on the crane.

EN 1991-3 gives as the principle governing the evaluation of fatigue that the loads causing fatigue shall be determined taking into account the distribution of loads lifted and the movement of the crane and how they effect the stress at a given fatigue detail. As an application of this principle EN 1991-3 provides a method of calculating a "fatigue damage equivalent load" by considering the number of crane cycles performed and the distribution of loads lifted over the lifetime of the crane. It is stated that this method of fatigue assessment is compatible with the crane design code (EN 13001) but is a simplified method which may be used in the design of gantry girders to compensate for incomplete information on the crane loads or movements.

The fatigue damage equivalent load is a normalisation of the crane loading to a constant amplitude load applied for two million cycles which would result in the same fatigue damage as the actual crane loading. The crane is classified into a fatigue class based on the number of cycles it will perform during its lifetime and the load spectrum.

The load spectrum kQ is determined by considering the amplitudes of the wheel loads as given in Equation 11.

$$kQ = \sum_j \left(\left(\frac{\Delta Q_{i,j}}{\max \Delta Q_i} \right)^m \frac{n_{i,j}}{\sum n_{i,j}} \right) \quad (11)$$

Where: $\Delta Q_{i,j}$ is the load amplitude of range j for wheel i ($Q_{i,j} - Q_{\min,i}$)
$\max\Delta Q_i$ is the maximum load amplitude for wheel i ($Q_{\max,i} - Q_{\min,i}$)
i is the number of the wheel
$n_{i,j}$ is the number of cycles carried out at load level j for wheel i

The number of cycles the crane performs relative to two million is expressed by ν, given in Equation 12.

$$\nu = \frac{\sum_j n_{i,j}}{N_i} \quad (12)$$

Where: N_i is the reference number of cycles (2×10^6)

The fatigue damage equivalent wheel load Q_e is then given by Equation 13.

$$Q_e = \varphi_{fat}\ \lambda_i\ Q_{\max,i} \quad (13)$$

Where: ϕ_{fat} is the fatigue dynamic factor, $\phi_{fat,1} = 0{,}5(1+\phi_1)$; $\phi_{fat,2} = 0{,}5(1+\phi_2)$, applied in the same manner as ϕ_1 and ϕ_2.
λ_i is the damage equivalent factor = $\sqrt[m]{kQ \times \nu}$
m is the slope of the fatigue S-N curve

The fatigue damage is then calculated assuming that the fatigue damage equivalent load passes over the relevant fatigue detail two million times.

The fatigue classification of the crane may be given in the specification documents of the crane or may be determined from a description of the cranes which fall into each fatigue class, given in a table in EN 1991-3. A further table is given which specifies the damage equivalent factors for each fatigue class for both normal and shear stresses.

11 BASIS OF DEVELOPMENT OF CRANE LOADING PROVISIONS IN SANS 10160-6

The updated crane load models will form Part 6 of the South African loading code SANS 10160. Various aspects of codified crane loading should be considered for the development of Part 6, in order to ensure that the crane loading provisions are consistent with the South African basis of design and applicable to the South African conditions.

The range of configurations of cranes whose loads may be described by the crane load models should be clearly defined. For example, it should be clearly stated whether the crane load models are applicable only for overhead travelling bridge cranes or if they may be applied to portal cranes or cranes with wheels at different levels. The applicability of the load models to cranes mounted on top of rails as well as under-slung cranes should be considered. The effects of different types of load lifting attachments (e.g. masts, grabs, magnets, ladles, hooks) should be examined and the applicability of the load models ascertained.

The assessment and selection of the load models from EN 1991-3 (ultimate and accidental limit states) and possible extension of these load models should be carried out on the basis of the following considerations:

1. Exhaustiveness: do the load models cover all the crane load situations that the support structure is likely to encounter during its lifetime?
2. Validity: are the underlying mechanics and method of calculation of the load model valid?
3. Applicability: are the EN 1991-3 load models applicable to South African conditions (e.g. construction tolerances) and crane manufacturing practices?

The load combination table provided in EN 1991-3 should be reviewed to ascertain whether all possible load situations have been included. The load combination table should be updated to include any omissions, additions or modifications of the load models.

The issue of crane support structure fatigue should be considered. EN 1991-3 provides a detailed method for obtaining fatigue loading whereas the current practice in the South African design standards is not to include fatigue in the loading code but rather in the materials codes. Fatigue is a situation where the loads and resistances are more closely linked than in ultimate strength design and it is therefore debatable whether fatigue should be dealt with in the loading code. Three options have been identified with respect to fatigue design provisions in SANS 10160-6: a detailed method of calculating fatigue loads in line of that given by EN 1991-3, the current practice of addressing fatigue design in the materials codes rather than the loading code or providing basic guidance on which crane load combinations occur frequently enough to cause fatigue in the support structure. In the event of the election for a detailed fatigue load calculation method, the fatigue provisions in EN 1991-3 should be critically assessed for their application to support structures.

The load models in EN 1991-3 are more complex than those in SABS 0160 and may be considered over-complex for small, simple crane installations. The option of including simplified crane load models for application to these types of crane installations should be considered. The simplified models should be shown to always yield conservative loads and their field of application should be clearly defined.

The partial load factor which is applied to crane loads should be reviewed. The crane partial load factor in SABS 0160 is the imposed load factor of 1,6. In contrast EN 1991-3 applies a crane partial load factor equal to the Eurocode permanent load factor of 1,35. There is a clear difference in value and classification with either imposed or permanent loads between the two codes. There is also no reliability basis for the SABS 0160 crane partial load factor. This issue should be addressed for the updated loading code SANS 10160.

12 CONCLUSIONS

An overview of the crane load provisions in SABS 0160 has been presented. These load models are over-simplified in comparison to those in international standards. A further shortcoming of the SABS 0160 crane load provisions is their dissimilarity to the loads applied for design of the crane itself. The crane provisions are therefore to be updated in the forthcoming revision of the loading code, SANS 10160-6.

Four crane loading codes were considered to find a basis for the revision of the crane load models for SANS 10160-6. The Eurocode (EN 1991-3) was found to have advantages over the German code (DIN 15018-1), International Standard (ISO 8686-1:1989) and Australian code (AS1418.1-1994) in that it is based on the codes used for the crane design but formulated specifically for support structures. It was developed in close cooperation with the developers of the crane design code and the section of the steel design code on the design of crane support structures. The use of these three codes together would ensure uniformity and harmonisation of the crane – support structure design process.

The basis of the Eurocode crane design code is the same as the codes used in South Africa for crane design and in the event of the crane support structure design code being adopted, the harmonisation could be carried through into South African crane – support structure design. EN 1991-3 was therefore selected to form the basis of the crane loading section of SANS 10160.

An overview of the crane load models in EN 1991-3 was given where their greater sophistication and transparency, with respect to crane mechanics, compared to those in SABS 0160 was observed.

The basis for the revision of the crane load provisions in the South African loading code has been developed. The aspects of codified crane loading to be considered were identified and are mentioned again briefly. The range of crane configurations covered by the crane load models should be identified. The load models should be selected for inclusion in to SANS 10160 based on their exhaustiveness with respect to load situations considered, validity of the underlying crane mechanics and applicability to South African conditions.

The crane load combinations should be assessed as well as the provisions made for fatigue loading. The crane partial load factor should be reviewed in order to determine a more rational, reliability based partial load factor.

13 REFERENCES

AS1418.1-1994 (1994): *Cranes (including hoists and winches)* Part 1: *General requirements*. Standards Australia.

DIN 15018 Part 1 (1984): Deutsche Norm: *Cranes, steel structures, verification and analysis*. Beuth Verlag GmbH, Berlin.

Dunaiski, P.E., Barnard, H., Krige, G. and Mackenzie, R. (2001): *Review of provision of loads to structures supporting overhead travelling cranes*. In: Zingoni, A. (ed.), Proceedings SEMC, pp. 1321-1328. Elsevier.

EN 13001-2 (2004). *Crane safety – General design* – Part 2: *Load effects*. CEN/TC 147 European Committee for Standardisation.

EN 1991-3 (2002). Eurocode 1: *Actions on structures*, Part 3: *Actions induced by cranes and other machinery*. European Committee for Standardisation (CEN), Brussels.

ISO 8686-1:1989 (1989): International Standard: *Cranes – Design principles for loads and load combinations*, Part 1: *General*. International Organisation for Standardisation, Geneva, Switzerland.

Krige, G.J. (1998): *Overhead crane and machinery loads*, SA National Conference on Loading. SAICE & SAISC, 9 – 10 September 1998, Midrand, South Africa.

Sedlacek, G. and Grotmann, D. (1996): EC 1: *Actions induced by cranes*, IABSE Report Vol 74, IABSE Colloquium on Basis of Design and Actions on Structures, Background and Application of Eurocode 1, Delft.

SABS 0160:1989 (1990) (Amended 1990). *Code of Practice: The general procedures and loadings to be adopted in the design of buildings*. South African Bureau of Standards, Pretoria.

6-2 Revised Provisions for Crane Induced Actions

Dunaiski PE & Dymond JS

1 INTRODUCTION

The crane load models in the South African code for loadings on buildings, SABS 0160:1989 (Amended 1993), have been shown to be over simplistic in comparison with crane load models in international codes of practice, namely the German Standard (DIN 15018-1:1984), International Standard (ISO 8686-1:1989), Australian Standard (AS1418.1-1994) and Eurocode (EN 1991-3). Furthermore, the SA loading code (referred to as SABS 0160 for conciseness) load models do not reflect the loads applied to the crane in the design of the crane itself, though the loads imposed on the crane must be the same as those imposed by the crane on the support structure. Because of this, the crane load models have been reviewed for the revision of the South African loading code SANS 10160.

The four codes mentioned above were investigated in order to find a basis for the revision of the crane load models. The Eurocode (EN 1991-3) was selected as the most suitable reference code because it is based on the methods used for design of cranes but has been modified for specific application to the support structure. EN 1991-3 is the most recently revised crane loading code and has been subject to an extensive review process.

Verification of the EN 1991-3 crane load models is required to determine their suitability for inclusion into SANS 10160-6. The basis that was developed for their verification can be described by: validity – assessment of the underlying crane mechanics and method of calculation; applicability – suitability of load models to South African conditions and the scope of SANS 10160-6; exhaustiveness – assessment of whether all possible crane load situations are accounted for.

The implications of the revised load models will be investigated in terms of design effort, encompassing the effort required for the calculation of the loads as well as obtaining the information required for the load calculation and cost of the support structure.

A further issue to be considered for the revision of the crane load models is the partial load factor. The partial load factor applied to crane loads in SABS 0160 is the imposed load factor of 1,6. EN 1991-3 applies a partial load factor to crane loads of 1,35 which is the Eurocode permanent load factor. There is no reliability basis for the application of either of these partial load factors to crane loads therefore an investigation into the reliability of crane imposed loads is required. A reliability code calibration of the crane load models has been carried out and is reported in Chapter 6-3.

2 SCOPE OF THE CODE

The scope of the code in this instance refers to the range of cranes whose wheel loads may be expressed by the load models in the code. The scope of SANS 10160-6 is stated as "overhead travelling bridge cranes on runway beams at the same level", i.e. excluding portal or semi-portal cranes and tower cranes. The scope of EN 1991-3 is stated as "cranes on runway beams" which is similar to SANS 10160 but without the proviso that the runway beams should be at the same level. EN 1991-3 was based on EN 13001-2: Crane Safety – General design – Part 2: Load effects, which considers all types of cranes, which should be borne in mind when assessing the applicability of the EN 1991-3 load models for incorporation into SANS 10160-6.

3 CRANE CLASSIFICATION

EN 1991-3 classifies cranes into four hoist classes in order to distinguish among cranes based on their hoisting characteristics, specifically the dynamic response to hoisting a load. Unlike SABS 0160 where most of the crane wheel loads depend on the crane class, in EN 1991-3 the hoist class affects only the dynamic factor for hoisting a load. The EN 1991-3 crane classification system has been incorporated into SANS 10160-6 by including as an informative Annex, the table of cranes and hoist classes from EN 1991-3, modified to cover the types of overhead travelling cranes typically encountered in South Africa. It has been shown that cranes tend to fall into corresponding classes in SABS 0160 and EN 1991-3, and hence SANS 10160-6.

4 VERIFICATION OF CRANE LOAD MODELS

The crane load models used in EN 1991-3 have been assessed by using the criteria for validity, applicability and exhaustiveness as discussed in Chapter 6-1. Based on this assessment, models from EN 1991-3 have been selected for incorporation into SANS 10160-6.

Comparative calculations have been carried out for a 40t crane which can be considered representative of cranes in South Africa (Warren et al. 2004) for the crane load provisions in SABS 0160 and SANS 10160-6, considering the crane as class 3 and hoist class 3. The characteristic crane wheel loads for each load case are shown schematically for both codes.

4.1 Vertical loads

4.1.1 Static vertical loads

The load model in EN 1991-3 for calculating vertical loads is static equilibrium. The assumption is made that the vertical force for each end carriage is divided equally among the wheels on that end carriage. This load model has been incorporated into SANS 10160-6. If the assumption about the distribution of loads among the wheels on an end carriage is clearly not valid for a particular crane, then the vertical wheel loads may be obtained from the crane manufacturer.

Figure 1 shows the static vertical wheel loads from SABS 0160 and SANS 10160-6 respectively.

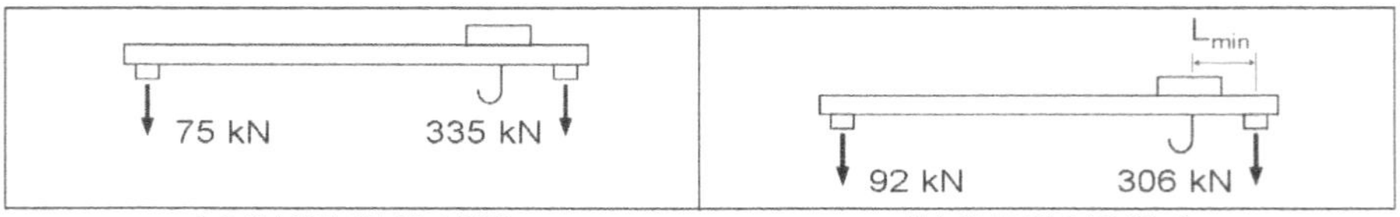

(a) SABS 0160:1989 (b) SANS 10160-6

Figure 1 Comparison of static vertical wheel loads

4.1.2 Basis for dynamic vertical wheel loads

Whereas SABS 0160 provides only a general impact factor, the magnitude of which depends on the class of the crane, EN 1991-3 distinguishes between dynamic effects caused by hoisting, sudden release of a load and travelling over uneven rails.

4.1.3 Dynamic effect of hoisting a load

The dynamic effect of hoisting a load off the ground is modelled by EN 1991-3 separately for the crane self weight and hoist load.

The factor applied to the self weight is given in EN 1991-3 as: $0{,}9 \leq \phi_1 \leq 1{,}1$ to reflect the upper and lower values of the vibration. The lower value is not significant for overhead bridge cranes but is of more interest for tower type cranes for which stability may be critical. Only the upper value of 1,1 has been incorporated into SANS 10160 for compatibility with the scope of SANS 10160-6.

The dynamic factor for the hoist load ϕ_2 is dependent on the crane hoist class and the steady hoisting speed and was incorporated into SANS 10160-6 unchanged.

4.1.4 Dynamic effect of releasing a load

The load situation of sudden release of the hoist load is applicable to grab and magnet cranes which routinely release part or the whole of the hoist load during normal operation. A dynamic factor ϕ_3 is applied to the hoistload to model the dynamic effects of sudden release of a part of the hoistload. The dynamic factor ϕ_3 will always be less than one and is therefore not critical for overhead travelling bridge cranes, but rather for cranes for which stability is critical. This load situation and dynamic factor have, however, been incorporated into SANS 10160-6 in order to represent in more detail the full range of crane behaviour and load situations. This load case may also be relevant for fatigue calculations.

4.1.5 Dynamic effect of travelling on uneven rails

EN 1991-3 specifies that if the rail tolerances in EN 1993-6 are observed then $\phi_4 = 1{,}0$ otherwise reference is made to a method of calculating ϕ_4 in EN 13001-2 (2004), which is the same as the method given in ISO 8686-1:1989. The same approach has been adopted for SANS 10160-6, but in the South African context the rail tolerances are given by SANS 2001 CS1 and the method for calculating ϕ_4 as given in ISO 8686 has been included in SANS 10160-6 as informative Annex B.

Figure 2 shows the vertical impact forces from SABS 0160 and the vertical dynamic forces due to hoisting a load off the ground according to SANS 10160-6. The dynamic effects due to releasing a load are dependent on the fraction of the load released and are not critical for the ultimate limit state verification and the dynamic factor for travelling on uneven rails is 1,0; therefore the loads for these two cases have not been depicted.

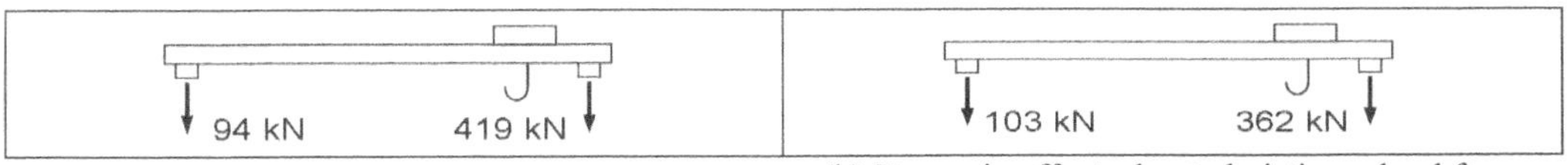

(a) Impact forces from SABS 0160:1989

(b) Dynamic effects due to hoisting a load from SANS 10160-6

Figure 2 Comparison of dynamic vertical loads

4.2 Horizontal wheel loads

4.2.1 Horizontal wheel loads due to acceleration of the crane bridge

The SABS 0160 load model for acceleration of the crane bridge considers the longitudinal loads only whereas EN 1991-3 considers the eccentricity of the crab causing the crane to skew, which is a more rational consideration of the complete crane behaviour under this operating condition. The method of calculating the drive forces in EN 1991-3 corresponds to the method of design of the wheel drives as carried out by the crane manufacturers. In EN 1991-3 the dynamic effects of the drive motors are modelled by considering the motor behaviour comprising smooth changes, sudden changes or backlash, whereas no provision for dynamic effects of the drive motors is made in SABS 0160.

The EN 1991-3 load model for acceleration of the crane bridge is rational, takes the smoothness of change of drive forces into account and calculates the magnitude of the drive force in a way that is compatible with the crane manufacturing process and has therefore been incorporated into SANS 10160-6. EN 1991-3 provides a range of options for the value of the dynamic factor modelling the dynamic effects of the behaviour of the drive, but no guidance is given on which value to choose. In order to simplify the calculation process, only the maximum value of the dynamic factor for each drive behaviour has been included in SANS 10160-6. Figures 3 shows the horizontal wheel loads calculated from SABS 0160 and SANS 10160-6 respectively.

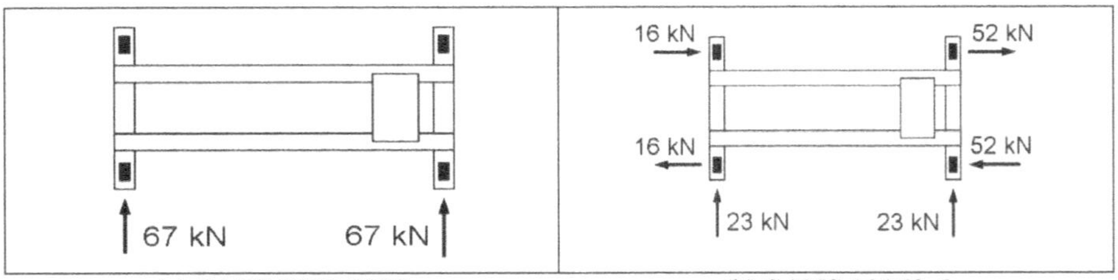

(a) SABS 0160:1989 (b) SANS 10160-6

Figure 3 Horizontal loads due to acceleration of the crane

4.2.2 Horizontal wheel loads due to skewing of the crane bridge in plan

The load model for skewing from EN 1991-3 is based on an empirical expression for the friction factor for a wheel rolling and slipping on a rail due to the crane running at an angle to the line of the rails. Coefficients are provided by EN 1991-3 for the calculation of the skewing loads based on four wheel configurations. Wheels are considered as either "independent" or "continuous" and either "fixed" or "movable". Independent wheels are not coupled in any way with the wheels on the opposite end carriage whereas continuous wheels are coupled either electrically or mechanically. Movable wheels allow lateral displacement of the wheels along the axle and fixed wheels cannot move laterally. Consultations with the leading crane manufacturers in South African determined that it is current practice to use only independent, fixed wheels, therefore only the coefficients relating to this wheel configuration were incorporated into SANS 10160-6.

Figure 4 shows the wheel loads due to skewing from SABS 0160 and SANS 10160-6 respectively.

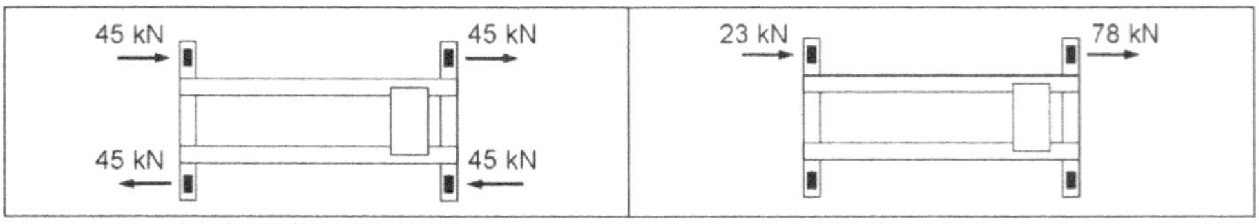

(a) SABS 0160:1989 (b) SANS 10160-6

Figure 4 Comparison of forces due to skewing

4.2.3 Horizontal wheel loads due to acceleration of the crab

EN 1991-3 states that the load situation of acceleration or braking of the crab may be assumed to be covered by the buffer forces related to movement of the crab. The buffer force is an accidental load that is used only for verification of the accidental limit state whereas acceleration and braking of the crab is a load situation that occurs frequently during normal operation of the crane and should therefore be taken into account for verification of the ultimate limit state and fatigue.

The forces due to crab acceleration are unlikely to be critical for verification of the ultimate limit state but may play a role in the assessment of fatigue depending on the operational procedure of the crane.

Furthermore, the partial load factor differs from 1,6 for the ultimate limit state considering regular loads to 1,0 for the accidental limit state which brings into question the validity of assessing a regular, frequently occurring load situation by means of an accidental load. Therefore the load model as provided for the crab buffer force in EN 1991-3 was incorporated into SANS 10160-6 as the load model for acceleration and braking of the crab as a regular, frequently occurring load.

Figure 5 shows the wheel loads due to acceleration and braking of the crab as given in SABS 0160 and SANS 10160-6 respectively.

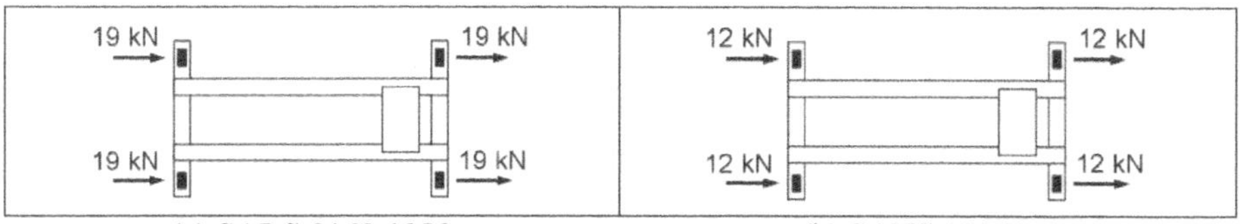

(a) SABS 0160:1989 (b) SANS 10160-6

Figure 5 Comparison of forces due to crab acceleration

4.3 Test Loads

The load tests that are performed on cranes are specified in SANS 4310:2002 Cranes – test code and procedures and the Occupation Health and Safety Act and Regulations 85/1993. Two sets of load tests are carried out on cranes: a 25% overload test performed during installation and commissioning and after refurbishment or upgrading and a 10% overload test performed annually.

According to SANS 4310:2002 the purpose of the 25% overload test is to verify the structural competence of the crane and components. The test is performed in such a manner as to maximise the rope loads, bending moments and axial forces in the major crane components. The load is lifted only 100 – 200 mm from the ground without any other movement of the crane. In order to achieve maximum bending moment in the crane bridge, the load is positioned at midspan of the crane bridge.

The purpose of the 10% overload test is the verification of the crane mechanisms and brakes. During this test the crane is controlled in the manner in which it will normally operate, i.e. moving all the drives, full travel of the crab along the bridge.

EN 1991-3 provides for a "static" test of 25% overload and a "dynamic" test of 10% overload which correspond to the load tests described above. The 25% overload test, however, will not be critical for the support structure as the load is lifted in the centre of the bridge and not next to one end carriage. The dynamic test with 10% overload is therefore the critical case for the support structure under test loads. Both test load situations have been included in SANS 10160-6 in order to provide the designer with a full description of the test procedure. The designer should however be aware of the position of the crab for the 25% overload case so that unnecessary conservatism is not included in the design of the support structure.

A dynamic factor ϕ_6 is given to model the dynamic effects of hoisting the test load off the ground and is calculated as: $\phi_6 = 0.5(1 + \phi_2)$. It is reasonable that the dynamic factor for lifting the test load is less than that for lifting a normal operating load because the test loading is carried out under more controlled conditions. The dynamic factor was included in SANS 10160-6.

The forces resulting from the test loads according to SANS 10160-6 are shown in Figure 6; SABS 0160 does not have a load model for test loads.

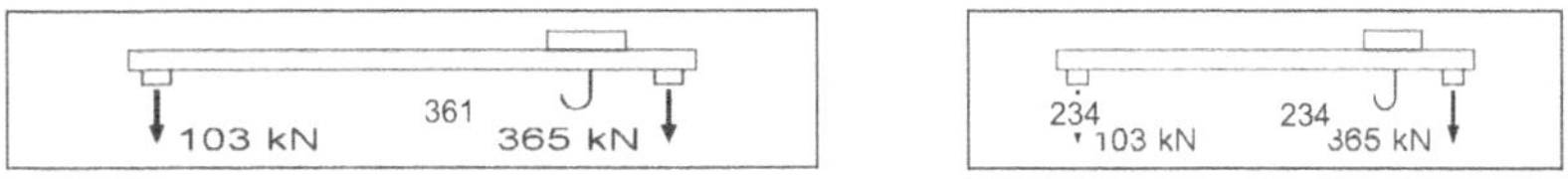

(a) Dynamic (b) Static

Figure 6 Forces due to test loads according to SANS 10160-6

4.4 Accidental loads

4.4.1 End stop forces due to crane movement

This load situation occurs when the crane runs into the end stops at the extreme of the runway. SABS 0160 provides two alternatives for the calculation of the end stop forces: the simplified method which applies the combined weight of the crane bridge and crab to each end stop and the more detailed method which recommends considering the resilience of the end stop and buffer but without providing guidance on how this is to be carried out. EN 1991-3 provides a method of calculating the end stop forces taking into account the resilience of the crane buffer in a rational manner.

A significant difference between the two codes is the manner in which the hoist load is treated. SABS 0160 states that if the hoist load is free to swing, it may be excluded from the calculation of the end stop forces whereas EN 1991-3 includes the weight of the hoist load even if it is free to swing. Experimental investigations have shown that the hoist load does indeed have an effect on the magnitude of the end stop forces, however, not in proportion to its weight relative to the crane self weight (Kohlhaas 2004). The EN 1991-3 load model is therefore conservative in considering the full hoist load but the load model in SABS 0160 is unconservative in excluding the hoist load. Some conservatism in the end stop forces is desirable because of the low cost of safety measures to withstand these loads (end stops, bracing and continuity plates) compared to the cost of the industrial building and crane; and the potentially high consequences of failure in terms of risk to life, damage to the support structure, crane and equipment and possible interruption of the industrial process.

The SABS 0160 simplified method often results in loads larger than EN 1991-3 but is based on a less rational calculation method. Whether the SABS 0160 end stop forces are larger than those calculated according to EN 1991-3 depends on the relative weights of the crane and the hoist load, the long travel speed and the stiffness of the buffers. In view of its increased rationality and conservatism, the EN 1991-3 load model for end stop forces due to crane movement has been included in SANS 10160-6. Figure 7 shows the end stop forces from the simplified method in SABS 0160 and SANS 10160-6 respectively.

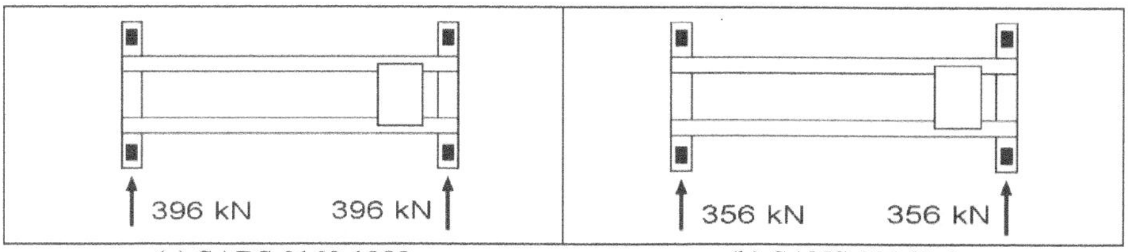

(a) SABS 0160:1989 (b) SANS 10160-6

Figure 7 Comparison of forces due to crane movement

4.4.2 End stop forces due to crab movement

The buffer forces related to movement of the crab have been discussed in the section on acceleration and braking of the crab. The EN 1991-3 load model for end stop forces due to crab movement has been included into SANS 10160-6, i.e. the load model for crab acceleration and crab buffer forces are the same.

4.4.3 Tilting forces

EN 1991-3 mentions tilting of cranes equipped with rigid lifting attachments, such as masts, briefly and states merely that the resulting loads should be determined. This approach has been carried over into SANS 10160-6.

5 EXHAUSTIVENESS OF CRANE LOAD MODELS

A further step in the verification of the crane load models in EN 1991-3 is the assessment of whether all possible load situations that a crane may experience during its lifetime have been accounted for.

A load model that is included in SABS 0160:1989, but does not appear in EN 1991-3, is one to provide for the effect of misalignment of the crane wheels and gantry rails. The misalignment load model in SABS 0160 considers the situation where the wheels or rails are misaligned in a toe-in or toe-out manner such that the crane exerts forces either pushing the two rails apart of pulling them together. The model is based on friction between the wheel and rail head with friction factors given as dependent on the class of the crane. This load model does not take into account the situation where the wheel flanges come into contact with the rail, as this would not occur if the crane wheels and rails were aligned within specified construction tolerances and therefore on a crane installation forming part of the scope of the code.

The misalignment load model was developed based on a description of assumed crane behaviour rather than experimental work or observation of cranes operating in practice. Similarly, the friction factors are based merely on the class of the crane in the absence of any experimental or theoretical work.

In order to decide whether this is a valid load situation which is not described at all by the load models in EN 1991-1, the behaviour of a crane should be carefully considered. For the simple case where the crane is travelling along a runway and encounters a misaligned rail on one side of the runway, the crane is then travelling at an angle to the misaligned rail. This behaviour is described by the skewing load model in EN 1991-3. Misalignment of the crane wheel is similarly represented by the skewing model. Extending this behaviour to the case where wheels or rails on either side of the runway are misaligned, the wheels or rails would need to be misaligned at the same angle; in the case of the rails, the misalignment of both rails would need to start at the same place along the runway.

This is a very unlikely situation and it can thus be concluded that any misalignment of the crane wheels or rails will cause skewing of the crane.

Based on this consideration of crane behaviour, and the lack of supporting experimental observations for the misalignment model in SABS 0160, it was concluded that the SABS 0160 model does not represent a realistic mode of crane behaviour which needs to be taken into account in SANS 10160-6. Therefore it is concluded that the loads caused by any misalignment of crane wheels or gantry rails are described by the skewing load model in EN 1991-3.

A load situation which does not appear in either SABS 0160 or EN 1991-3 is that of storm wind on an exposed gantry overriding the crane brakes and driving the crane into the end stops. An exposed gantry may be an open gantry or a crane installation under construction where the cladding has not yet been completed. There is much evidence to suggest that this is a critical load situation if sufficient anchorage has not been provided for the crane.

Provision has been made in SANS 10160-6 for this load situation by stating that wind forces may exceed braking forces which would result in movement of the crane if the crane is not equipped with clamping devices. It is recommended that the forces resulting from the crane being blown onto the end stops should be calculated using the load model for buffer forces, where the initial speed on impact of the crane is calculated in an appropriate manner, i.e. taking acceleration due to the wind force into account. This load situation is classified as an accidental load situation.

6 CRANE LOAD COMBINATIONS

Crane load combinations are vertical and horizontal wheel loads which are grouped together and considered as one characteristic crane action for combination with other loads, such as wind or imposed loads. EN 1991-3 provides a table specifying the crane load combinations with the relevant dynamic factors. This table has been modified for inclusion into SANS 10160-6 to take into account the extra load cases and to exclude the in-service wind which is covered by the load combinations recommended in SANS 10160-1. The load combination table as it is in SANS 10160-6 is given in Table 1.

Table 1 Groups of crane loads to be considered as one characteristic crane action

	Symbol	Groups of loads											
		ULS								Test load	Accidental		
		1	2	3	4	5	6	7	8	9	10	11	12
Self weight of crane	$Q_{c,k}$	ϕ_1	ϕ_1	1	ϕ_4	ϕ_4	ϕ_4	ϕ_4	1	ϕ_1	1	1	1
Hoist load	$Q_{hl,k}$	ϕ_2	ϕ_3	-	ϕ_4	ϕ_4	ϕ_4	ϕ_4	-	-	1	1	1
Part of hoist load	$\eta Q_{hl,k}$ (1)	-	-	-	-	-	-	-	1	-	-	-	-
Acceleration of crane bridge	H_T, H_L	ϕ_5	ϕ_5	ϕ_5	ϕ_5	-	-	-	-	ϕ_5	-	-	-
Skewing of crane bridge	H_S	-	-	-	-	1	-	-	-	-	-	-	-
Acceleration or braking of crab	$H_{T,3}$	-	-	-	-	-	1	-	-	-	-	-	-
Misalignment	H_M	-	-	-	-	-	-	1	-	-	-	-	-
Test load	Q_T	-	-	-	-	-	-	-	-	ϕ_6	-		-
Buffer force	H_B	-	-	-	-	-	-	-	-	-	ϕ_7	-	-
Tilting force	H_{TA}	-	-	-	-	-	-	-	-	-	-	1	-
Storm wind		-	-	-	-	-	-	-	-	-	-	-	1
1) $\eta Q_{hl,k}$ is the part of the hoist load that remains when the payload is removed, but is not included in the self weight of the crane													

The characteristic load combinations from SABS 0160 and SANS 10160-6 have been determined for the 40t crane and are presented below.

SABS 0160 does not provide guidance on the crane load combinations further than that the three sets of horizontal transverse forces need not be considered to act simultaneously. This results in three ultimate limit state combinations and one accidental load combination which are shown in Table 2.

Table 2 SABS 0160:1989 crane load combinations

Load combination	Vertical wheel loads	Horizontal wheel loads
ULS 1 Vertical with impact + Crab acceleration + Crane acceleration	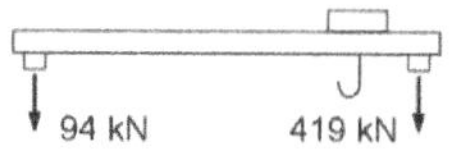	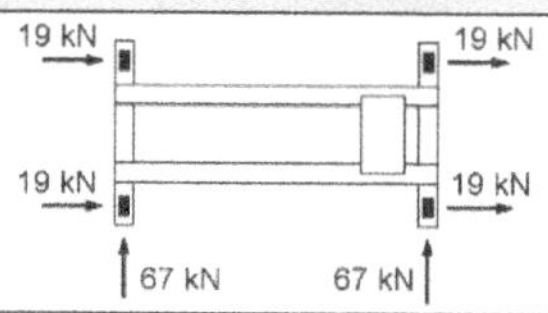
ULS 2 Vertical with impact + Skewing + Crane acceleration	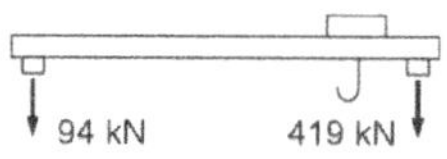	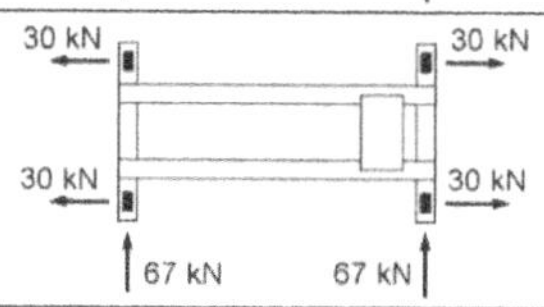
ULS 3 Vertical with impact + Misalignment + Crane acceleration	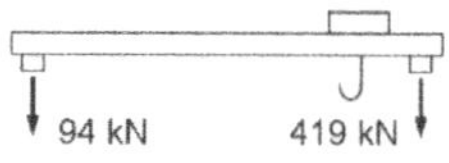	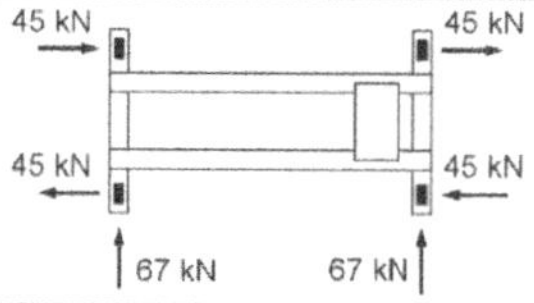
Accidental 4 Vertical without impact + End stop forces	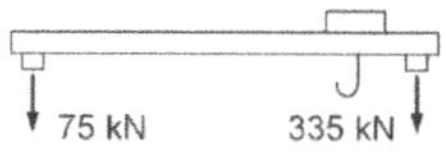	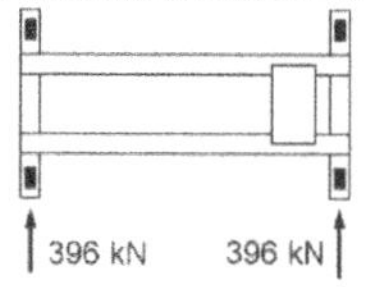

The load combinations in Table 1 result in eight ultimate limit state combinations, one test load combination and two accidental combinations. Load combination 2 as given in the table has not been calculated because it depends on the operating conditions of a crane that regularly releases a part of its hoist load. Load combination 8, for which the permanent portion of the hoistload remains, was calculated using 20% of the maximum hoist load as permanent. The load combinations are shown in Table 3.

Each load combination models a particular crane operating scenario, e.g. ULS load combination 1 describes the crane hoisting a load and accelerating and ULS load combination 5 describes the crane travelling at a constant speed and skewing.

It can be seen from inspection of Tables 1 & 3, that only combinations 1, 5 and 7 will be critical for verification of the ultimate limit state for the support structure. The remaining combinations are included for the purposes of providing a complete picture of the behaviour of the crane for fatigue purposes.

Table 3 SANS 10160-6 crane load combinations

Load combination	Vertical wheel loads	Horizontal wheel loads
ULS 1 ϕ_1 (Self weight) + ϕ_2 (Hoist load) + ϕ_5 (Crane acceleration)	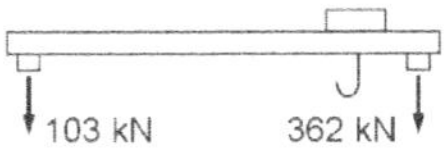	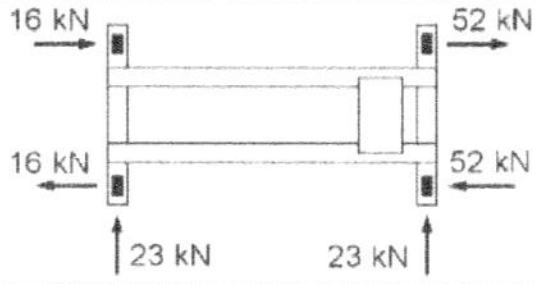
ULS 3 Self weight + ϕ_5 (Crane acceleration)	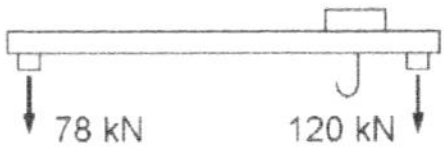	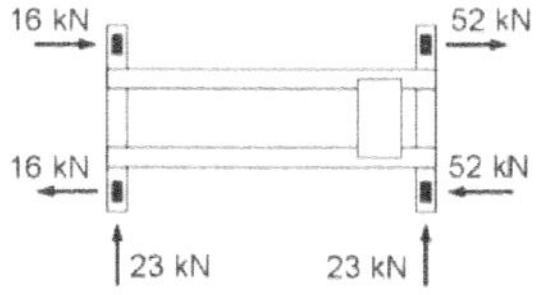
ULS 4 ϕ_4 (Self weight) + ϕ_4 (Hoist load) + ϕ_5 (Crane acceleration)	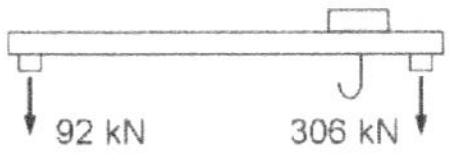	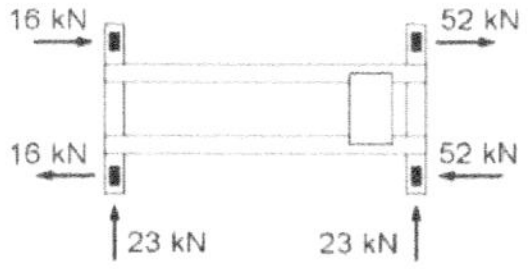
ULS 5 ϕ_4 (Self-weight) + ϕ_4 (Hoist load) + Skewing	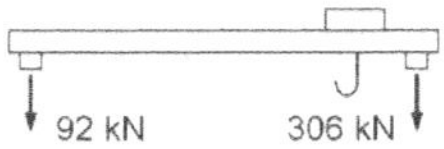	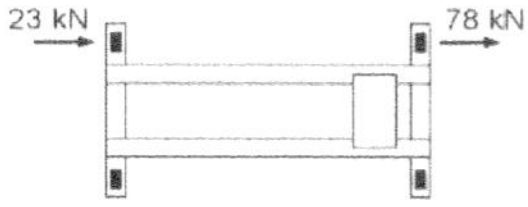
ULS 6 ϕ_4 (Self weight) + ϕ_4 (Hoist load) + Crab acceleration	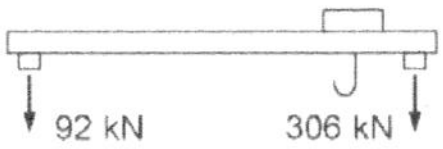	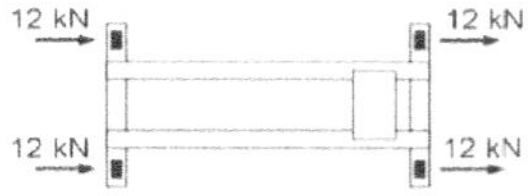
ULS 7 ϕ_4 (Self weight) + ϕ_4 (Hoist load) + Misalignment	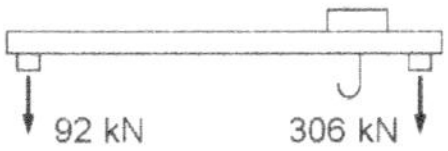	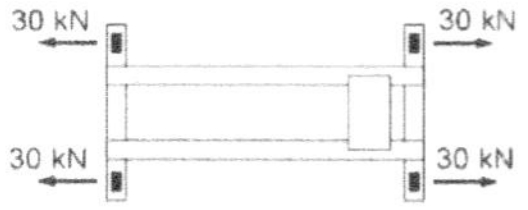
ULS 8 Self weight + η(Hoist load)	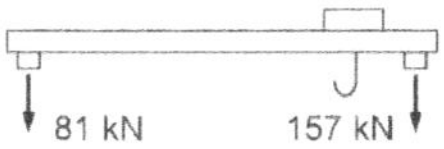	
Test 9 ϕ_1 (Self weight) + ϕ_6 (Test load) + ϕ_5 (Crane acceleration)	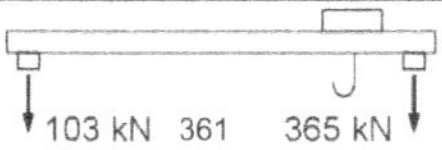	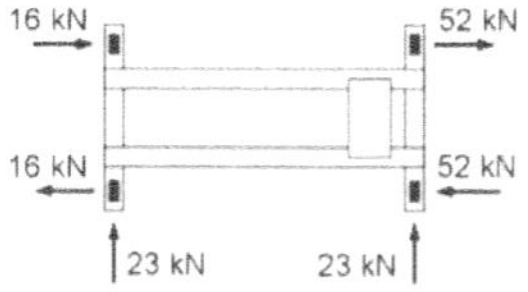
Accidental 10 Self weight + Hoist load + ϕ_7 (Crane buffer force)	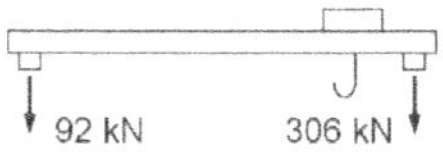	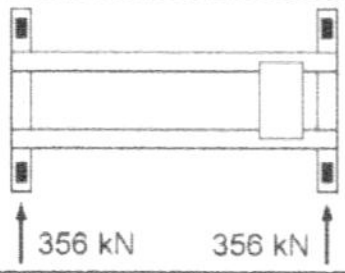

7 FATIGUE

EN 1991-3 provides a method for assessing fatigue in crane support structures by considering the number of cycles the crane will perform and its intensity of loading. A "fatigue damage equivalent load" is calculated which is a constant amplitude load for two million cycles. This method of normalising the fatigue loads was derived from the design method for the crane itself and it is uncertain whether it is applicable to crane support structures.

One indication of the lack of adaptation of the crane fatigue method to the support structure is that the wheel load amplitude has been considered for the stress range. This is true for the crane because during a cycle of lifting a load, travelling, setting down the load and travelling back, the crane experiences a stress range which may be represented by the change in wheel load from minimum load (crab furthest away, no load) to the maximum load experienced during the cycle (crab at a point along the bridge, a given load lifted).

The same is not true for the support structure. For example, considering the tensile stress in the bottom flange of a simply supported beam, before the crane moves onto the beam the stress is that caused by the own weight of the beam, when the crane passes over the beam the stress increases until it is a maximum due to the crane, and then reduces to the initial value again. In the case of a continuous beam, the stress may range from compressive to tensile as the crane passes over.

A further point to be considered is that the EN 1991-3 fatigue method assumes only one stress range per crane cycle. One crane cycle may mean one pass of the loaded crane over a particular fatigue detail followed by a pass of the unloaded crane travelling back into position to lift the next load. Furthermore one crane pass over a particular structural detail represents two or more wheel passes which may result in multiple stress ranges per pass of the crane.

In view of the uncertainty about the validity of the EN 1991-3 fatigue method in application to the support structure, it was not included in SANS 10160-6.

Traditionally the South African loading code has not specified fatigue loads. However, it was considered that it would be helpful to a crane support structure designer to provide some guidance on loads that are likely to cause fatigue. This is therefore the approach taken for SANS 10160-6. It is stated that load combinations 1, 2, 3, 4, 6, 7 and 8 should be considered for verification of the structure for fatigue and that the number of stress cycles should be determined taking the intended use and design life of the structure into account. The load combinations stated above are ultimate limit state combinations which would be factored by the crane partial load factor for verification of the ultimate limit state but considered as serviceability loads (i.e. partial load factor = 1,0) for fatigue verification.

8 IMPLICATIONS OF NEW CRANE LOAD MODELS IN SANS 10160-6

The implications of the new crane load models in SANS 10160-6 can be classified in terms of additional information required for the load calculations, the design effort and the cost of the resulting support structure. The first two points will be discussed here. The effects on cost of the support structure has been discussed extensively by Dymond (2005), indicating that the revised load models result in a reduction of about 10% in vertical loads but that horizontal loads are so dependent on design parameters that no clear trend can be discerned.

8.1 Additional crane information

The increased sophistication of the crane load models in SANS 10160-6 requires that more crane parameters are taken into account for the calculation of the wheel loads than for SABS 0160. The additional crane parameters and the relevant load case have been summarised in Table 4.

One of the problems associated with the additional information required is that crane manufacturers are currently not accustomed to providing this information. The second problem is that preliminary design of the support structure is often carried out before procurement of the crane has taken place and the parameters of the crane are therefore not known. The additional parameters required for calculation of the loads in this case translates to additional parameters which must be estimated, thereby increasing the uncertainty in the preliminary design.

Table 4 Additional information required to calculate loads according to SANS 10160

Load Case	Additional information
Static vertical wheel loads	L_{min} – minimum distance between hoist and rail
Dynamic effect of hoisting a load	v_h – steady hoisting speed
Dynamic effect of releasing a load	type of hoist – grab/magnet Δm – portion of hoist load released
Acceleration of crane bridge	number of wheel drives behaviour of drive changes – smooth/sudden/backlash
Skewing of the crane bridge	clearance between wheel flange and rail width of rail head
Crane buffer forces	S_B – spring constant of buffer ξ_b – degree of plasticity of buffer (energy – force – displacement buffer diagrams) v – long travel speed

In an effort to overcome these two problems, consultations with crane manufacturers are required to obtain their participation in the process of supplying the additional information. Preliminary design is often carried out with reference to tables of standard crane parameters supplied by the crane manufacturers which could possibly be updated to include the additional information in a standard form.

8.2 Design effort

The design effort required for the design of a crane support structure can be separated into the effort required for calculation of the loads and that required for the subsequent design of the support structure.

The load models in SANS 10160-6 are more sophisticated, and therefore more complex, than those in SABS 0160, and as such a greater effort is required in the calculation of the loads. A rough estimate of the relative amount of work required can be obtained by the number of pages of hand calculations required to determine the crane load combinations – 4 pages for SABS 0160 compared to 10 pages for SANS 10160-6. In practice, load calculations would be carried out using a computer with a software program or spreadsheet and the additional effort would therefore be a once off effort only, consisting of becoming familiar with the load models and setting up the software.

The subsequent design of the support structure could, however, be larger using the crane load models from SANS 10160-6 for which there are three potentially critical crane load combinations compared to two critical combinations from SABS 0160.

9 CONCLUSIONS

The crane load models from EN 1991-3 were assessed for inclusion into SANS 10160-6 on the basis of the validity of the crane mechanics and calculation method, their applicability to South African conditions and their exhaustiveness in describing the full range of load situations a crane is likely to experience during its lifetime. With the exception of fatigue, the crane load models from EN 1991-3 are more rational models which describe the crane behaviour in more detail than SABS 0160.

The EN 1991-3 load models for the ultimate limit state, test load and accidental limit state were incorporated into SANS 10160-6. Some minor modifications have been made to the crane load models to fit them to South African conditions and practice, e.g. including only the independent fixed wheel case for skewing as this is current crane manufacturing practice.

The validity of the fatigue load calculation from EN 1991-3 in application to the crane support structure is uncertain and it was therefore decided to omit this from SANS 10160-6. The approach taken to fatigue was instead to specify which load combinations are relevant for fatigue and to leave the onus on the designer to assess the number of crane cycles and movement of the crane in the building.

Two crane load situations that are not taken into account by EN 1991-3 but that were identified as of importance were misalignment of the crane wheels or rails, and storm winds on open gantries. Load models for these two conditions were included into SANS 10160-6.

The implications of the new crane load models were investigated. The increased sophistication and therefore complexity of the SANS 10160-6 models include more crane parameters in the load calculations and therefore more information is required from the crane manufacturer, often in the preliminary design phase where such information is difficult to obtain. Crane manufacturers should be consulted in this regard in order to determine a method of supplying the required information in the preliminary design phase, e.g. in tables of standard cranes.

The SANS 10160-6 load models require more calculation effort to obtain crane load combinations than SABS 0160. The use of computer programs and spreadsheets should reduce this effort to a once-off set up cost and it is therefore not significant. The design effort required for the subsequent design of the support structure could also be greater for SANS 10160-6 because there are three potentially critical load cases compared to two for SABS 0160.

10 REFERENCES

AS1418.1-1994 (1994): *Cranes (including hoists and winches)* Part 1: *General requirements*. Standards Australia.

DIN 15018 Part 1 (1984): Deutsche Norm: *Cranes, steel structures, verification and analysis*. Beuth Verlag GmbH, Berlin.

Dymond, JS (2005): *Reliability based codification for the design of overhead travelling crane support structures*, PhD Thesis, Department of Civil Engineering, Stellenbosch University.

EN 1991-3 (2002). Eurocode 1: *Actions on structures*, Part 3: *Actions induced by cranes and other machinery*. European Committee for Standardisation (CEN), Brussels.

EN 13001-2 (2004). *Crane safety – General design* – Part 2: *Load effects*. CEN/TC 147 European Committee for Standardisation (CEN), Brussels.

ISO 8686-1:1989 (1989): International Standard: *Cranes – Design principles for loads and load combinations*, Part 1: *General*. International Organisation for Standardisation, Geneva, Switzerland.

Kohlhaas S (2004): *Impact forces on end stops for overhead travelling crane support structures*. Technical Report, University of Stellenbosch.

Occupational Health and Safety Act (Act 85 of 1993) and regulations, Compiled by Lex Patria Publishers, South Africa.

SABS 0160:1989 (1990) (Amended 1990). *Code of Practice: The general procedures and loadings to be adopted in the design of buildings*. South African Bureau of Standards, Pretoria.

SANS 4310:2002/ISO 4310:1981: *South African National Standard: Cranes – Test code and procedures*, Standards South Africa, Pretoria.

6-3 Reliability Assessment of Crane-induced Actions

Dymond JS, Retief JV & Dunaiski PE

1 INTRODUCTION

The partial load factor applied to actions induced by electric overhead travelling cranes in SABS 0160:1989 is the same as the imposed load factor of 1,6 which has been calibrated for floor loads in office buildings. No reliability assessment has been carried out on crane induced loads in order to ascertain whether this partial load factor is suitable for crane loads.

The Eurocode standard for crane induced loads, EN 1991-3: *Actions induced by cranes and machinery*, recommends a partial load factor for crane loads of 1,35. This is the same partial factor used for permanent actions in Eurocode. In spite of the fact that the Eurocode reference level of reliability is more conservative than SA design practice, as discussed in Chapter 1-1 and Chapter 1-2, a more conservative value for the crane load partial factor [1,6] is used in South African practice, as compared to that of EN 1991-3 [1,35].

There is no literature available on the method by which the crane partial load factor in EN 1991-3 was decided on. It can therefore be assumed to be judgement based, rather than derived from reliability considerations. Similarly in the case of SABS 0160:1989 there is no background material available to justify the applied partial factor, other than the implication that crane loads should be treated as variable loads.

In general, there is therefore a need for a reliability assessment of crane induced loads in order to determine a suitable rational partial factor. Specifically, there is a need to determine a proper value for the procedures for crane induced actions specified in Part 6 of the revised SA loading code SANS 10160, which refers to EN 1991-3, as outlined in Chapter 6-2. This chapter presents the reliability assessment that has been used to determine the partial factor for crane loads on support structures as specified in SANS 10160-6. The assessment is done in accordance with the general requirements for structural reliability as specified in SANS 10160-1 *Basis of structural design.*

2 DESIGN OF CRANE SUPPORT STRUCTURES

Crane support structures are required to withstand a range of load situations over their lifetime. The ultimate limit state and serviceability limit state consider all the general operating conditions of the crane as well as the load tests that the crane is required to perform. Combinations of ultimate limit state vertical and horizontal crane loads are combined with other loads, such as wind or roof imposed loads, for verification of the design of the industrial building which houses the crane installation. The crane support structure is subject to the accidental load situation of the crane running into the end stops at the extreme of its travel on the gantry. The support structure is also subject to fatigue due to the cyclic nature of crane loading. Furthermore many industrial buildings contain more than one crane in each bay or in adjacent bays.

The reliability analysis has been carried out for combinations of crane loads with wind or roof imposed loads, fatigue and the accidental limit state, and has been reported extensively by Dymond (2005). The discussion here will be confined to the ultimate limit state, considering crane loads alone, with reference to a single crane.

3 RELIABILITY BASED CODE CALIBRATION

The reliability assessment of crane induced actions was carried out in the form of a reliability based code calibration. The procedure followed for the reliability code calibration was that given by Faber & Sørensen (2002). Each point will be discussed in the subsequent sections:

(i) Definition of the scope of the code
(ii) Definition of the code objective
(iii) Definition of the code format
(iv) Development of limit states equations
(v) Development of stochastic models
(vi) Determination of optimal partial load factors
(vii) Verification of partial load factors
(viii) Sensitivity assessment of reliability modelling and partial load factors

3.1 Definition of the Scope of the Code

The scope of the code refers to the range of cranes whose wheel loads are represented by the crane load models in SANS 10160-6. The scope thus provides for electric overhead travelling bridge cranes with wheels at the same level excluding portal or semi-portal cranes. This is similar to the scope of SABS 0160:1989.

The code calibration was carried out on specific selected example crane installations. In order to form a sufficient basis for the code calibration these cranes and support structures should be representative of the scope of the code. This would be ensured if the range of parameters of the crane and support structure covers the range of the scope of the code, i.e. the range likely to be encountered in practice. A description of the selection of the representative cranes has been given by Warren *et al.* (2004) and will only be summarised here.

The crane parameters affecting crane loads in SANS 10160-6 were identified and their range and distribution determined in consultation with a leading crane manufacturer with reference to over 500 cranes supplied in South Africa. The governing parameters were found to be the crane capacity and bridge span. The example cranes were selected from crane installations in practice to cover the range of capacities and the most likely range of bridge spans, and can thus be considered as representative of overhead travelling cranes in South Africa. The main parameters of the representative cranes are given in Table 1.

Table 1 Parameters of representative cranes

Crane Capacity (SWL)	Crane bridge span	Weight of bridge	Weight of Crab	Number of wheels	Outer wheel spacing	Hoist type
t	m	kN	kN	#	m	
5	19,2	55	6,4	4	3	hook
40	23,8	298	98	4	4,4	hook
260	28,5	1970	1298	16	12	ladle

The range of crane support structure configurations was investigated and it was confirmed that the support structures of the representative cranes were typical of small, medium and large cranes and thus covered the range of the configurations likely to be found in practice. A summary of aspects of the support structures for the three representative cranes is given in Table 2.

3.2 Definition of the Code Objective

The calibration of the code is performed to support its reliability objectives. The primary objective is for structures designed according to the code to achieve a level of reliability which exceeds a specified minimum value. The complementary objective is to achieve economic design solutions, which can be expressed in terms of consistency of reliability across the scope of applicability of the code.

Table 2 Support structures for the representative cranes

Crane Capacity	Girder configuration	Girder section	Column
5t	2 span continuous	Hot rolled I section with top flange channel cap	Steel columns. Girders supported on corbels
40t	simply supported	Welded mono-symmetric I girder	Concrete columns. Girders supported on chairs on top of columns.
260t	simply supported	Welded symmetric I girder with surge plate	Steel laced columns with two I section legs. Girders supported on top of inside leg.

A table of 50 year target reliabilities based on a cost – benefit analysis is given in JCSS (2001) and is shown in Table 3. This scheme implies that the optimum target level of reliability should consider both the consequences of failure of the structure and the related cost of improving its reliability.

Table 3 50 year target reliabilities (JCSS 2001)

Relative cost of safety measures	Consequences of failure		
	Small	Moderate	Large
Large	1,7	2,0	2,6
Moderate	2,6	3,2	3,5
Small	3,2	3,5	3,8

Although Eurocode allows the level of reliability to be set nationally, the default target reliability of $\beta_T = 3{,}8$ is used for the reference level. This corresponds to small cost of safety measures and large consequences of failure. The approach taken can however be interpreted as a level of reliability to be reached *on average*, rather than viewing the target reliability as the absolute minimum value (EN 1990:2002). South African codes are calibrated, for ductile failure modes, to a *minimum* target reliability of $\beta_T = 3{,}0$ corresponding approximately to small cost, small consequences or moderate cost, moderate consequences. A reliability of $\beta = 3{,}0$ was the implicit level of reliability of existing practice at the introduction of limit states design into South Africa (Kemp *et al.* 1987, Milford 1988). In order to maintain consistency throughout the set of South African standards, the target reliability for the crane load calibration was selected as $\beta_T = 3{,}0$. See Chapter 1-2 for a more detailed discussion for the level of reliability set in SANS 10160-1.

3.3 Definition of the Code Format

Typically the code format refers to the number and type of partial load factors, characteristic values of loads, load combination rules and the number and type of load combination factors (Faber & Sørensen 2002). The load combination rules and type of load combination factors have already been defined for South African codes and will thus not be considered further (see Chapter 1-2).

Characteristic values are defined in the Eurocode basis of design (EN 1990:2002) as the "principle representative value of an action", which may be determined on a statistical basis as is the case for imposed loads and wind loads. In the case of cranes the principle representative value would be the nominal value, be it of the crane self weight, hoist load or geometry.

In further discussion, the code format may be interpreted as the number and type of partial load factors. Four code formats were considered, which take into account the different statistical properties of the crane self weight, hoist load and vertical and horizontal wheel loads. Different code formats were investigated in order to assess the manner of achieving a consistent reliability over a range of design situations. The four code formats, together with the relevant partial load factors, are defined below.

1. **One partial load factor:** The factor (γ_C) applied to the characteristic wheel loads. This is the practice followed in both SABS 0160:1989 and EN 1991-3.
   $E_{V,H} = E(\gamma_C Q)$
2. **Two partial load factors:** Separate partial factors are applied respectively to the crane self weight (γ_{Csw}); and to the hoist load (γ_{Ch}); for the calculation of the design values of the wheel loads.
   $E_{V,H} = E(\gamma_{Csw} Q_{sw}; \gamma_{Ch} Q_h)$
3. **Two partial load factors:** One partial factor (γ_C), applied to the characteristic wheel loads, as for code format 1; but with an additional partial load factor (γ_H) applied to the horizontal wheel loads.
   $E_V = E(\gamma_C Q)$; $E_H = E(\gamma_C \gamma_H Q)$
4. **Three partial load factors:** Separate partial factors for crane self-weight (γ_{Csw}) and hoist load (γ_{Ch}), as for code format 2; with an additional partial load factor (γ_H) applied to the calculated horizontal wheel loads.
   $E_V = E(\gamma_{Csw} Q_{sw}, \gamma_{Ch} Q_h)$; $E_H = E(\gamma_H \gamma_{Csw} Q_{sw}, \gamma_H \gamma_{Ch} Q_h)$

Where: E_V, E_H are vertical and horizontal crane wheel load effects
Q_h is the weight of the hoist load
Q_{sw} is the self weight of the crane

3.4 Development of Limit State Equations

Limit state equations for reliability analysis can be presented in the form of the probabilistic performance function:

$$g(X) = R(X_R) - E(X_E) \quad (1)$$

Where: X is the vector of random or *basic* variables
$R(X_R)$ is the true resistance of the structure as expressed in terms of basic variables X_R
$E(X_E)$ is the true action effect on the structure in terms of basic variables X_E

It is common practice in calibration of the Eurocodes to separate the resistance and loading because of the wide scope of design situations provided for in the codes. Calibration of the crane load models is concerned with a limited range of structures, i.e. crane support structures, therefore the resistance and loading can be considered more readily together. By using the existing design codes for structural resistance, this calibration procedure would require only consideration of the partial load factors.

The reliability assessment was carried out in two steps: Firstly the configuration of the structural element was obtained by designing it to just comply with code requirements to obtain an economically designed element. Secondly a reliability analysis was performed of the economically designed element modelled in terms of the basic variables, using the nominal values to determine the parameters of the basic variables. A distinction should therefore be made between the probabilistic limit state equation (Equation (1)) considered for the reliability analysis and the design equation (Equation (2)) used to size a structural element in the economic design.

$$R_d \geq E_d \tag{2}$$

Where: R_d is the codified factored resistance of the structure
E_d is the codified factored action effect

The purpose of a code calibration is to set the minimum level of reliability inherent in the provisions of the code under consideration by adjustment of the partial factors applied to the codified design procedure. Many designs carried out according to the relevant code would have a higher reliability than this minimum level due to rounding up of member sizes for practical requirements. Since calibration is concerned with the lower *limit* of the code design, it is essential that no conservatism is included in the design of the member whose structural reliability is to be assessed, as this would result in inflated reliability values for the code. Therefore the economic design was carried out so as to *exactly* satisfy code requirements, i.e. the design equation was treated as:

$$R_d = E_d \tag{3}$$

In the limit state equation given in Equation 1, $R(X_R)$ and $E(X_E)$ represent the actual (probabilistic) load conditions and resistance of the structural element. It is often not possible to model the actual resistance and loading truly, so recourse is made to codified methods of calculation with the inclusion of modelling factors that express the probabilistic characteristics of the difference between calculated values and actual values. This was the approach taken for the calibration of the crane load models.

The loading refers specifically to crane loading which was determined from the crane load models in SANS 10160-6. Modelling factors were added to the basic variables in converting the code load models to the representation of $E(X_E)$. The resistance in the design equation was determined from the South African materials codes SANS 10162-1:2005 and SABS 0100-1:1992. The resistances in the limit states equations $R(X_R)$ were determined in accordance with the materials codes models for the relevant structural elements, with the inclusion of modelling factors.

Calibration of the ultimate limit state was carried out on the structural elements most directly loaded by the crane loads: the crane girder and crane columns. In setting up the limit states equations, the resistance of the member was expressed in terms of its *critical* failure mode, i.e. the failure mode which governed the size of the member.

4 DEVELOPMENT OF STOCHASTIC MODELS

The variables that were considered as random or basic variables were selected from the limit states equations on the basis of their displaying a significant variability. Basic variables can be classified into two groups in terms of the type of uncertainty displayed: aleatory or epistemic. Aleatory uncertainty refers to inherent variability and randomness and dominates the uncertainty in material properties, geometric properties and some loads, e.g. wind. Epistemic uncertainty is due to a lack of knowledge, imperfect representations of reality or imperfect predictions and characterises modelling uncertainties and some loads, e.g. self weight.

The stochastic models for material properties, geometric properties, dead load and modelling factors for resistances were obtained from literature. Two fundamental sources of information were consulted: the seminal work that was carried out for the development of the American National Standard A58 reported by Ellingwood *et al.* (1980) and the more recent probabilistic model code developed by the Joint Committee on Structural Safety, JCSS (2001).

No suitable literature was available on the stochastic models for loads lifted by cranes or modelling factors for the calculation of crane vertical and horizontal forces. New stochastic models were therefore developed for the code calibration.

4.1 Stochastic Model for Loads Lifted by Cranes

Stochastic models for granular loads lifted by grab cranes were developed by Köppe (1981) and used in reliability analyses on crane loading by Pasternak *et al.* (1996). These load models were not suitable for hook cranes in South Africa, so new load models were developed. The development of the stochastic models for loads lifted by cranes has been described by Warren *et al.* (2005) and will be reviewed briefly here.

Cranes were divided into four classes as described by Dunaiski *et al* (2001), as expressed in terms of the Safe Working Load (SWL):

1. **Light:** Crane rarely lifts the SWL and normally only small proportions of the SWL
2. **Medium:** Crane fairly frequently lifts the SWL, normally moderate proportions of the SWL
3. **Heavy:** Crane normally lifts high proportions of the SWL
4. **Very heavy:** Crane regularly lifts the SWL

Both "arbitrary point-in-time" as well as extreme hoist load distributions were developed in terms of crane class for the purpose of the reliability assessment of crane loads.

4.1.1 "One cycle" hoist load distributions

The "arbitrary point-in-time" models were referred to as "one cycle" distributions because they model the probability distribution of the size of the hoist load for one crane cycle, i.e. one load lifted. Beta distributions were used and the shapes of the distributions were developed from histograms and descriptions of loads lifted obtained from crane operators as well as the description of the crane classes.

The upper tail of the load lifted was identified as an important parameter in determining a model for the one cycle hoist load distribution. The upper tail was related to the SWL and the degree of control of limiting the hoist load in terms of the SWL during all possible operational conditions. Two scenarios were considered for the upper tail of the distribution:

1. **Good control:** The overload limit switch restricts the maximum load a crane lifts to less than 1,10 times the SWL.
2. **Poor control:** Insufficient supervision results in the overload limit switch being turned off and the crane lifting up to 1,25 SWL, taken from the specified static test for the crane hoist system. This scenario represents situations which are known to occur in practice where overloading of the crane does take place.

Figures 1 & 2 show the "one cycle" distributions for good and poor control respectively.

4.1.2 Extreme hoist load distributions

The extreme hoist load distributions model the probability distribution of the largest load a crane will lift in a given number of cycles N_{extr}. The extreme hoist load distributions were developed by simulating N_{extr} hoist loads from the "one cycle" distribution and selecting the largest load. This was repeated until a set of largest (extreme) loads was obtained, with a sufficient number of repetitions to provide stable statistical parameters. Polynomials were fitted to the values for the mean and standard deviation with respect to N_{extr} up to 2×10^6 cycles, in order to obtain analytical expressions for these parameters. The extreme hoist load distributions are therefore very versatile and may be applied to any limit state or load combination by selecting the appropriate value of N_{extr}.

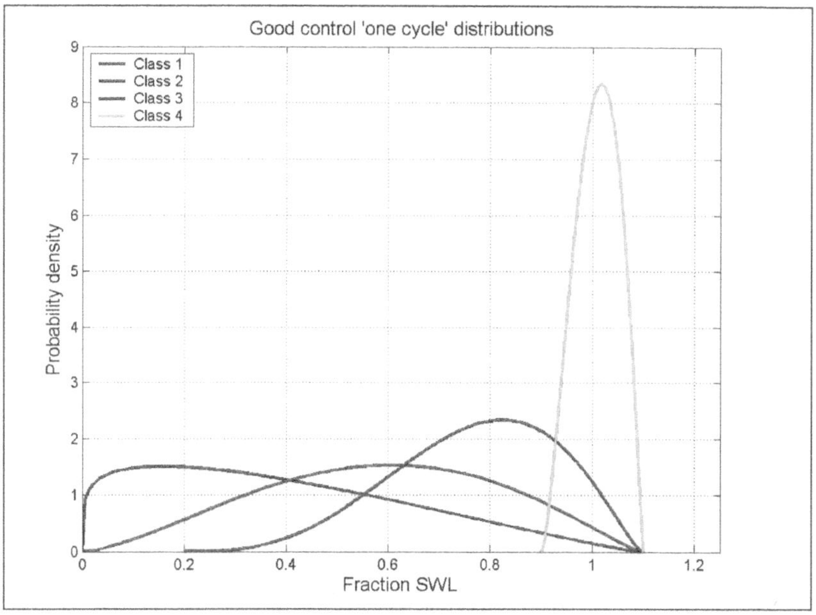

Figure 1 **"One cycle" distributions for *good* control scenario**

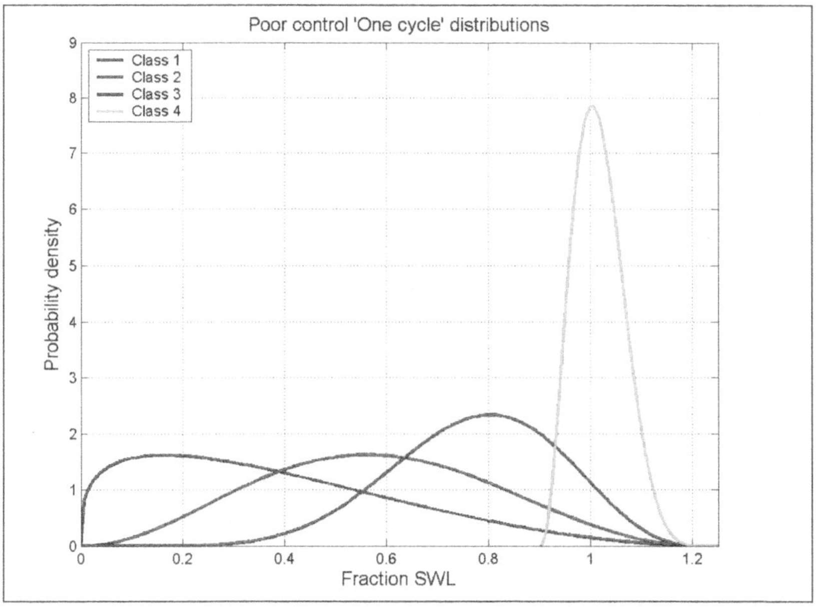

Figure 2 **"One cycle" distributions for *poor* control scenario**

The trends in the mean values and standard deviations are shown in Figures 3 & 4 respectively. The expected value clearly approaches the upper limit of the hoist load with increasing N_{extr}, whilst the values for the standard deviation become relatively small.

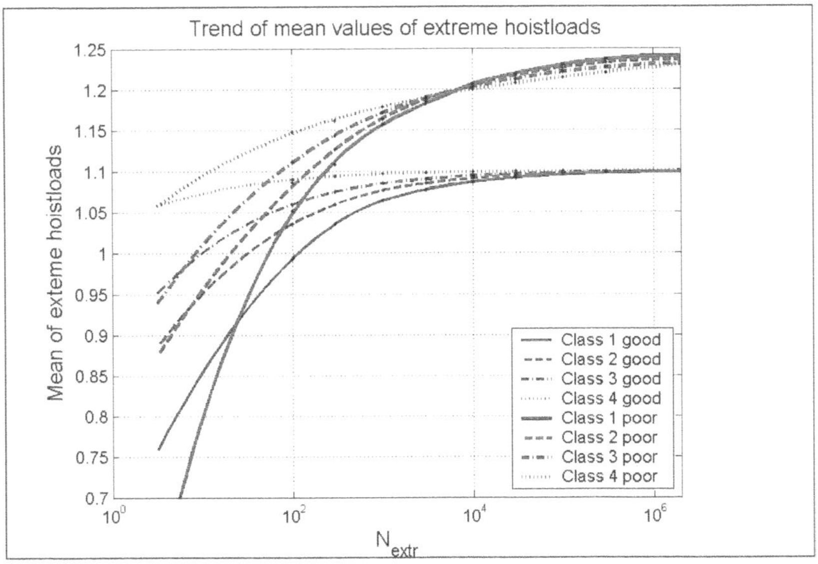

Figure 3 **Trend of mean values for extreme hoist load distributions**

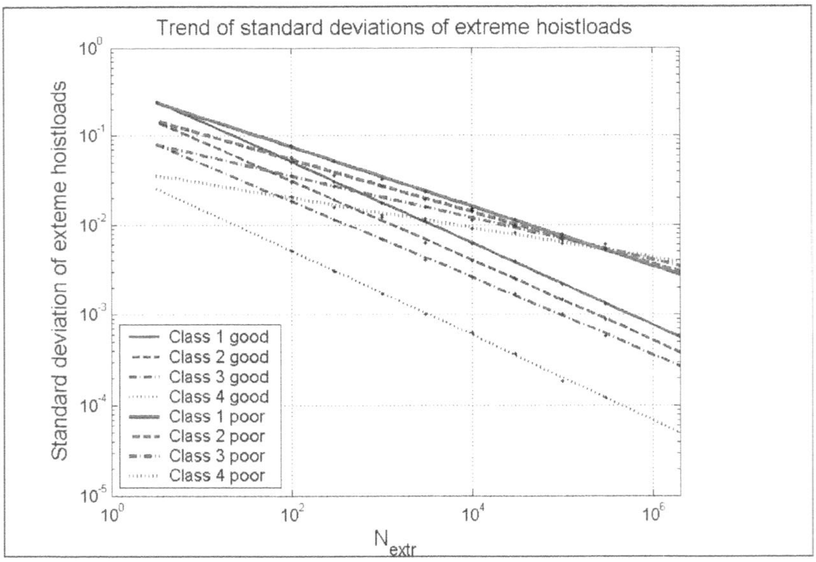

Figure 4 **Trend of standard deviations for extreme hoist load distributions**

4.2 Modelling Factors for Crane Imposed Loads

Modelling factors take into account the difference between results yielded by calculation models and reality, expressed in terms of the statistical properties of the ratio of actual to calculated values. Two parameters are defined: the bias λ gives the average value of the ratio, which reflects for example, whether the model is systematically conservative ($\lambda < 1$) and the dispersion expressed as the coefficient of variation δ of this ratio.

Three levels of information are employed in the selection of the statistical parameters for modelling uncertainties:

1. The ideal situation is when experimental data are available and the bias and coefficient of variation may be calculated from measurements of the phenomenon.
2. In the case where no such information is available, engineering judgment can be employed; this could include consultations with experts.
3. Nominal provision may be made for modelling uncertainties by applying a factor with no bias and a reasonable *upper* bound of the coefficient of variation.

Generally little information is available on modelling factors for crane induced loads, requiring the application of rather basic representation of this source of uncertainty. The approach taken was to break down the calculation process and assign model factors to its respective components.

In addition to the parameters of the model factor, an appropriate probability distribution needs to be established. A lognormal distribution is typically used for modelling uncertainties in the absence of data to the contrary (JCSS 2001).

There were no experimental data available for the vertical and horizontal crane wheel loads. A limited numerical simulation was carried out for the vertical wheel loads to investigate the effect of one wheel lifting off the rail, which may occur in the case of girders with unequal spans. The bias was selected as less than one because the vertical loads include dynamic factors which model the largest load that will occur, whereas the mean will be less than this load. A coefficient of variation of 5 % was selected based on the numerical investigation and the fact that the calculation model is based on straightforward static equilibrium.

The horizontal wheel loads were more problematic to estimate because, unlike gravity loads e.g. floor loads, it is not possible to make visual observations, and due to lack of experimental results nominal provision was made for the modelling uncertainty in a bias of 1,00 and a coefficient of variation of 15 %. The horizontal wheel loads were judged to be more uncertain than the vertical loads due to the increased complexity of the load models, the general lack of clear rational mechanical models or available literature and the difficulty of obtaining reasonable estimates.

The values of bias and coefficient of variation for each of the vertical and horizontal crane loads are shown in Table 4.

Table 4 Modelling uncertainties for crane wheel loads

Load	λ	δ
Vertical wheel load	0,95	5%
Horizontal wheel load	1,00	15%

5 DETERMINATION OF OPTIMUM PARTIAL LOAD FACTORS

The reliability assessment was performed on individual structural elements of the support structure in the two-step manner described earlier, *viz.* economic design followed by a reliability analysis of the economically designed member.

The First Order Reliability Method (FORM) using the Rackwitz-Fiessler procedure of equivalent normal distributions in a matrix formulation as described by Nowak & Collins (2000) was used for the reliability analysis. The reliability analysis procedure was programmed in MATLAB.

5.1 Parametric Range of Calibration

The calibration was carried out over a number of parameters, which are defined along with the practical range in Table 5 below as an optimisation process. The critical case which resulted in the lowest level of reliability was identified for the representative crane installations and structural element; the crane class, hoist load ratio $Q_h/(Q_{br} + Q_{cr} + Q_h)$, N_{extr} and degree of control were varied parametrically. For each code format a set of partial load factors was determined so that the required level of reliability was achieved for *the critical case of combination of parameters*.

The critical case for the ultimate limit state was found to be the 5t crane column in combined axial compression and bending. Reliability decreases with increasing N_{extr} and a value of 1×10^6 cycles was selected as a reasonable upper limit. A Class 4 crane resulted in the lowest reliability for the good control situation and a Class 1 crane for the poor control situation. Class 1 cranes (maintenance and general workshop cranes) are more likely to work in a low control environment than the higher class cranes and it is therefore reasonable that this should be the critical case for the poor control scenario.

Table 5 Range of calibration parameters

<table>
<tr><th>Parameter</th><th colspan="4">Range</th></tr>
<tr><td>Representative crane</td><td>5t</td><td colspan="2">40t</td><td>260t</td></tr>
<tr><td>Structural configuration:
▪ Girder</td><td>Rolled I-girder with channel capped top flange</td><td colspan="2">Mono-symmetric welded I-girder</td><td>Symmetric welded I-girder with surge plate</td></tr>
<tr><td>Structural configuration
▪ Column</td><td>Rolled I-column</td><td colspan="2">Concrete column</td><td>Laced steel column</td></tr>
<tr><td>Ratio of horizontal to vertical load effects</td><td colspan="4">Normally occurring ratios in structural elements considered</td></tr>
<tr><td>Crane class</td><td colspan="4">Class 1 – 4
▪ Load models for each class applied to each representative crane</td></tr>
<tr><td>Hoist load ratio:
$Q_h/(Q_{br} + Q_{cr} + Q_h)$</td><td colspan="4">Consider the range 0,30 – 0,85
▪ Standard cranes cover full range (most likely Class 1 & 2)
▪ Process cranes cover only 0,30 – 0,65 (most likely Class 3 & 4)</td></tr>
<tr><td>Number of cycles for N_{extr}</td><td colspan="4">Up to 1×10^6</td></tr>
<tr><td>Degree of control:
▪ Hoist load upper limit</td><td colspan="2">Good control – 1,10 SWL</td><td colspan="2">Poor control – 1,25 SWL</td></tr>
<tr><td>Code format</td><td colspan="2">Hoist load and crane weight:
▪ Combined
▪ Separate</td><td colspan="2">Additional partial load factor for horizontal loads:
▪ No provision
▪ Included</td></tr>
</table>

5.2 Calibration Steps

For code formats with multiple partial load factors (see Section 3.3) calibration was carried out stepwise in the order: γ_C, or γ_{Csw} followed by γ_{Ch}, then γ_H, in each step selecting a value to be applied in determining the subsequent partial factor:

- The full range of hoist load ratios was considered for calibration of γ_C .
- The hoist load ratio was set to zero for calibration of γ_{Csw} after which the full range was considered for calibration of γ_{Ch} .
- The calibration of γ_C or γ_{Csw} and γ_{Ch} was done in the following manner:
  - for Code Formats 1 and 2 it was carried out using the load situation which resulted in the largest ratio of horizontal to vertical load effects,
  - for Code Format 3 and 4 the vertical load effects only were considered, in order to eliminate effects from horizontal loads.
  - The horizontal partial load factor γ_H was then calibrated considering the maximum ratio of horizontal to vertical load effects.

5.3 Results

Graphs are presented in Figures 5 & 6 of the reliability as a function of the ratio of hoist load to total crane weight for both the good and poor control scenarios. Figure 5 shows Code Format 1 where one partial load factor is applied to the calculated wheel load and Figure 6 shows Code Format 2 where the crane self weight (γ_{Csw} = 1,62) and hoist load are factored separately before calculation of the wheel loads in the process of obtaining the optimal design.

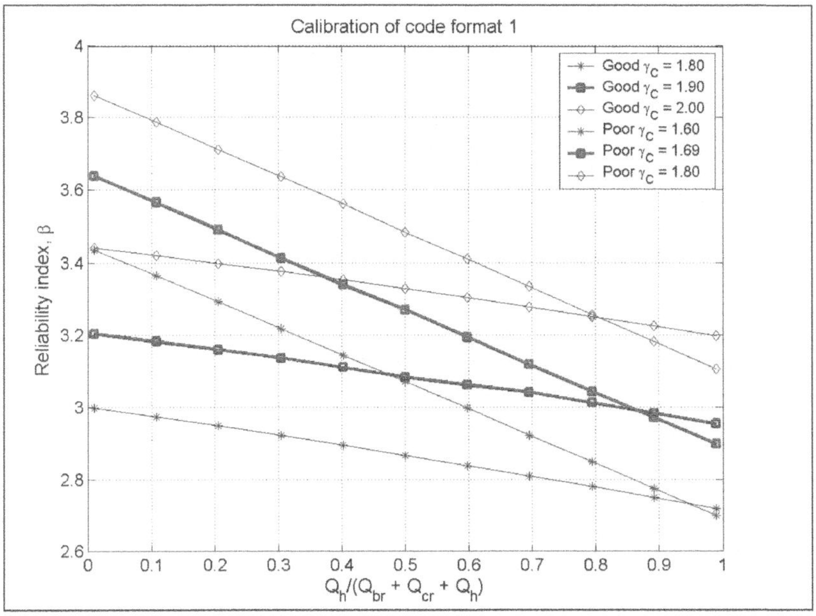

Figure 5 ULS calibration results for Code Format 1

It can be seen from Figure 5 for Code Format 1 that the reliability decreases as the ratio of hoist load to total crane weight increases. The same trend is observed for Code Format 3. This is due to the increasing effect of the mean value of the hoist load being modelled to be larger than the SWL. As can be expected this effect is more pronounced for the poor control hoist load model. This effect is eliminated if the hoist load and self weight of the crane are factored separately, as in Code Formats 2 and 4. For this situation it can be seen from Figure 6 that it is possible to select values of the partial load factors that result in a constant reliability over a range of hoist load to total crane weight ratios.

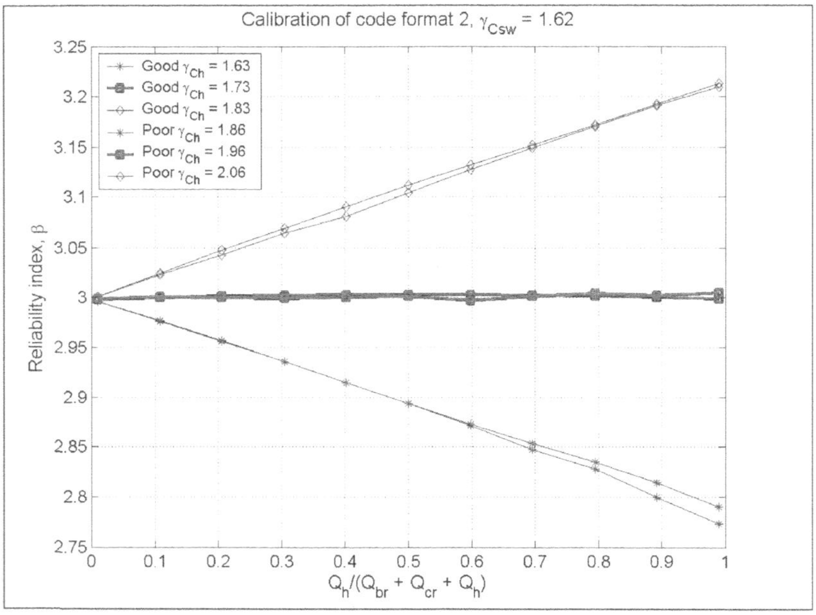

Figure 6 ULS calibration results for Code Format 2

The same phenomenon is observed with different ratios of load effects caused by horizontal and vertical loads, where the reliability decreases with an increasing ratio of horizontal/vertical load. This effect is ameliorated by the use of the additional partial load factor for horizontal loads in Code Formats 3 and 4.

6 SENSITIVITY ASSESSMENT OF CALIBRATION

The reliability results and consequent calibrated partial factors are based on a number of developed and selected factors. A sensitivity assessment of the results determines the degree to which the results are dependent on these factors. It therefore provides an indication of the degree to which the results are reasonable, both in terms of providing sufficient reliability and conversely not being unnecessarily conservative. The sensitivity assessment also provides an indication of areas which could fruitfully be explored further.

6.1 Crane Load Model

Figure 7 shows a graph of the reliability index as a function of N_{extr}, for both good and poor control for all four crane classes. The maximum practical hoist load ratio of 0,85 was used because this is the value that results in the largest sensitivity to N_{extr}.

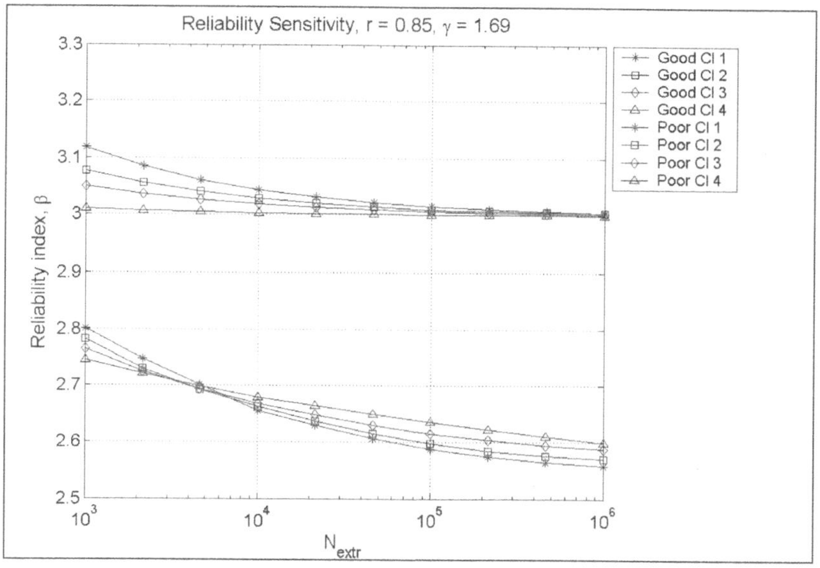

Figure 7: Sensitivity of reliability to class of crane and N_{extr}

The graph shows reliability results for Code Format 1 with the calibrated partial load factor of γ_C = 1,69 for good control. This case is representative of the sensitivity of the other code formats to the values of the respective calibration parameters. N_{extr} was considered in the range of 10^3 - 10^6 cycles as reasonable lower and upper limits; with 10^3 cycles representing a crane performing 1 cycle every 2 weeks for 25 years and 10^6 cycles representing a crane performing 120 cycles per day (approximately 8 cycles an hour,16 hours a day) for 25 years.

Design codes make provision for design of structures based on acceptable practices of construction, maintenance and use of the building. Overloading of cranes is an abuse of the structure which does occur in practice: Figure 7 highlights the reliability implications of such overloading when the design of the support structure is based on the premise of good control. At $N_{extr} = 10^6$, the reliability drops from $\beta = 3{,}0$ for good control to $\beta = 2{,}56$ if the control is actually poor. This results in a *significant decrease* in reliability, which relates to an almost fourfold increase in probability of failure.

The general trend is for the reliability to increase as N_{extr} decreases due to the change in statistical parameters of the "N_{extr} cycle" distributions. The degree to which the parameters of the "N_{extr} cycle" distributions change with N_{extr} can be explained by considering the upper tails of the "one cycle" distributions in Figures 1 & 2. The steeper the upper tail, the less sensitive the parameters are to N_{extr}. This explains why reliability for poor control is more sensitive to N_{extr} than for good control and also why the lower class cranes for good control are more sensitive to N_{extr} than the higher class cranes.

Reliability for good control is relatively insensitive to N_{extr}, most significantly for the dominant calibration case of Class 4 where the reliability ranges from 3,00 to 3,01 for N_{extr} of 10^6 and 10^3 respectively. N_{extr} is therefore not a significant parameter for calibration of partial load factors for good control. For poor control however, reliability is more sensitive to N_{extr}, with the reliability index ranging from 2,56 to 2,74 for N_{extr} of 10^6 and 10^3 respectively, causing it to be a more significant calibration parameter in this case.

The critical crane class for calibration of good control was shown to be Class 4. However, it can be clearly seen in Figure 7 that for high N_{extr} the reliability is not sensitive to the class of crane; even at $N_{extr} = 10^3$ cycles, the difference in reliability between Class 4 and Class 1 is only 0,11.

The reliability for poor control is more sensitive to the crane class at 10^6 cycles, although the difference in reliability between Class 1 and Class 4 is only 0,04, which is not significant. Inversely, for situations not expected in practice, Class 1 cranes are critical at a large number and Class 4 cranes at a low number of cycles, with the cross over at $N_{extr} \approx 4\times10^3$.

In summary, for good control neither the class of crane nor N_{extr} are significant parameters in the calibration and no substantial conservatism is included in considering the critical case for the determination of the partial load factors.

The reliability for poor control was not significantly influenced by the class of crane but was sensitive to N_{extr} with a difference in reliability of 0,2 between 10^3 and 10^6 cycles.

6.2 Basic Variables

Figure 8 shows the sensitivity factors for the significant basic variables in a reliability analysis on the 5t crane column. This case can be taken as generally representative of the relative importance of the various basic variables to the calibration results across the parametric range and the code formats respectively.

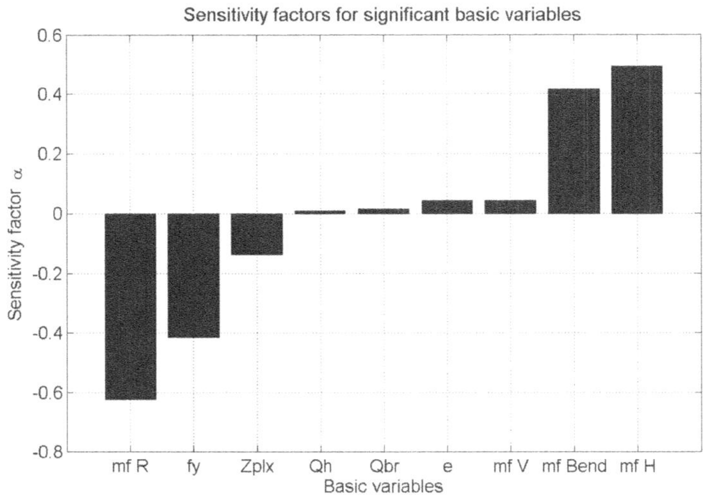

Where:

mf R	modelling factor for resistance
fy	yield strength of steel
Zplx	plastic section modulus
Qh	weight of hoist load
Qbr	weight of crane bridge
e	eccentricity of the load
mf V	modelling factor for calculation of the vertical wheel loads
mf Bend	modelling factor for bending moments obtained from an analysis program
mf H	modelling factor for calculation of the horizontal wheel loads

Figure 8 Sensitivity factors for 5t crane columns

The modelling factors, on both the resistance and loading side, are the dominant basic variables. This was true for all the structural elements and is generally true for many code calibration situations.

The concern is the dominance of the modelling factor for calculation of the horizontal crane wheel loads. This was the modelling factor for which there was the least information available and was therefore only nominally provided for. In view of its dominance in the reliability analysis, this stochastic model should be reviewed and refined. Experimental and numerical investigations into crane wheel loads are being undertaken (Dymond *et al* 2006), the results of which may be used to update the statistical properties of the modelling factor for horizontal wheel loads. Until such a time, the partial load factors obtained in the code calibration results are based on the best information presently available.

7 CALIBRATION RESULTS

Values of the partial load factors calibrated to achieve the target level of reliability for the four code formats are summarised in Table 6 in terms of the level of operational control. These partial factors represent the values required across the full range of crane classes, hoist load ratios, hoist load modelling parameters, and failure of the critical structural elements for the representative cranes and support structures.

Table 6 Calibrated ULS partial load factors

Code Format	Good Control				Poor Control			
	γ_C	γ_{Csw}	γ_{Ch}	γ_H	γ_C	γ_{Csw}	γ_{Ch}	γ_H
1	1,69				1,90			
2		1,62	1,73			1,62	1,96	
3	1,43			1,26	1,59			1,26
4		1,35	1,44	1,27		1,35	1,62	1,27

The main features of the respective results are the following:

- The partial factors resulting from the calibration are generally large in comparison to the value of 1,6 for crane wheel loads presently used in SABS 0160:1989. This result can largely be ascribed to the effects of horizontal loads on the support structure. Indications of this effect are given by the relatively large sensitivity factor for horizontal load modelling uncertainty (Figure 8) and the large effect of separate treatment by an additional partial factor for horizontal loads for the Code Format 3 & 4.
- The partial factors for the case of poor control are consistently higher than those for good control for the corresponding code format, with an increase of ~ 0,2 of the partial factor related to the hoist load, which implies an increase of ~ 12% in the design hoist load. The parametric assessment also indicates that this case is more sensitive to the respective parameter values, indicating inconsistency of reliability across the range of practical situations.
- Differentiation in treatment of self-weight and hoist load does not result in significantly different partial factors for the respective load types, although it improves the consistency of reliability across the range of hoist load ratios. As indicated in Figures 5 & 6 this is just an improvement in consistency for the range of reliability index values of 3,0 – 3,2 achieved by Code Format 1 for good operational control. For poor control the improvement is much higher (from values of 3,0 – 3,6), but as indicated above this is more of an indication of the sensitivity of this case to the various situations and parameters.
- A separate multiplicative partial factor for horizontal loads as applied in Code Format 3 & 4 reduces the partial factors for vertical loads significantly, to almost halfway between the SANS 10160 value of 1,6 and the EN 1991-3 value of 1,35. However this would require an increase of about 27% of the horizontal design load, either through the separate partial factor or by adjusting the load models for horizontal load.
- Code Format 4 comprising separate factors for the hoist load and self weight and an additional factor for the horizontal load best fulfils the calibration objective of maintaining a consistent reliability by accounting for the statistical differences between the crane self

weight, hoist load, vertical and horizontal loads. This improved reliability performance comes of course at the price of an elaborate set of three partial factors.
- Generally good consistency of reliability across the crane classes is achieved, as is shown by Figure 7. Although there are indications of higher levels of reliability for lower classes of cranes at a lower number of cycles, which is somewhat opposite to what a proper trend should be, any possible adjustment is not warranted by the present study. A refined assessment would be required to consider differentiated reliability treatment of support structures as a function of crane classes.
- The calibration is primarily directed here towards the relation between partial load factors and the reliability which is achieved. The importance of structural resistance, as reflected by the values of resistance related factors shown in Figure 8 should be noted.

When considering the result that Code Format 4 was shown to best fulfil the code calibration objective of maintaining a consistent level of reliability, it is necessary at this point to consider a further practical issue involved in code calibration, namely code complexity. Theoretically it would be possible to develop a code format which would result in consistent reliability over a range of materials, loading conditions and structural configurations by applying a plethora of partial load factors; however, this would not result in a practical design code. Consistent reliability and code simplicity are often opposing calibration objectives and optimisation is therefore required.

One option in striving to meet the conflicting requirements of consistent reliability and code simplicity would be to adopt Code Format 3, which consists of one partial load factor applied to the calculated wheel loads and incorporation of the additional load factor into the horizontal load model. The reduction of transparency of the reliability measures in the code would not be significant since it could simply be applied by an adjustment of empirical models, whilst retaining the existing code format. Due to the nominal information base on modelling uncertainty for the determination of horizontal crane wheel loads, such a measure would require additional investigation before it were applied.

A further decision which would have to be made for adoption of the partial load factors into a design code is whether the poor or good control scenario partial load factors should be included. The issue of whether a design code should allow for misuse of a structure which is known to occur (e.g. overloading of the crane) or whether the solution is stricter supervision, is not only contentious, but is part of a general dilemma with which the structural designer is confronted.

8 CONCLUSIONS

The reliability assessment and calibration of crane induced loads on support structures was performed, based on consideration of representative installations, structural elements and their failure modes, across a practical range of design parameters. From the results, the following issues need to be considered regarding the reliability of the stipulations for crane induced loads in SANS 10160-6:

8.1 Recommended Code Format

The main purpose of an elaborate code format which applies multiple partial factors to the respective design parameters is to improve the consistency of reliability across the practical range of design variables. Although the alternative formats investigated provided some insight into the relative importance of the various factors, it is concluded that the application of a *single partial factor* provides sufficiently consistent reliability for the design process.

Horizontal loads are the most significant candidates for separate treatment. However, adjustment of empirical factors of horizontal load models based on improved information of its reliability performance could achieve the same effect without complicating the code format.

8.2 Level of Reliability

The reliability assessment and calibration is based on the requirements given in the present SA loading code SABS 0160:1989, and maintained in the basis for structural design given in Part 1 of the revised SA loading code SANS 10160.

Based on the premise of the study, the present single partial factor of 1,6 for crane induced loads of SANS 10160-6 does not result in achieving sufficient reliability. The nominal provision made for the performance of models for horizontal crane loads that were applied, in absence of proper information, is the main source for such insufficient reliability. Although the models for skewing from EN 1991-3 and misalignment between the crane wheels and rails from SABS 0160:1989 both represent established practice, they are of a quasi-mechanical and empirical nature and lack information on their reliability characteristics. Research on these topics is certainly needed (Dymond *et al* 2006).

The reliability calibration indicates that there is sufficient justification to *at least* maintain the present practice of applying a partial factor of 1,6 to crane loads, as compared to the recommended value of 1,35 given EN 1991-3.

8.3 Upper Limit of Hoist load

It should be noted that characteristic crane induced loads are based on nominal values of crane self-weight and SWL specifications. The assessment of achieving sufficient reliability, as discussed above, is conditional on only nominal exceedance of the specified SWL, as represented by the reliability model for good operational control. The calibration clearly demonstrates that overloading, which is modelled here as poor control, significantly reduces the reliability of the design.

Although this is known to occur in practice, no provision can be made for *out of specification* conditions. At best a warning can be provided of the implications of such practice.

8.4 Differentiated Crane and Reliability Classes

The reliability assessment and calibration also indicate that required partial load factors were not significantly sensitive to the crane classes. The only justification for differentiated partial factors in terms of crane classes would be when differentiated levels of reliability are set as a function of crane class. Although SANS 10160 Part 1 provides the basis for differentiated reliability classes, this issue was not considered in the present investigation, nor is it provided for in Eurocode EN 1991-3.

An extensive range of operational conditions, which are not necessarily correlated with crane classes, clearly applies in practice to crane installations and their buildings. This provides justification for executing a proper assessment of a scheme of reliability classes for the design of crane support structures, in order to utilise the option provided in SANS 10160-1 (and EN 1990) for differentiated reliability treatment.

9 REFERENCES

Dunaiski PE, Barnard H, Krige G & Mackenzie R (2001): *Review of provision of loads to structures supporting overhead travelling cranes*. In: Zingoni, A. (ed.), Proceedings SEMC, pp. 1321–1328. Elsevier.

Dymond, JS (2005): *Reliability based codification for the design of overhead travelling crane support structures*, PhD Thesis, Department of Civil Engineering, Stellenbosch University.

Dymond JS, Dunaiski PE, Retief JV & De Lange JH (2006): *The future of crane support structure design in the South African context*. SAISC Steel-50 Conference, Johannesburg.

Ellingwood B, Galambos TV, MacGregor, JG & Cornell CA (1980). *Development of a probability based load criterion for American National Standard A58*. NBS Special Publication 577, National Bureau of Standards, US Department of Commerce, Washington DC.

EN 1990 (2002). *Eurocode – Basis of structural design*. European Committee for Standardisation (CEN), Brussels.

EN 1991-3 (2002). Eurocode 1: *Actions on structures*, Part 3: *Actions induced by cranes and other machinery*. European Committee for Standardisation (CEN), Brussels.

Faber MH & Sørensen JD (2002). *Reliability based code calibration*. Joint Committee on Structural Safety, Workshop on Code Calibration, Zurich.

JCSS (2001). Joint Committee on Structural Safety - *Probabilistic model code*. JCSS internet publication: http://www.jcss.ethz.ch/

Kemp AR, Milford RV & Laurie JAP (1987). *Proposals for a comprehensive limit states formulation for South African structural codes*. The Civil Engineer in South Africa, vol. 29, no. 9, pp. 351–360.

Köppe U (1981). *Nutzlastkollektive von Kranen*. Hebezeuge und Fördermittel, vol. 21, no. 2, pp. 36–39.

Milford RV (1988). *Target safety and SABS 0160 load factors*. The Civil Engineer in South Africa, vol. 30, no. 10, pp. 475–481.

Nowak AS & Collins KR (2000). *Reliability of Structures*. McGraw-Hill.

Pasternak H, Rozmarynowski B & Wen Y-K (1996). *Crane load modelling*. Structural Safety, vol. 17, pp. 205–224.

SABS 0100-1:1992: *Code of Practice, The structural use of concrete*, Part 1: *Design*. South African Bureau of Standards.

SABS 0160:1989 Amended 1990: *Code of Practice, The general procedures and loadings to be adopted in the design of buildings*. South African Bureau of Standards.

SANS 10162-1:2005: *Code of Practice, The structural use of steel*, Part 1: *Limit states design of hot-rolled steelwork*. South African Bureau of Standards.

Warren JS, Dunaiski PE & Retief JV (2004). *A comparison between crane induced load effects from SABS and Eurocode*. In: Zingoni, A. (ed.), Proceedings SEMC. Balkema.

Warren JS, Retief JV & Dunaiski PE (2005). *Reliability models of overhead traveling crane loading for code calibration*. In: Proceedings Ninth ICOSSAR. Millpress.

LIST OF CONTRIBUTORS

Peter Day [(1)]
Champion SANS 10160 Part 5
Chairman: Jones & Wagener (Pty) Ltd, Consulting Civil Engineers
Past Chairman ISSMGE TC23 on Limit State Design

Prof Peter Dunaiski [(1)]
Champion SANS 10160 Parts 2, 6, 7 & 8
Vice Dean: Teaching, Faculty of Engineering, Stellenbosch University
Chairperson: SABS SC 59I *Basis of structures design and actions.*

Dr Juliet Dymond
Research Fellow
Department of Civil Engineering
Stellenbosch University

Prof Milan Holický
Deputy Director: Klokner Institute
Czech Technical University, Prague
Member Eurocode CEN TC 250

Dr Adam Goliger [(1)]
Champion SANS 10160 Part 3
Chief Engineer
CSIR Built Environment
SA Representative: Int Assoc Wind Eng

Andries Kruger
Specialist Scientist,
Climate Service
South African Weather Service

Prof Johan Retief [(1)]
Champion SANS 10160 Part 1
Department of Civil Engineering
Stellenbosch University
SABS Representative to ISO TC98 *Basis for design of structures*

Prof Jan Wium [(1)]
Champion SANS 10160 Part 4
Member: SAICE WG on Loading Code
Chairperson: SABS SC 59A Working Group on Revision of SANS 10100

(1) Members of the SAICE Working Group on the Revision of the South African Loading Code

Made in United States
Cleveland, OH
05 November 2026

47673667R00147